W9-AHH-597

WOOD

ALSO BY HARVEY GREEN

The Uncertainty of Everyday Life 1915–1945

Fit for America

The Light of the Home

CRAFT • CULTURE • HISTORY

WOOD

HARVEY GREEN VIKING

VIKING

Published by the Penguin Group

Penguin Group (USA) Inc., 375 Hudson Street, New York, New York 10014, U.S.A • Penguin Group (Canada), 90 Eglinton Avenue East, Suite 700, Toronto, Ontario, Canada M4P 2Y3 (a division of Pearson Penguin Canada Inc.) • Penguin Books Ltd, 80 Strand, London WC2R 0RL, England • Penguin Ireland, 25 St. Stephen's Green, Dublin 2, Ireland (a division of Penguin Books Ltd) • Penguin Books Australia Ltd, 250 Camberwell Road, Camberwell, Victoria 3124, Australia (a division of Pearson Australia Group Pty Ltd) • Penguin Books India Pvt Ltd, 11 Community Centre, Panchsheel Park, New Delhi—110 017, India • Penguin Group (NZ), Cnr Airborne and Rosedale Roads, Albany, Auckland 1310, New Zealand (a division of Pearson New Zealand Ltd) • Penguin Books (South Africa) (Pty) Ltd, 24 Sturdee Avenue, Rosebank, Johannesburg 2196, South Africa

Penguin Books Ltd, Registered offices: 80 Strand, London WC2R 0RL, England

First Published in 2006 by Viking Penguin, a member of Penguin Group (USA) Inc.

10 9 8 7 6 5 4 3 2 1

Copyright © Harvey Green, 2006
All rights reserved

Illustration credits appear on pages xi–xvii.

ISBN 0-670-03801-6

Printed in the United States of America
Designed by Carla Bolte • Set in Granjon with Centaur

To woodworkers everywhere,
to those who appreciate their work,
and to those who preserve,
conserve, and carefully use the forests

CONTENTS

ILLUSTRATIONS

Unless otherwise noted, photos are by the author or from his collection.

Introduction

Where I live—south-central New Hampshire—many people heat their houses either wholly or partly with wood. Piles or neatly stacked rows of cordwood are tucked between trees or under blue tarpaulins in the backyard or, rarely, on the front lawn. A few enterprising or lucky wood burners store their supplies out of the weather. Most obtain their supplies of firewood in the usual manner, buying it cut and split. A small number own enough wooded land and have the tools and the inclination to harvest their own firewood. They hew and limb their trees, cut them to the proper length, and split the pieces, either by hand or with pneumatically driven wood splitters. After aging them for about a year, the wood is ready for the woodstove, furnace, or in rare instances, the fireplace.

A few of us, however, manage this process slightly differently, in a sort of middle ground between the do-it-yourself loggers and those who have the truck show up with the fuel ready to burn. I get logs cut in four-foot sections. Then I cut them to the lengths required by my wood stove. These days I rent a splitter (for which my elbows are grateful), and once the machine has worked its effortless magic, I stack it in the woodshed for the winter.

A few years ago I was hotly in the midst of splitting some big hunks of tree trunk when I made a profoundly disturbing discovery. I knew that the big tree—now sawed into two-foot lengths—was maple. What I discovered only after cutting it all up and beginning to split it was that it was one of the woodworker's prizes—tiger, or curly, maple. I salvaged some short pieces from the logs and have them still, but the damage was done.

Curly maple has been a choice wood for fine furniture and gun stocks in North America for about three centuries. Its dramatic striped grain glows and shimmers when finished. "Choice curl'd maple" is specifically identified in eighteenth-century ledger books, furniture orders, and inventories of household furnishings. It not only looks good, it is also exceptionally hard and durable. The hardness allows for closer tolerances and more precise machining or shaping. Of course it is also that much more difficult to shape and smooth it.

Part of its appeal may have lain in the difficulty of identifying it in the forest. There is disagreement among experts about the conditions that produce the curled or striped appearance in maple and other woods, among them cherry, oak, sycamore, and ash. It may be the result of a tree twisting as it grows. Any number of situations in the forest can induce this, such as other trees in the canopy blocking access to the sun or a tree attempting to straighten itself after partial dislodging, perhaps caused by a storm or another tree falling against it. The complication for the would-be worker of this wood (at the outset, since there are plenty of complications once the stuff is in the workshop) is that the interior grain pattern in most cases leaves no discernible marker on the bark. In this case the first layer of wood under the bark did reveal a wave pattern on the surface. Had I removed some bark while the logs were in four-foot sections (with a pointed tool known as a bark spud), I might well have had the brains to take the logs to a sawmill to salvage some reasonably sized boards from the firewood pile.

Furniture-grade lumber was an important part of colonial American trade activity from the onset of the eighteenth century. Ships plying the Atlantic and the Caribbean brought both straight and figured mahogany, rosewood, and other "exotics" to cabinetmakers and their aspiring and socially conscious patrons in the fine houses in cities and the countryside. Imported wood was expensive, but it was an important status marker or symbol for the well-heeled. (I can find in words associated with wood no linguistic equivalent to the equation of fine leather shoes with wealth in general; no one appears to have been called "well-wooded" or "well-lumbered.") Fine furniture with visually complex grain, smooth surface, and a deep patina conveyed wealth, taste, and refinement. Wood in this highly "artificial" form (so called because human artifice had been employed to transform the forest log into furniture) was testament to perhaps even greater levels of refinement than were shoes, boots—and heels.

Maple, however, had an appeal that the noticeable stripes of rosewood or the flamelike patterns of crotch mahogany lacked. (Wood sawed from the joint of a trunk and a major limb of a tree—called the crotch—generally possesses swirling flamelike patterns that when properly matched and positioned take on the appearance of flames.) Unlike mahogany, it was a native wood, and like birch (whose occasionally dramatic figuring could also be positioned on a piece of furniture to resemble flames) it was less expensive to acquire. Its nativity came to matter in the latter eighteenth century, when American independence from Great Britain became a political, economic, and cultural issue.

In the lumberyards in which figured maple can be found today, it and its cousin, bird's-eye maple, command a premium price. (Bird's-eye maple typically has a multitude of small round spots and grain swirls throughout, the aberration the result of branches that fail to grow or, in some cases, diseases of the wood.) Even with modern machinery both grain patterns can be devilish to shape and smooth. To some extent this explains why some modern furniture factories discard figured woods they receive in large loads of raw materials, reasoning that the wear on expensive machines is not worth any possible premium in the price of the finished product. Thus does mechanization homogenize.

For millennia wood has been a familiar and ordinary substance, common in people's lives, used for heat, shelter, food preparation, light, tools, weapons, toys, storage, land and sea vehicles, decoration and design. It is in fact *so* familiar that we often dismiss it as a "common" substance—even as we spend money replicating it artificially—and yet its use and appearance send a strong signal of class, status, authenticity, or something less tangible or definable. Today wood has ceased to be the jack-of-all-materials, superseded by (usually) manmade substances in everything from ships to golf clubs. But it persists not just for reasons of utility and convenience but for its affective—even Romantic—presence. That is unlikely to change, no matter what new materials or flashy goods become available to consumers. I believe it is the cultural power of wood—and by extension the forest—that informs this reality. Another way of thinking about it is that wood seems to us somehow more historical and hence more "real."

Part of the continuing lure of wood is that it is entirely organic, associated with nature in its wild and domesticated forms. While some conservationists would argue that this is exactly why wood should be superseded as a material and fuel, most consumers demonstrate that these associations are what makes wood more desirable. That these distinctions are in fact not quite so easily drawn is beside the point.

Ordinary people assume that they know or can readily discover how to work with wood, whereas working with metals and plastic is mysterious, dangerous, or requires tools that are too expensive and too large for most people to acquire or use. Metal is hard, difficult to cut when cold, and can be sharp enough to seriously wound a person. Most metals require great heat before they can be manipulated, and metalworking (other than jewelry making) is associated with soot, the heat of the forge, and muscular smiths sweating at the anvil. Plastics are more easily made malleable, but they can give off toxic fumes when heated, as well as benign odors that merely smell dangerous. Both are manufactured materials and, like plywood or other "engineered" wooden materials, the products of huge factories and advanced science and technology. Wood that comes from boards and not sheets by contrast seems a genteel and unthreatening material with few secrets and dangers.

Laminated wooden materials—though not manufactured "sheet goods," as modern plywood is termed—are not exactly new in human history and were not viewed with even mild skepticism when they first appeared. In the early 1840s, the New York City furniture maker John Henry Belter patented a new method for bending laminated rosewood furniture in two directions simultaneously, thus making possible his intricately carved, lightweight, and extraordinarily strong chairs and sofas. He became one of the most important producers of expensive "French-style" furniture in the pre–Civil War United States. No one seems to have complained about his transformation of wood.

Belter machined and engineered wood to produce decorative furniture. Using wood decoratively, however, was a much older tactic in the ornamentation of furniture. For centuries before Belter's innovation, furniture makers had dressed up their wares by covering less expensive and less visually interesting woods with a thin sheet of fancier wood (*veneer*) or with intricate inlaid designs of small pieces of these thinly sliced woods (*marquetry*). Some species are naturally colored or can be tinted in such a way as to make the color possibilities almost endless, and hence the decorative options as broad as can be imagined. Veneering and marquetry work, long practiced in the Islamic world, became popular in Europe in the latter half of the seventeenth century, especially in the imperial capitals of Stockholm, Copenhagen, and Amsterdam.

These three great mercantile empires traded extensively with Africa and the East Indies, where exotic wood species were abundant.

Anyone who has endured meetings of one kind or another has run across the modern age's combination of lamination and the sort of industrial production that seems to be a counterpoint to wood's affective qualities. This incarnation is the ubiquitous folding metal table with a brown plastic "walnut" surface attached to some sort of composite material. There are "practical" arguments for this common, yet bizarre surface choice. Plastic is impervious to water and easier to clean. It does not stain or need refinishing, as does wood, but it is difficult to repair should the surface be chipped or cut or begin to separate from its substrate. But the inescapable visual fact remains: Manufacturers and their customers want the table to *seem* to be wood because it is still the material of choice and of status. The plastic replica, wherever it is used, is inescapably the mark of the economically disadvantaged or socially and aesthetically unaware. Captains of industry get real wood in their boardrooms and offices. Foot soldiers do not.

In some ways the proponents of practical plastic are correct. Wood has formidable disadvantages. It rots; it burns; it is eaten by insects and chewed by other creatures. Trees take decades to grow to maturity, or at least to usability. They are huge, and they require dangerous work to be removed from the forest and processed into useable forms. Trees are inconsistent in quality and resistant to most forms of human effort to regulate and manage their growth. They burn up in wildfires and die in floods. Great winds fell them. Because wood is a plant material, it is subject to the damaging effects of heat, light, and humidity; it expands, contracts, changes color, cracks, and breaks as environmental factors change. Sanding and sawdust endanger a worker's lungs, and a few woods are poisonous or are allergens. Veneer layers delaminate from their base materials. Some metals oxidize in harmful or inconvenient ways, but most wood finishes either scratch easily or are subject to staining, darkening, or bleaching.

Why, then, do we keep coming back to wood? In part this is because other materials have their own drawbacks. Stone is difficult to obtain, because of both its weight and its location. Manageably sized rocks are scattered on the landscape; large deposits are generally underground. Rock is difficult to manipulate into useable forms. It is also "cold," a characteristic that can make it undesirable as a building material. Like metals and some plastics, it conducts heat—thus it is quickly affected by the lack of it. Wood, on the other hand, is more easily acquired on the landscape (by axe, saw, and gravity) and is more easily transformed by human artifice. It is "warmer" by virtue of its lack of heat conductivity. It is easier to build with wood than with stone.

Wood also endures because it is now thought of as a traditional, even (ironically) preindustrial material. Thus wood carries with it the connotation of age and the "simplicity" of nostalgia. Because wood was the main building and furnishing material of the ancient and immediate past, it is ever present in the most familiar material elements of history: historic houses and museums, both of which are filled with what previous generations called "old lumber" and what contemporary observers call antiques.

Before we get too nostalgic for the "wooden age" in human history, however, we should remember that the Industrial Revolution in the West began with wood as its major material. Before 1850 most machines—spinning wheels, looms, plows, rakes, shovels, hoes, churns—were made almost entirely of wood. Even at the outset of the age of steel, machines for home and factory production were made mostly of wood, with iron or steel fittings attached at areas of greatest friction or where cutting took place. But iron and steel require fuel (wood, charcoal, coal, coke) to smelt the ores and melt the metals. Wood requires no further transformation of its substance in the wild. Metals and plastics may be the materials of industrialism today, but wood made the revolution possible.

Modern "engineered" wood in some ways may be superior to wood in its original form or composition. Plywood does not warp or twist in the same way that most solid woods do. Laminated beams may in some cases be superior to their solid-wood counterparts as building supports. Particle boards, chip boards, and oriented strand boards make use of previously discarded parts of trees and thus can play a part in the wiser use of them. For all the advantages of modern materials, however, post-and-beam builders can still find clients who are unconcerned with the cracks that inevitably appear in large pieces of wood as they go through periodic drying, rehydrating, twisting, and bending. People still erect log buildings, though their appearance is more like an ordinary house than the crude structures first erected in northern Europe and Russia (where the log "cabin" was first developed) and later in North America.

The Industrial Revolution had the effect of transforming wood into just another machined mass-produced commodity, or so we might think if we visit a forest-products factory or a large lumber mill that turns logs into boards and beams. The uniformity developed in the building trades smacks of both the standardization of goods of all sorts and of the sly tinkering with things that characterizes capitalism and probably all economic systems, come to that.

Consider what the trade calls "sheet goods." These are the various wood-and-glue materials that come in four-by-eight-foot sheets (occasionally four by ten) in quarter-, half-, and three-quarter-inch thicknesses. The dimensions correspond to the recommended spacing of vertical studs in walls, or between rafters or floor joists—sixteen or (less commonly) twenty-four inches. Standardized distances make considerable sense, especially if one has to do renovations or repairs or expand a structure.[1] Dimensions of solid wood products have also been standardized. Framing lumber in the United States is usually spruce and produced in multiple widths with a standard thickness; it is called "two-by" lumber because the pieces were originally a uniform two inches thick, and the widths are available in increments of two inches—2 × 4, 2 × 6, 2 × 8, and so forth. Sheathing or other building lumber is usually sold in three-quarter-inch thickness and in similar width increments. Much wider boards than framing lumber sizes are available, although they are less common today as the old-growth forest disappears. In most of the rest of the world, lumber comes in metric sizes of roughly the same dimensions.

But in reality they aren't the dimensions we think they are. Novices are shocked when they discover that two-by-fours are at most 1½ × 3½ inches, two-by-sixes are 1½ × 5½ inches, and so on. Three-quarter-inch plywood is no longer that dimension; it is more like eleven-sixteenths inches thick, while half-inch plywood is seven-sixteenths; quarter-inch is about three-sixteenths. Buying boards thicker than the standard three-quarter stock gets one into another Lewis Carroll–like world. Inch-thick boards are called five-quarter, as if they were 1¼ inches thick, and the pattern continues. Where did that quarter-inch go? What about the missing half-inch in the thickness of framing lumber? This fudging of the size is part of a pattern in a consumer society in an economic system in which the pressure to cut costs and increase profit margins is immense. Consumers get that runaround all the time in other goods. Packages of food get smaller while the price remains the same or increases only a little, as if we are not going to notice. The "fast ones" the manufacturers pull fool only the fools; consumers just accept them because their options are limited in a mass society of big machines and big companies. They would like to think that in the old days people might have confronted the local tradesman or merchant who tried to give them the shaft. They may be right, but that is merely conjecture and probably just nostalgia or romanticism at work.

These changes have to do with maximizing profits for the supplier of goods while, in the case of lumber or sheet goods, maintaining the effectiveness consumers had come to expect. In truth it probably made no difference for purposes of framing to shrink the stud thickness from two to one and one-half inches. Three-quarters of an inch is probably plenty thick for sheathing lumber. These reductions save lumber, just as new methods of joining small pieces of lumber to make one large piece make use of more of the tree, and some sheet goods use what was once burned as waste or merely dumped. The amount of waste from the lumbering practices of previous centuries would appall even the grumpiest moss-back today. It is, thus, for the good of the majority that many of these innovations and standardizations have occurred in the handling of wood, even if some of the methods employed to fabricate building materials may have been harmful to the consumers' or woodworkers' health.

The irony of industrialization and standardization lies not in dimensions smaller than one expects; it is not about industrialists large and small being too clever by half (an inch) or about being generally shifty and furtive. The irony lies in the will of people to continue to see wood as organic and natural, and manufactured goods to be artificial and somehow tricked up, even as industrialization seems to homogenize wood. Manipulated organic materials *are* artificial—the products of human art and artifice. People used to refer to the trades as the mechanic arts, meaning that artisans' hands transformed the natural—the rough, the curvilinear, the asymmetric, the coarse—into the artificial—the smooth, the straight, the symmetrical, and the refined—by virtue of their skills and knowledge. This was considered a good thing, part of the process of elevating humans above the animals. But the artificial, in its most skilled incarnation a tour de force that instilled awe in the beholder, has become a sign of the alleged oppression by the mechanical and the industrial. It is not the material that has changed; it is we who have done so. What previous generations took for granted or even viewed with hostility—the wild, dark, and threatening forests where danger lurked—we now value as it disappears before our eyes.

Industrialization began with wood—as a fuel to make other products and as a material from which to make the machines that transformed human life wherever mechanization and industrialization appeared. By the middle of the nineteenth century, wood ultimately provided a substance—pulp for paper—on which the transmission and preservation of knowledge have since become dependent.[2] In the end, the Industrial Revolution did not replace wood nor in all ways supplant it.

———

This introduction has raised some of the ironies and complexities of examining wood in human history, particularly the linkages between wood and other materials generally credited with the onset of the Industrial Revolution in the West. Chapter 1 introduces readers to the material—its physical qualities, how it can differ from species to species—and offers basic information on the physical structure of wood as well as characteristics such as grain and figure that give wood its appearance. The second chapter begins a functional analysis that takes up most of the book, starting with wood's sheltering role. Chapter 3 examines the ways people have traditionally worked with wood and solved the problems encountered in transforming logs into boards, planks, curves, and veneers, and the ways in which people have joined board and beam to create structures and other useful artifacts of everyday life. The fourth chapter describes the ways in which wood was for centuries the material of empire, particularly the naval powers that dominated much of the world since at least the time of the Persian Empire. Chapter 5 analyzes the aesthetic, artistic, religious, and decorative aspects of wood, including its use in sculpture and carving, musical instruments, and furniture. From faith and the arts I then examine, in chapter 6, one particular and important form of construction—the box—in a spectrum of manifestations that include everything from barrels and crates to chests of drawers and caskets. The seventh chapter is devoted to an investigation of the little things made of wood, such as toothpicks, golf tees, smoking pipes, canes, clothes pins, matches, chopsticks, and toys, that have had a continuing impact on human life. The eighth chapter is an analysis of wood in war and play—activities that are as dissimilar in their intentions as their artifacts are similar in form. The ninth chapter investigates wood burning for heat, preservation of foods, and ceremonial and other sweating, and concludes with a discussion of the historical patterns of deforestation. I close with some observations about the continuing fascination with wood as an element with which to build buildings and to fashion works of art and function.

―――――――

Once in the history of North America boards and planks of great width were not rarities, though they were difficult to produce without heavy machinery. Huge trees of nearly every species were abundant on the landscape. Gigantic trees grew in the western part of the continent, as they still do in those tropical regions remote enough to have yet avoided the burning and logging that every day decreases their number. North American redwoods and sequoias, now for the most part protected be-

FIG. I
"The Wawona Tunnel, Tree and Surrounding Forest, Mariposa Grove, Yosemite Nat. Park, Calif." Keystone View Co., ca. 1930. The road could more easily have gone around the tree.

FIG. 2
"A monster Sequoia just felled in grove at Converse Basin, Cal." Underwood and Underwood, ca. 1910.

cause of their majestic size and their declining numbers, once provided wood so immense in size that loggers and businessmen could not wait to bring them down. Photographers recorded not only the size of the trees but the humorous (to them) antics of loggers and assorted wits who tunneled through the heartwood of a giant so that a car could drive through the tree, or who felled a tree and built a house of the wood on the stump. [FIG. 1] The self-satisfied look of a couple of tree cutters as they lounge against such a tree that they have brought to earth represented the pride in "subduing nature," of bringing the big tree down to earth and down to their size. [FIG. 2] This is not to say that cutting down one of those behemoths was easy, or free of danger, or that the cutters were somehow strange or purely exploitive. What strikes the modern eye as so odd about these images is that they suggest that those involved seemed so completely oblivious to the long-range effects of their work, even as they knew that it took centuries to grow such a specimen. By 1900 the U.S. government had a reasonable idea of the amount of treed landscape in the country, and of the extent of the old-growth forests. It was as if there were no tomorrow for wood, as if it were still something of a nuisance, as it had been for the first farmers trying to clear land for tillage. One president in the late twentieth century actually wondered if trees were not still some sort of obstacle in the path of economic progress.

We live in a different world now. The diversity of building materials and the shrinking of the earth's supply of wood raise questions about how and when we should use it. Synthetic materials extruded or pressed into complex shapes—moldings, window sash, clapboards complete with "grain," doors—have the advantage of resisting swelling and shrinkage, thus providing consistently tight joints and easy installation. It makes sense to argue that using solid wood for a surface we intend to paint may not only be obsolete but also irresponsible. On the other hand some of these new materials will never decompose, and the waste by-products of their manufacturing process may also cause further environmental degradation. Wood's "disadvantages"—rot and combustibility—

might well be seen as advantages compared with some of the drawbacks of inorganic alternatives.

As resources dwindle, wood has now accumulated other cultural values that it did not necessarily possess previously. The values we now link to the organic, the natural, the pure, and even the primitive are not necessarily those of nostalgia, of reverie, or of protest against a mechanized society. Wood was enmeshed with industrialism, but that relationship is now forgotten, washed away by the values we now associate with the steel, plastics, and synthetics that have overtaken wood. Put another way, the laboratory and the factory (or "plant," as it was once termed) have taken over the nightmarish or frightening aspects people once reserved for the deep and dark forest, also called "the woods."

We face well-documented threats to the earth's well-being that have come to pass because of the notion that the planet was some sort of cornucopia. The mathematics of forest resources and human consumption have been altered forever by technological advances in the harvesting of wood resources, a situation only partly ameliorated by the development of new materials to replace wood. But the production and disposal of many of these materials are potentially harmful to the air and water supplies humans must have. Moreover, wood still exerts a cultural attraction on many people, so the "solutions" to the problem of managing wood resources and the related problems of maintaining biodiversity, erosion control, and water resources promised by substituting manufactured substances are limited. By attempting a cultural history of wood it is my hope that the human species can use the past to find a clearer path to a balanced and more appreciative understanding of the place of this amazing material.

Into and Out of the Wood(s)

1

J ust about everyone in the world over the age of six or seven years old probably knows that wood comes from trees, although they might not be certain how the tree becomes the board and the house or the piece of furniture, or why the various species look and feel different. Wood's physical and visual characteristics and differences are broad enough to make it a material that has for centuries served people in a myriad of ways. While trees themselves are not my primary concern here, knowing something about how they grow is important to under-standing why wood for millennia was the most important material humans have used for shelter, heat, and light, or beautiful, sacred, and mundane things.

Trees grow just about anywhere on the planet where there is water, soil (and in a few places where there seems to be none), and where the annual climate is warm enough to support photosynthesis and fluid movement for enough time to complete the growth process. There are no trees in Antarctica or at elevations so high that there is either no soil or too little warmth. This is a discernible zone of no arboreal growth termed the *tree line,* above which greenery is in the form of shrubs, mosses, lichens, and other diminutive plants, some of which are too small to be seen by the naked human eye. Conversely, some species can grow in the mud at the bottom of bayous, swamps, and other places regularly or constantly inundated with water without decaying. Many of these woods, such as bald cypress (*Taxodium distichum*), also known commonly as southern, swamp, yellow, or gulf cypress, are resistant to water-induced rot and are commonly used for piers and expensive outdoor furniture. Teak (*Tectona grandis*), on the other hand, which is also used on boats and for furniture and decking, needs well-drained soil to thrive. Red maple (*Acer rubrum*), also called swamp maple because it grows in wet sites, is nonetheless intolerant of weathering, though it is more dimensionally stable than its harder and more famous relative, sugar maple (*Acer saccharum*).

Coniferous trees are in general more tolerant of harsh climates, as the great pine, spruce, and other evergreen forests of wintry climates attest. As the term *evergreen* implies, conifers in general do not lose their greenery in winter, as broadleaf, or deciduous, trees do in climates of significant seasonal change. In temperate and tropical climate zones the forests can be dense and the trees relatively large in diameter if there is a high rate of annual rainfall. In desert areas large trees are rare. In general a great wide belt of woodlands circles the central area of the earth on either side of the equator (except in desert areas such as in northern Africa), their heartwoods often of a darker and more varied hue than the relatively light heartwoods of the northern climates. There are nonetheless a few dark northern woods, black walnut and black ash being the most common in North America.

But how do trees actually live and grow? It is generally easy enough to figure out if a tree is alive. A live tree has needles throughout the year or leaves in warmer seasons. In ancient societies in areas where winter forces dormancy on the trees and strips them of their leaves, people thought the seasonally changing trees and other plants either went to sleep or died in the cold months, only to be reborn magically in spring, hence the great celebration when the snows melted and the trees leafed.

In 2001 an unusually snowy and cold winter began in southern New Hampshire when a big coastal storm dumped about eighteen inches of snow and ice in mid-November. It got cold and stayed that way until mid-April. There was no January thaw or any other premature warming that winter, and the snow kept piling up. Roofs collapsed; people got even more ornery than usual. While the deep snow protected the sensitive plants, it also protected the voles, mice, and other little creatures that burrowed beneath the snow and in the ground frozen only a couple of inches deep. One or two of those varmints ate the bark and the tender layers of wood just beneath it around the circumference of my lone sour cherry tree, girdling it. In the spring I saw the damage and knew the tree was finished. It gamely sent out leaves and a few blossoms that summer. It is now a tiny pile of sawn branches probably headed for the smoker. Those adorable fellows from *The Wind in the Willows* and *Peter Rabbit* had executed my tree.

The hungry rodents had broken the vital connection between my tree's roots and its stem and superstructure. The elementary botany most schoolchildren learn tells us that tree roots absorb moisture from the soil, and that this water and dissolved trace minerals are conducted upward to the branches and leaves through the sapwood (*xylem*) to the leaves, where photosynthesis (literally "light-making") converts the liquids to compounds essential for the tree to grow. The nutrients are then sent back to all areas of the tree through microscopic vessels in the *phloem,* a tissue in the tree's inner bark. Each year a tree adds a layer of wood to its outer edge, just beneath the bark, expanding the diameter of the trunk and its branches a tiny amount (hence the expansion cracks and other fissures in the outer bark). Technically, trees are thus classified as perennial vascular plants with the capability of adding growth to preexisting growth, as opposed to plants that send up shoots, grow, flower, and die back to their roots (*perennials*) or completely (*annuals*). This process occurs throughout the tree structure, in the stem (or, in more common terms, the trunk) as well as in the branches that make up the crown of the tree.

The microscopically thin *cambium layer* beneath the bark is the business end of the growth process. In this region the tree's cells are living (they have protoplasm), as opposed to nearly all of the rest of the cells in the stem, branches, and twigs, which are relatively hollow and in part serve to conduct moisture and sap throughout the tree. Cambium cells produce growth in the tree by dividing into two types of cells: one on the outer edge that soon begins to turn into a bark cell (phloem); the other becomes an inner cell of the sapwood (xylem). These cells produce each year's new growth at the periphery of the stem. Remove even a relatively narrow band of the cambium layer around the circumference of the tree, and lateral growth will soon stop and, like my cherry tree, the plant will die. There are, of course, some growth areas at the tips of branches that provide for the expansion of the crown upward, and thus for more leaves for photosynthesis. Most cells within a tree, therefore, only get one chance to grow. Sapwood cells transport sap throughout the tree. In the case of the maple tree, however, humans have found that they can liberate some of the sap for use as a sweetener. Maples are the most famous source of boiled sap sweetness, but other trees can be used as well. Scandinavians tap birch trees, and Finns produce a quite potable birch liqueur.

There are a few other cells that maintain their protoplasm within the tree's structure. Unlike the great bulk of the tree cells, these renegades are more or less horizontally oriented. That is, they are not cells of elongated shape whose long axis is parallel to the tree stem. These cells, called *rays* or *medial rays,* radiate from the core, or pith, of the tree stem and are visible to the naked eye as spots of different colored wood or lines radiating from the pith, especially in trees such as oak and beech. Careful or lucky sawing of a log produces *ray flecks* that seem to dance upon the surface of smoothed and polished boards, adding to the visual character and commercial value of the wood. You can make them longer, shorter, rounder, or more or less dense depending on the angle and orientation of the log at the sawmill.

When a tree is felled, the crosscut stem reveals its *annual rings,* a series of roughly concentric circles of alternating colors, light and dark. Each represents one year's growth. In trees grown in climates in which winter slows or stops growth, the inner band of light *early wood* is laid down in the spring and the outer band of *late wood* as winter approaches. The rings are the traces of the cambium layer reproductive cells doing their annual work. [FIG. 3]

FIG. 3
Enlargement of a cross-cut piece of red oak. Each combination of light (early) and dark (late) wood constitutes an annual ring. The black dots in the early wood are the ends of the pores that have been exposed by cross-cutting the wood. The light diagonal lines are medial rays.

In more temperate or tropical climates, the distinction between early and late wood is minimal in many cases and therefore the differences in color that most people think of as the grain are less distinct or even barely visible. Tropical woods such as Honduras mahogany (so called because the earliest shipments of the wood originated in Honduras, though it is native across a wide stretch of territory from Brazil northward to Mexico), and jatoba (*Hymenaea courbaril*), an orange to reddish-brown hardwood that is native to about the same area, have a more even color than do hardwoods of wintry climates. Climate, however, is but one of the factors instrumental in determining the growth-ring distinctions in color. Cell structure and arrangement, in particular the size and density of cells that conduct liquids up to the leaves and back down to growth areas in the tree, can also provide color distinctions.

As long as the wood cells are actively conducting sap, water, and nutrients, the wood remains a lighter color in most trees and is appropriately called sapwood. This lighter colored wood usually forms a band around the periphery of the tree stem, encompassing the *heartwood,* a central area in which the cells are no longer needed to conduct liquids throughout the tree. [FIG. 4] The scattered rays and the cell walls of the

FIG. 4
A cherry board showing light-colored sapwood and heartwood. At lower left is a small area of bark and the cambium layer.

disused center of the stem combine with minerals and other compounds to produce the distinctive color of the heartwood. A cherry board cut from the entire width of a tree trunk, for example, will have borders of light sapwood on its longitudinal edges and the characteristic reddish brown color between them. Variations in color differences between sapwood and heartwood are common, from the nearly indiscernible difference in white ash to the dramatic difference in the appropriately named purpleheart. Different species have characteristically different sapwood widths. Black cherry sapwood usually comprises about ten to twelve annual rings, while sweet birch and magnolia normally contain sixty to eighty rings.[1] Makers of wooden products in which strong visual quality is desirable tend to reject sapwood, preferring the more consistent and characteristic coloring of heartwood.

Hardwood and Softwood

If you press your thumbnail into a piece of white pine, it leaves a discernible indentation in the surface. Try the same act on a piece of ebony and you experience something quite different. Not only will there be no mark, but you might break your fingernail. This little test merely confirms what people throughout the world have known by experience for millennia. Woods have different physical as well as visual characteristics, and once understood, these can be put to a variety of useful purposes, from flotation to building structures meant or required to take considerable stress or load. Unfortunately and inconveniently, the terms *hardwood* and *softwood* do not accurately describe the actual physical characteristics of density and hardness, much as one might wish that they would. Instead these terms actually relate to two different ways the trees reproduce.

It's all in the seeds. Seeds without an exterior shell produce the softwood trees, which include pine, cedar, and others with foliage that resembles needles or scales. For the most part these are conifers, more commonly termed evergreens or coned trees. Pine, spruce, Douglas fir, and larch trees—all conifers—also contain resin canals, tubes larger than and parallel to the tree stem cells and which contain the sticky

FIG. 5

"Tapping Pine Trees for Turpentine," Keystone View Co., ca. 1900. Long-leaf pine, which grew abundantly from North Carolina to Florida, was the source of the sap that was distilled into turpentine, rosin, pitch, and tar.

substance so treasured by early shipbuilders. The resin is produced by cells on the inner walls of the canals. Newly sawed resinous softwoods often show droplets of the resin (mistakenly called sap) on their surfaces, as do trees gashed, slashed, or pruned. Harvesting the resin from pines, particularly southern yellow pine, has traditionally been accomplished by slashing the bark of the tree and setting a bucket under the cut to collect the dripping pitch. [FIG. 5]

Hardwoods, on the other hand, are trees whose seeds are covered with some sort of exterior casing or shell, such as acorns, hickory nuts, apples, peaches, and other such nuts or fruit. The cells of hardwoods are also quite different from those of softwoods. They are usually larger in diameter than softwood cells and arranged end to end, with little or no separation between one cell and its longitudinal neighbor. Hardwood cells tend to resemble miniature straws running parallel to the entire length of a tree stem. In some hardwoods the *pores,* or the ends of the tubes exposed by crosscutting a log, are concentrated in the early wood of the annual ring and can be quite large in comparison to other hardwood species. Dip an end of a piece of straight red oak in bubble-blowing soap and blow on the other end near the bark: bubbles will form on the soapy end.[2] Red oak and other such woods are termed *ring porous* because the tubular structure is found in the early wood. But it is not quite that straightforward. Some ring-porous woods do not have larger earlywood cells and smaller, denser late-wood cells, and the arrangement is not quite like the "straws" of red oak. White oak, for example, is unlike red oak because it forms thin obstructions to flow as the sapwood becomes heartwood. These minuscule twisted membranes rigidify during the change and block passage of liquids and air.

Other hardwoods, such as cherry and birch, have vessels of more even diameter and distribution across the yearly growth. Called *diffuseporous* woods, they hold liquid in—or keep it out—and are thus more useful for making cups, bowls, and other vessels. Presumably the ancients who first tried to make barrels or buckets out of red oak were disappointed when their creations leaked, even if their joinery was perfect, while those who by chance used white oak were rewarded if their coopering skills were sound.

Makers of dugout watercraft probably looked first for buoyancy and ease of working the wood when they embarked on the laborious task of hollowing out a large log. And they were lucky to have done so. Hollowing out an oak log would have been a nightmare of long hours hacking at an inhospitable tree stem, but then they might well have produced an elegant boat that would have drawn water from its ends, because red oak is ring porous. But softwoods such as cedar, pine, or larch, which are *nonporous* (because their cells are structurally different), or woods that are diffuse porous were successful choices. The ends of these boats would not conduct water into the vessel even though cutting had sheared off the longitudinal cells of the tree, exposing them to water.

Thus does the English language cloud rather than clarify. One might be tempted to think that a porous wood would be structurally weaker and less dense than a nonporous wood, but such is almost never the case. Hardness, flexibility, and load-bearing capabilities are associated with density (weight per unit of volume) and cell structure rather than porosity. Moreover, measuring the weight of identical volumes of hardwoods and softwoods reveals some startling results, demonstrating the complexity of understanding strength in woods. As R. Bruce Hoadley demonstrates in *Understanding Wood,* American black cherry (*Prunus serotina*), readily identified as a hardwood by those who have tried to work it, is about the same specific gravity (a comparison of a substance's density compared to that of water) as Douglas fir (*Pseudotsuga manziesii*), a softwood, and southern yellow pine is about the same specific gravity as American black walnut (*Juglans nigra*). Maybe the thumbnail test is not such an accurate one after all.

Variations in wood species are considerable. Lignum vitae (*Guaiacum officinale*), also called iron wood, is heavier than an equal volume of water (so it will not float) and is about twelve times heavier than balsa (*Ochroma pyramidale*), which certainly does float. Brazilian rosewood (*Dalbergia nigra*), one of at least a dozen varieties of a popular tropical wood native to Africa, South America, South Asia, and the Caribbean, is almost exactly the same specific gravity as water. Most well-known Western Hemisphere hardwoods weigh between 40 percent and 70 percent as dense as water.[3]

Why does this matter? Strength and load-bearing capacity are important if we want to build a house, barn, or bridge. Density, flexibility, and tensile strength matter when we are building the skeleton of an airplane or the hull of a ship, as the British found out when they launched all those cannonballs at the USS *Constitution* about two centuries ago. Oak is very difficult to work, especially after it has dried out, but I will feel better about it as the ridgepole (that long beam at the point of a roof, running from one end to the other) of my barn than I will about a beam made of Eastern white pine, which is unlikely to hold up as long. The strength of oak in large part comes from thicker-walled fibrous cells in the spaces between the thin-walled cells that conduct the sap through the tree. Some softwoods are less resistant to breakage than hardwoods since they are composed primarily of very long thin-walled fibrous cells and are often riddled with resin canals and in some cases, pockets for resins.

Growing and Felling Trees

For much of human history, people cut and used trees for a variety of purposes by harvesting them wherever they found them; they then moved on to new forests as they depleted what was near. Better and more efficient transportation methods eased this effort, but cultivation of trees, first for the foods they produced for human consumption and eventually for their lumber and for decorative and other functions (such as a windbreak or for privacy), has been around for millennia. Evidence from both archaeology and dendrochronology (measuring a tree's age by counting annual rings) suggests that humans have been growing and using trees for at least six thousand years. The Roman writer Virgil discussed silviculture approximately two thousand years ago in his *Georgics*. In North America it took fears of lumber shortages and the establishment of forest preserves in the early twentieth century before most lumber companies began to replant forests. In a few European countries and in Japan, where the forests had long since nearly disappeared, replanting schemes had been tried since the seventeenth and eighteenth centuries. In the midst of the Great Depression of the 1930s and the destructive dust and wind storms that helped create the American Dust Bowl, the federal government embarked upon a tree-planting program known as the Shelter Belt Project, in which hundreds of thousands of trees were planted in rows to provide some protection from the prevailing westerly winds across the Great Plains.

People have been cutting timber to meet these needs since they first figured out how to attach a stone or metal axe head to a handle. Harvesting trees had been a gradual process throughout most of the world until the great expansion of population that began in China in about 700 CE took down forests for agricultural land and fuel supplies. The European oceangoing exploration and imperial expansion into East Asia and Africa that began in the fifteenth century also gobbled up trees, a process that later industrialization accelerated.

Virtually all of North America, save the prairies, the Great Plains, those areas in which Native Americans had burned the woods for farming, and the arid intermountain West, was covered with dense forest when Europeans first set foot on the continent. The mass clearing of the North American continent began in earnest in the nineteenth century, a result of the confluence of several important factors.[4] First, the decline of the European imperial presence in what is now the United States unleashed Americans who had for decades presumed that the land to the west of their coastal settlements would be theirs to exploit. Second, the opposition of the indigenous inhabitants on those lands in the end proved ineffective to the European-Americans' designs. Third, changes in the technology of timber harvesting enabled this truly massive effort to succeed. The scale of what was possible in cutting and transporting wood increased as the machines and the engineering skills of the timber cutters advanced, exponentially increasing the amount of wood produced, or at least the number of trees cut. Finally, the vast size of the North American continent and its forests led people to believe that the timber supply was inexhaustible, a conclusion that suggests either limited numeracy (to put it in polite terms), a firm faith that the Divinity would replace or in some way provide for wooden needs once the trees were gone, or more perversely, that the end of cheap and high-quality wood was a problem later generations would have to solve for themselves. It was much the same in colonial India, Australia, and New Zealand as it had been on Easter Island and in the Norse settlements of Greenland and Iceland, although the conservation ethic began in India a few decades earlier than it did in North America.[5] Until the twentieth century, massive timber clearing proceeded more slowly in the tropical and semitropical rain forest areas, but in areas around ancient cities in Africa, India, South and Central America, and Southeast Asia timber clearing had been ongoing for centuries.

The qualities of thrift and stewardship—which are supposed to be central or at least important in the Protestant theology that allegedly informed the American "national character" and British colonial adventuring throughout the world—seem to have been left behind in the voracious consumption of the woods that marked the nineteenth century. This consumption ethic—rather like a drunk on a weekend bender—soon reached the rain forests of the equatorial regions of the earth, where wood now, as it had been centuries before to the north and east, is both a resource and a pest to be eradicated for the glories of agriculture. There is no doubt that farming is important—trees for the most part do not make such great fodder for humans and most animals—but there is enough cleared land extant at this moment to feed the entire world and probably has been for decades prior to the recent cutting and burning of rain forests.

One way to begin to balance the demand on wood with the ability of trees to regenerate probably was discovered by accident. Some tree stumps, such as pine, rot and die; others, such as maple, oak, birch, pin cherry, alder, and hazel soon generate several branches simultaneously (called "spring") to recover from their beheading, thus producing a new crown of leaves and small branches necessary for photosynthesis. Some species, such as black cherry and aspen, also generate new stems, but from underground, directly from the roots of the tree. Coppicing, or deliberately cutting and recutting to produce multiple new branches, involves lopping off the topmost part of a tree stem, either at ground level or at about six to nine feet above the ground (called pollarding) to produce a multitude of small new branches at the cut point. These branches, which normally begin as lateral shoots parallel to the ground, quickly turn upward to grow toward the sun, thus producing the curved wood favored for tool handles, supports for climbing plants, hurdles, fuel, binders in plaster, gates, and fencing. In some cases where the wood is very pliable when young, such growth is woven to make baskets.

The new growth is a desperate attempt by the tree to generate new crown material. The spring grows at a much more rapid rate than do branches in an untouched crown, which is usually too high off the ground to reach with any ease. It is a renewable process that can be repeated in anywhere from three to ten years, depending on the species. There are some trees in England that have been pollarded for thousands of years.[6] The disadvantages are primarily visual, for those who criticize the practice. Coppiced and pollarded woodlands either appear to be dense thickets of spindly poles or areas of gnarly stumps in need of a serious haircut.

Seeing beauty or ugliness in the woodscape has more to do with neo-classical or romantic aesthetics than with function, and critiques of the appearance of coppiced or otherwise manipulated trees are centuries old. This criticism arose in part from the social ambitions of wealthy European landowners who by the sixteenth century had set aside and replanted forests for their particular brand of hunting, which had long been a leisure pursuit of royalty. At about the same time builders in cities and of great country estates embraced trees as part of the decoration of avenues, promenades, and open spaces, making these areas more like what we think of as parks. By the eighteenth century the wilderness had caught the attention of Romantic poets, essayists, painters, and other artists who celebrated trees as "grand" and "sublime" and portrayed the wildness of the forests as both an antidote and a relief from the pressures of urban life as well as a potential clue to recapturing an allegedly "lost" state of primitive grace and well-being.[7] When the countryside ceased to be seen as merely functional—that is, as painterly aesthetics began to include carefully constructed visions of the forest as somehow Edenic and uplifting, rather than terrifying or merely dark and dense— coppicing and pollarding came to be seen as defiling what nature had created. The critique, however, was embedded in the class system of Europe, one in which most of the critics of the practice were relieved from worry about whence came their fuel and raw materials for farming.[8]

As in most instances in which one is transforming some natural material for human use, technology and technological change are critical components not only in *how* a job is accomplished but also in what sorts of jobs *can* be accomplished. In any case, felling trees is a difficult and dangerous task, but it is more difficult with a stone axe than with a metal-edged axe, and with the latter it is much more taxing than with a human-powered saw, and all are inferior to a modern gasoline or electric chain saw.

The process of harvesting trees seems simple enough. Fell the tree with axe or saw, cut the logs to the length of the fireplace or, by the eighteenth century, wood-burning stove, then split the big pieces into manageable size for the fire. Splitting also helps dry the logs, thereby making the burning more efficient. The entire process could be completed in the woods, although in many areas large draft animals—horses, oxen, elephants—hauled logs from the forest to the home. For wood as a building material, the job is more complicated. Whole logs must be removed from the woods to a mill or some other area in which they can be sawed and split more conveniently into parallel lengths of desired thickness.

Even with modern tools, dropping a tree, especially a big one, is not an easy job. Ideally a standing tree leans in a direction that provides a clue to where it will probably fall. The first cut, always less than one-half the radius, should be on the "downhill" side of the stem, opening up a V-shaped notch for the tree stem to close up as it falls. The logger then makes a larger cut on the "uphill" side of the stem, slightly above the first cut. About halfway through the stem, gravity takes over, the tree leans (the small piece uncut in the middle of the tree acting as a hinge), falls, and its accumulating velocity snaps the hinge.

This rosy—and dangerous—scenario happens most of the time. But many complications can arise. Trees usually grow in groups, and the falling tree sometimes catches in the branches of a neighbor, producing what tree cutters call a "hanger," or more ominously, a "widow-maker." A strong wind can then dislodge the felled tree from its tenuous support. Such a situation has become even more dangerous in the past century because chain saws and the heavy equipment commonly used in the woods mask the noise that the wind or a crashing tree produce.[9] Loggers who continue to work near a widow-maker never hear it coming down until it is too late.

The axe—both head and handle—was until the twentieth century the subject of continual research, tinkering, and alteration, probably from the time a tree feller first figured out that clobbering a tree for long periods of time was tiring, and that the process was faster and more efficient if the head were sharp. At some point people also discovered that the shape of the head might make a difference. In the same way, an accident of growth in a tree limb used for a handle might well have been the inspiration for curving the handle to make the swing a little easier. The job was so difficult that all sorts of people worked at trying to make it easier. After patenting laws went into effect (in England in the late seventeenth century, in the United States in 1790), dreams of wealth from the registration of an idea or improvement on an existing form spurred inventors, hucksters, and lawyers to record and fight over their innovations, much to the benefit of future historians. The U.S. patent and trademark office is loaded with nineteenth-century patents for improvements in axe heads, as it is with improvements in other laborious tasks.

As all that effort was being exerted to develop the perfect axe, in some parts of the world a parallel effort using sheet metal or iron made it possible to produce a competing tool for bringing down trees. Saws are essentially rectangular or trapezoidal sheets of metal with a handle on one or both short edges and one long side filed to produce a series of sharpened teeth. Such saws had been around since about the eighth century BCE, and the Romans and the Chinese seem to have developed a different form, the *bow saw,* at about the beginning of the Christian era in the West. Bow saws are instruments with very narrow strips of metal (less than one inch wide, for the most part) that are held in tension in an H-shaped wooden frame. The blade links the bottom of the vertical elements of the H, and a wound doubled string attached to the top of the H holds the blade in tension. To increase tension you simply insert a small length of wood between the strings, wind it up, and keep the ten-

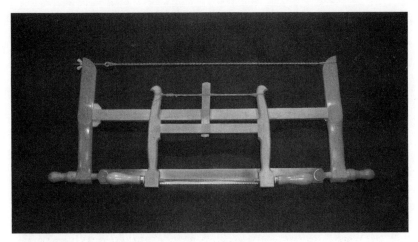

FIG. 6
Bow saws. Beech and pine. Ulmia Manufacturing Co., Sweden, ca. 1980.

sion by bracing the wood piece against the central bar of the H. The *frame saw* is a similar form, an open wooden frame with the saw blade in the middle, held in tension by wedges where the blade is attached. [FIG. 6]

Large frame saws were the best tools for sawing boards from logs, and bow saws were fine tools for cutting smaller pieces of wood, but felling the tree required something different. The improved metallurgy that was developed in many parts of the world after 1000 CE enabled saw makers to produce the wider yet still flexible blades needed for hewing big trees. With teeth properly filed, two men could make short work of a tree—even a large one. The key, of course, was in the preparation and maintenance of the saw teeth. When most people think of a saw, they envision a row of continuous little triangles with their points up. A closer look down the length of a crosscut saw blade (so named because it is designed to cut across the grain) reveals that the points are bent slightly and alternately out from the plane of the larger sheet of metal that is the entire saw blade. The points on each side cut through the cell walls of the wood more efficiently because saws so bent, or *set,* create a slightly wider channel (called a *kerf*) than the thickness of the blade and slide more easily through the piece to be cut than those without set. (Some saws for fine work have little or no set, but they are the exception.)

How efficient were these saws? Consider that many of the giant redwoods and sequoias in the western United States were sawed by hand. Trees of ten feet or more in diameter were dropped by two men with saws of what seem now to be of enormous length. What we moderns think of as enormously difficult—even miraculous—acts of labor were the stuff of daily life for many workers in the past, and hand tools of efficiency surprising to modern people were routinely developed. Finding a substitute for human power to cut wood was not the primary challenge; by the late eighteenth century lumbermen knew that the power of falling water could be harnessed to do that. But a stream or river is immobile; you have to haul the logs to it. The challenge—largely unmet until the railroads had been built up in the middle of the nineteenth century—was bringing alternative power into the woods.

By the late nineteenth and early twentieth centuries the advent of the internal combustion engine and efforts to miniaturize it eventually resulted in a gasoline-powered saw with a continuous chain of teeth. Andreas Stihl is generally credited with inventing the chain saw in the 1920s, and it was widely used by the 1940s. Succeeding incarnations of the chain saw have refined and made safer the original concept without altering its basic idea.

The nature of big lumber operations in the woods is still a mystery to ordinary people. Unlike their predecessors with axe, saw, and draft animals, industrial timber harvesters of the nineteenth century built open-air "factories" on site, using cables and winches attached to existing trees. They built long log trestles across ravines and burned "unusable" branches and logs to power the steam engines they hauled into the woods. Big operations even laid their own railway tracks into the woods, much as logging companies now build roads for trucks and skidders (vehicles that remove logs from their felling place).[10] [FIG. 7]

FIG. 7
"In the Lumber Regions of Washington—A Walking Dudley," Keystone View Co., ca. 1898. On the back of the card: "To exhaust the supply [of standing timber in Washington] would require at the present rate of cutting from 700 to 1,000 years."

What is most striking about these operations, however, is not the technology but the sheer size of the logs being sawed or moved. Until white people landed in the Western Hemisphere and in most of the other tropical areas of the world, most of the forest was still what we now call old-growth, that is, never harvested or burned in a systematic way or for profit. Nearly all trees in these environments went through their normal life course of sprouting, competing for space on the forest floor, and growing until they were attacked by predatory insects, fungi, or disease, or felled or damaged by weather. Many therefore grew to sizes with which we are not familiar—several feet in diameter—save for deliberately preserved areas or freak survivals, such as the allegedly nine-thousand-year-old yew tree in Fortingall, Scotland, the kauri trees of northern New Zealand, or the giant sequoias in the American West.

The demand for lumber in the industrializing world was enormous, and the productivity of the hitherto untouched forests made cheap lumber available for the houses, factories, and all other sorts of buildings and uses to which wood could be put. What many thought would take centuries to remove and use up disappeared within decades. In the United States the progress of the great clearing mowed down the pine forests of the Great Lakes states in the nineteenth century, the hardwood forests of the Appalachian Mountains, the yellow pine and cypress woods of the South, and then turned to the great coniferous forests of the far western states. Canadians took slightly longer to saw through their northern and western woods, probably because there were fewer people there and the north woods were in a more forbidding climate. In India the British and Indian lumbermen and colonial bureaucrats first began tentative steps toward management of timber resources in the mid-nineteenth century. In the United States the first real steps in that direction came from the hunter-conservationist president, Theodore Roosevelt, who set aside millions of western forest lands as national forests by executive decree on his last full day in office in 1909, much to the chagrin of the Congressional representatives of the western states in which the woodlands were situated.

Today our demand for cheap lumber is unabated, but sawn wood is available in forms that would have made nineteenth-century lumbermen laugh. Consumers now can obtain their cheap wood in the form of composite materials, made of chips and glue or sheets of veneers (also glued together) or "real" wood pieces joined together with glued sawtooth joints. Quickly grown softwoods such as pine or spruce are the most common fare in the lumberyard, usually loaded with small to medium-sized knots, the remains of little branches whacked off before they could grow in diameter. "Clear" boards—devoid of knots (or nearly so; the standards keep "evolving")—are dear, and a near treasure if visually unusual. Hardwoods are a small part of today's lumber trade.

Transporting Logs

A tree felled in a forest does not do much good for a builder in Tokyo, Calcutta, London, or Cairo. Getting logs out of the woods and someplace to mill them was at least as big a job as getting the tree down. Early wood harvesters used their domesticated animals to move logs to a running body of water; from there they were floated to mills often powered by the same current. [FIG. 8] Managing that journey of logs downstream was a tricky and dangerous business at best. Logrolling and log walking

FIG. 8
*"Great Chained Log Rafts...
On the Columbia River, Washington," Keystone View Co.,
ca. 1900. On the back of the
card: "For days and days these
huge rafts glide down the Columbia River towards the Pacific.... There are great forests
of redwood, fir, cedar, larch,
spruce and hemlock. Some of the
trees are easily 300 feet high and
30 feet in circumference."*

are in contemporary times by and large limited to lumbering athletic
contests, but in the heyday of hand felling and limbing they were skills
necessary for the dangerous job of guiding (or trying to guide) big heavy
timbers. The water was often dangerously cold, and the logs were so
heavy that two crashing or merely pushing against the human body
would crush the life out of a worker. Rivers and streams did not always
behave as lumber workers wanted them to, and logs sometimes piled
upon one another or arranged themselves in ways that jammed up the
transit downstream. In those cases workers often had to scamper from
log to log to try to lever the jam apart, hoping that the breakup would
not take them or one of their limbs with it.

Armed with chain saws, tree cutters in commercial operations today
ride off-road vehicles into the woods, in the company of gasoline- or
diesel-powered vehicles to bring the logs out once they are ready for the
mill. In some cases machines have been developed that grasp a tree stem,
slice it across grain with a large saw blade, lift the sawn trunk up, trim
it of branches, and place it onto a flatbed trailer, which is then pulled out
of the forest on logging roads.

Of course it saves a lot of effort and time if you don't have to saw logs
into boards. This was an obvious benefit of a log building, but you still

had to cut notches at the corners to make the structure stand up and then fill the spaces between the logs with mud or some other *chinking*. Both were tedious, but it beat swinging the broadaxe while standing beside or atop the log in order to produce square beams. Even in houses of the well-to-do, to whom saving labor or capital was not necessarily a concern, but to whom refinement mattered, unsquared and barked logs were used as rafters, under the roof and out of sight of those in the re- fined areas such as the parlor or dining room. But whatever their status as cultural icons, log cabins were coarse, and most people abandoned or disguised them as quickly as possible.

More "artificial" cladding—clapboards and shingles—replaced or covered logs on the outside of finer houses. These were usually riven, or split, from logs with straight grain, rather than sawn. The tool for making shingles is a *froe*. It has a thin blade eight to ten inches long, with a handle set at ninety degrees to it. You hold the handle in one hand and drive the blade into the top of a short piece of a log; this should crack it along the grain. Then you twist the handle sideways, and a shingle pops off. Clapboards were longer and required several thin wedges to be driven along the length of a long, small-diameter straight log. Once split off, they could be smoothed using a two-handled tool called a *drawknife,* so named because you used it by drawing it toward you.

Many modern occupants of older houses find comfort or aesthetic reward in exposing the beams and upper floor joists of their houses, thereby showing what they think are tool marks. This would amuse the original dwellers of the house since the rough-hewn marks of the broad- axe were meant to be hidden, given that they were coarse, unlike the refined surfaces of smoothed interior trim wood and furniture. Smooth- ness was a mark of status, manners, wealth, and civilization; refined manners, refined sugar, smooth furniture, and fine textiles were desir- able alternatives to roughness in manners, goods, food, and clothing. Silk was better than skins or homespun, though the latter form of cloth- ing was endowed with the ideological qualities of "honest virtue" by people all over the world in times of political turmoil and resistance.

The "leather shirts" of backwoods Virginia virtue and the "homespun" citizenry of the American Revolution have their counterparts in Gandhi's choice of simple clothing as well as that of nationalists in Africa who sought political independence and legitimacy by wearing traditional clothing as their uniform. Romantic nationalism appears often among the world's people, often celebrating nostalgically the very material goods or conditions in which vaunted ancestors placed little or no conscious value.[11]

Milling

Before the advent of powered mills, sawyers cut logs into planks, either with a long-bladed saw held fast in a wooden frame or with a long two-handled saw that resembled a felling saw. This was a lengthy, laborious

FIG. 9
Pit sawing, Bhutan. Photograph by Yoshio Komatsu, in Athena Steen, et. al., Built by Hand. *Courtesy of Yoshio Komatsu.*

FIG. 10

Animal-powered mill. Åland, Finland, ca. 1850–80. A draft animal—horse or ox, usually—walked in a circle around the central post, the toothed large gear wheel engaged a smaller gear wheel above, turning the shaft that carried circular motion into the barn or mill. Here the wheels turned grinding stones to make flour, or drove machines connected by leather belts.

process that involved two workers, one standing above or atop the log and the other below it, either in a pit or on the ground below the elevated log and top man. Obviously the latter had the better job. In remote or impoverished parts of the world, pit saws and large bowsaws are still in use. [FIG. 9]

Powered sawmills date at least as far back as the early fourteenth century in Europe. The first such operations simply replaced the pitman's muscles with animal or water power, without changing the up-and-down action of the blade. Horses or other large domesticated animals generated power by walking around a shaft to which they were tethered. Toothed gear wheels transferred power to the mill or workshop by meshing with smaller wheels and shafts. [FIG. 10] Large wooden wheels harnessed the power of flowing water by transferring their circular mo-

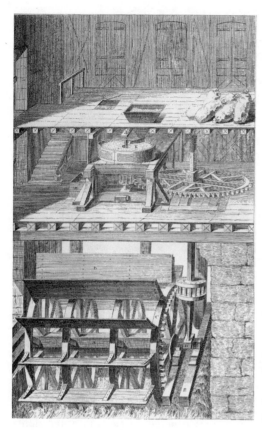

FIG. 11
"Moulin à Eau" (Water mill). Plate 6, volume 1 of Denis Diderot, L'Encyclopédie *(1751). Water flowing into the mill at the bottom of the engraving sets the apparatus in motion.*

tion to the wheel shaft, which in turn transformed that motion into up-and-down action by means of a cam—a teardrop-shaped device bored through at its center that was slipped over and attached to the shaft. When the shaft turned, the uneven shape of the cam produced a repeating up-and-down motion in the vertical arm against which the cam slid. When this arm is attached to a saw blade, the water and its machine will do the work. [FIG. 11] Additional gear assemblies were developed to advance the log on its bed, sliding it into the blade at a consistent rate. Eventually inventors figured out that they could save energy and complication in the mill if they simply took the circular motion as a given and changed the nature of the saw blade from a rectangle to a circle. The in-

vention of the circular saw blade, often credited to the American Shakers, transformed large-scale sawmilling. In the twentieth century, small electric motors and circular saw blades brought the sawmill to the home workshop and the job site. Later mills used steam engines to do the work and resembled the factories that made other consumer goods—except many were in the middle of the woods. [FIG. 12]

Modern sawmills differ from their nineteenth-century antecedents mainly in the methods they use to move the logs, in the diversity of lumber products they produce, and most significantly, in the accuracy of cutting and the uniform sizing of the boards produced. Computer-controlled saws and movement of the goods throughout the mill limit waste and heavy and dangerous labor, but are capital-intensive and in some cases do not allow for the exceptional piece of wood to be separately milled. Indeed, such variation in logs is often regarded as a problem in some automated mills.

Once a log is at the mill, the axis or axes on which it is sawed determine the sort of lumber to be produced. If we think of a log as a series of cylinders (one for each year's growth) to be transformed from the round into the rectangular, we can see that the positioning of the log relative to

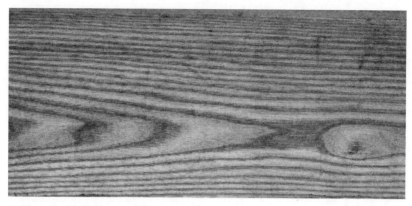

FIG. 13
Typical grain orientation of a flat-sawn board. The oval shape at the center is at the point where the saw intersected the annual ring tangentially. This board came from a cut near the edge of the tree.

the saw blade determines what patterns we will see in the sawn board. For example, if a saw blade cuts through the pith, the log will fall in two halves. When we look at the sawn surface, we see roughly parallel lines running along its length, revealing the width of the annual rings. Saw one of the halves again, parallel to the initial cut but a couple of inches toward the outside of the log (and away from the flat edge, thus making a two-inch-thick board), and the newly cut surface shows a greater distance between rings, especially in the middle of the board. As we move further from the center of the log, we gradually cut the annual rings more and more tangentially, slicing through more and more new wood and showing what appears to be more distance between the rings. These "flat-sawn" boards exhibit oval or flattened diamond patterns and arcs of annual rings on the ends of the boards. [FIG. 13]

To obtain nothing but boards like those first obtained from either half of the log, a miller quarters the log along its length and then saws narrower boards perpendicular to the growth rings. Quartersawing produces about the same number of square feet of boards as that produced by flatsawing, but the sacrifice is in both the number of cuts to be

FIG. 14
Coffer. English or American oak, ca. 1700. The wavy lines are where the saw intersected with the medial rays of the oak tree.

made (two to three times as many) and in the width of the boards produced. The advantage in quartersawing some woods is the often dramatic visual patterning that is revealed, usually in the form of ray flecks. Since the rays radiate out from the pith center, cutting in quarters intersects these rays at oblique angles and in unpredictable ways. In woods such as oak, beech, and sycamore, the flecks can be large and complex. [FIG. 14] Furniture of the European and American Arts and Crafts Movement of around 1900 made use of quartersawn oak in particular. This was a decorative decision consistent with the movement's goal of reducing machine-made ornament and emphasizing naturally occurring visual interest.

Hardness, Softness, and Strength

We have seen that the terms *hardwood* and *softwood* do not necessarily correlate with density, weight, or durability. But intuitively—or sensually—we know by touch that red or white oak is harder and more difficult to saw, shape, and smooth than are cedar or white pine, and that Brazilian cherry, a heavy and deeply toned reddish brown wood, is much heavier and a lot harder to work than are balsa or spruce.

The "feel" of a wood species does not help us much when we want or need to know about its other physical properties. Hoadley identifies four different ways to measure the strength of various woods commonly used in the United States. *Compression* refers to the ability of wood to withstand pressure applied to opposite ends of a piece, pushing inward. *Tension* is the measurement of the reverse application of force, the pressure to pull the wood apart from each end. *Shear* measures wood's strength when we try to separate it along its long axis. By applying force on half the board's width from one direction while pushing in the other direction from the opposite half of the other end, we can break the board lengthwise. Finally, he measures *static bending,* or the amount of deviation from straightness a board shows as you push in one direction from each end while exerting force in the other direction at the center of the board. (In more mundane terms, this is akin to two people holding a plank, one at each end. Someone sits in the middle of the board on the top. We measure the deviation—the bend—and increase the weight until we achieve breakage, usually in the form of splintering or snapping in two pieces.) Woods that can take a high degree of static bending without breaking—such as maple—are favored for flooring in which a certain amount of spring is desired, as in basketball courts or dance floors.

Hoadley's table of compression, tension, shear, and static bending of thirty common hardwoods and twenty-four common softwoods illuminates some extreme results and strengths but does little to correlate with what our thumbnail seems to tell us.[12] Hardwoods do not endure compression parallel to the grain much better than do softwoods, and a few softwoods, western larch (*Larix occidentalis*), for example, bear up under the strain better than do red or white oak. But almost all the hardwoods bear up under compression perpendicular to the grain better than do the softwoods. Maybe it has to do with the cells. Recall that softwoods have a different type of cell in the stem—long and fibrous, often with thinner walls and with less fibrous material between them—while hardwoods, porous though they may be, are more rigid, in part because their cells are of varying sizes and many are smaller in diameter than softwood cells.

Hoadley shows that white ash (*Fraxinus americana*) withstands considerable compression parallel to the grain—more than all but four other woods—and only four woods withstand more compression perpendicular to the grain. Similar comparisons are demonstrated with regard to tension, shear, and static bending, although the woods that exceed these qualities in ash are not the same in every category of measurement. With its light weight and relatively great strength, white ash emerges as a wood of great utility overall, making it ideal for striking materials (such as a baseball), especially if it is done perpendicular to the grain. Hence the child's first instruction (in the wooden bat era) was to point the label of the bat toward the sky so that the proper baseball swing meets the ball with a force perpendicular to the grain, since the bat companies usually placed the label on the flat-sawn part of the stick.[13] Shagbark hickory (*Carya ovata*) endures static bending well beyond the point that all other woods listed fail. It also outdoes most of the rest, save black locust (*Robinia pseudoacacia*), in withstanding compression, tension, and shear. Hence the use of that wood as a shaft for early golf clubs, which required both "whip" and strength, especially after rubber and harder materials were used to manufacture golf balls.

Test results for compression perpendicular to the grain do, however, get us somewhere closer to our thumbnail mystery, which so far has been unsolved by several scientific results. Only basswood (*Tilia americana*) and red alder (*Alnus rubra*) of the hardwoods (because they have covered seeds) have lower resistance to this sort of compression than do any of the softwoods. When you press your thumbnail on a board it is usually the application of compression perpendicular to the grain direction and you leave a mark. This is no surprise because our thumbnail is in fact a small edge that exerts considerably more pressure per square inch than if we applied the same force with the palm of our hand. Stiletto heels may look sexy but they are murder on a pine floor, since a woman's entire weight is concentrated on two tiny heels and the balls of

FIG. 15
Bird's-eye maple, curly maple, and lacewood.

her feet. The latter leave no mark because the toe of the shoe is relatively large, thereby distributing weight; the former do so because of their small area. Try writing with a ballpoint pen on a pine table without a pad and you will leave the text engraved on the table surface as well as inked on the paper.

Figure, Grain, Color

People confuse *grain* with *figure* all the time. Whereas the former refers to the pattern of the rings as revealed by sawing, the latter refers to oddities in growth and arrangement that produce curly, bird's-eye, blister, quilted, crotch, flame, feather, and other such aptly named formations in the wood.[14] [FIG. 15] Cabinetmakers who carefully select and saw their lumber can take advantage of visually arresting or comforting patterns in the annual rings of the trees or in the variant forms uneven and twisted tree growth can produce. A discerning client wanting a new chest of drawers or a desk might well be interested in the aesthetics of wood and in the opportunity to own a distinctive and showy piece. Thus both a cabinetmaker and client would probably reject pine, since it shows little difference between the new and the late wood in its annual rings, while cherry, mahogany, and oak show great differences in color and grain.

Figured wood is often the result of events in the life of a tree that change its annual-ring orientation. For example, when a tree goes about producing branches for its leaves, flowers, fruits, nuts, or seeds, the new growth produces strain and compression in the rings where the branching takes place. Rings at the crotch of the tree are often wavy, and when sawed or sliced show a feathery or wavelike effect. Tree stems that are bent in their growth patterns as they compete for sun or compensate for an uneven anchorage in the ground likewise show distorted ring patterns on the curve, a condition that can produce complex figure. Sliced across its length, a bent tree will show an uneven pattern in the growth rings, rather like ovals instead of circles. In softwoods the outer (lower) side of the curve is compressed and will tend to warp or separate when dried. In hardwoods leaning or bent trees form *tension* wood on the inside, or upper side, of the curve. In addition to having a greater tendency toward warping, these woods are harder to finish smoothly. Tension wood, however, is stronger than hardwood without bend and thus was a double bonus to shipbuilders, who took advantage of the shape of the curved tree for use in the curved sections of a hull. Since a silken finish was not necessarily an advantage on a watercraft, the furniture maker's annoyance was the boat builder's bounty.

Twisting growth in a tree stem, a fungal or other external attack, failed branch growth, or a multitude of other factors can produce uneven, wiggling growth in annual rings. When sawed, this appears to be striping in the grain—as in the curly maple I found in my firewood pile. Sometimes the cambium layer of a tree will grow in alternating swirls that interlock or with small pinwheel-like swirls, which produce bird's-eye figure. Tiny raised areas in the cambium layer can produce blister, or quilted, figure, if the bumps are elongated. Wavelike wiggles in long cells in hardwood, when cut radially from the log, produce curly, or tiger, figure.[15]

Figure was important to the wealthy and powerful because it was both unusual—and therefore more valuable to those seeking the status of rare things (what the wealthy called *virtu*)—and because it was a visual pun or trick, a clever demonstration of wit that the ordinary folk presumably did not or could not appreciate, if they ever got the chance to see it. Satinwood, for example, was so called because in the right light, especially that of candles, its polished surface looked like satin. Lacewood's dense flecks and ovoid figural shapes gave it the appearance of finely worked textiles, and crotch figuring often looked like feathers or flames. Figure was akin to patina, the evidence of age and old money that was so treasured by the rich and the knowing. Figure with the darkness and dust of age was probably the best of both worlds.

Woods get the characteristic colors of their species from the chemical composition of their heartwood. Genetic differences of the species determine the characteristic colors of heartwood and sapwood. Variations within a given species may also be a factor, as may the trace elements in the soil of a particular habitat. Light and air—especially oxygen—can alter the appearance of heartwood. American cherry, for example, darkens considerably and relatively rapidly when exposed to light. When freshly cut, cherry is a light yellowish brown. In a year the difference between exposed and covered cherry is dramatic. If left exposed to light it will eventually become dark reddish brown. Jatobá, a dense South American wood, is orange-red when first cut, but quickly turns a dark red brown. Mildly colored pine, fir, cedar, and other coniferous woods become gray when exposed to light and the elements and are oxidized to brown in a dry environment. In fact most woods oxidize over time, but the change is often little noticed until part of a piece covered for years is exposed and compared to the adjoining area that has darkened. Connoisseurs of antique furniture will look for the differences in oxidation for verification of the age of a piece, since that level of subtlety is seldom approached by forgers and absent from pieces originally produced as reproductions but later passed on as antique.

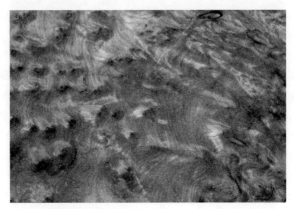

FIG. 16

A slice of burl. The swirling lines indicate that wood has grown around some disturbance in the wood—insects or failed branches, for example.

A tree's attempt to fend off fungi or other enemies often has the effect of transforming a relatively straight-grained wood into an oddity with great visual appeal. Insect infestation and the tree's defenses against it may leave irregular swirls in grain and discoloration where the insects have bored into it. Some beetles and ants will eat their way through parts of the tree stem, and their holes and tunnels seen as decorative, as in wormy chestnut. *Spalting* occurs when fungi, whose characteristic trail of degradation is a white rot, attack wood. As the decay proceeds, black or very dark brown lines border pockets of white rot. If found in time and dried, such wood often has attractive patterns, often with organic forms delineated by the dark borderlines. Burls are the bulbous growths found on the bark of trees, and they are usually filled with a multitude of swirled pinpoints of stunted or failed growth tips. [FIG. 16] Difficult to manipulate, burls were favored for food preparation (especially chopping bowls), tobacco pipe bowls (their ignition point quite higher than tobacco's), and as decorative thin sheets of complex figured material on small boxes and paneled furniture. The number of pieces of burl needed to cover a space and the difficulty in cutting, flattening, matching, and gluing up slices of burl made the goods expensive and also made for bold design and decoration.

Aromatics and Medicinals

People also value some wood for other qualities, such as aroma or their allegedly positive effects on human health and comfort. In Japan, China, and southward to Australia, camphor wood (*Cinnamomum camphora*) is a lightweight and soft yellow-brown to reddish brown wood best known for its aroma and use in chests and trunks in which textiles (and corpses) were placed for storage. It was also the source of a powerful preservative obtained by distillation, and camphor oil and camphor wood boxes were prized possessions among both Asian and Western consumers from the beginnings of the China trade in the eighteenth century through the early twentieth century, when synthetic camphor oil was developed.

In the West, some varieties of cedar found favor because of their aromatic qualities. Spanish cedar (*Cedrela odorata*) was used for cigar boxes. Some forms of North American eastern red cedar (*Juniperus virginiana*) are aromatic enough to resist moth infestation and thus provide safe storage for woolen materials. Distillations of pau rosa (*Aniba rosaeodora*), a yellowish wood native to Brazil, have been used in the perfume industry for centuries, while the sap of the North American sweet birch (*Betula lenta*) has been used in rural areas to produce a fermented birch beer. Early settlers from Europe made use of North American dogwood bark (*Cornus florida*) to make a concoction thought to reduce fevers. The balsamo tree (*Myroxylon balsamum*), a native of southern Mexico and the Pacific coast of South America, was one of the most prized aromatic woods. Native peoples used the tree's resin as a medicine for centuries before European contact, and it later found great favor first in Spain and then in the rest of Europe and North America, where people used it liberally, thinking it a near panacea. It is still used in salves and syrups and in the perfume industry.[16] This species is distinct from the North American balsam fir (*Abies balsamea*) tree, the aroma of which in the nineteenth century was thought beneficial in fighting or preventing tuberculosis. Various patent medicine hucksters promoted syrups that boasted of the salubrious effects of the balsam fir in a bottle, but their promises were for the most part hollow, if not utterly deceitful.[17]

The medicinal and aromatic qualities of wood are largely marketed in the form of distillates, or the liquids obtained from boiling the wood and then condensing the evaporated compounds in a cooled environment. Teas and other drinks can be made from the bark or roots of certain trees, such as sassafras (*Sassafras albidum*), a shrub or tree found in the eastern half of North America from Ontario to Texas, and were popular for about a century after 1850. On the whole, however, wood has found little medicinal use when compared to herbs and other plants.

Dangers of Wood

For all of its positive natural and organic associations, working with wood can be dangerous, even if we exclude the terrors of power-tool accidents. Recent research into the long-term effects of inhaling wood dust has made wearing a dust mask a requirement in most commercial operations. Moreover, pulmonary congestion is not the only threat with which to contend. Some species of trees contain toxic substances, in most cases in the bark and leaves, but in others the lumber itself is dangerous. Trace elements of strychnine, atropine, salicylic acid, protocyanide compounds, and other substances in the wood or sap of a tree make some species toxic, especially if heat (perhaps from a saw or confinement in an attic storage area) is present, since compounds can volatilize in such conditions. Spalted wood can be harmful in certain circumstances because of the chemical reactions that accompany the fungal degradation that causes its appearance. Allergies vary widely among those affected, but wood in some ways is no different than any other plant, though its negative effects may be more subtle or less investigated than more common afflictions, such as sensitivity to poison ivy. The perfume that envelops a woodshop and makes working with the material such a pleasure may also include irritants and allergens. Rosewood and walnut are special culprits in this regard.

Woods that have been chemically treated to resist rot and insects can present additional dangers. For years creosote, a tar derivative, and some oil-derived substances were used to treat evergreen woods used for utility poles, fence posts, and anything else that was going to be sunk into the ground or set in water. It would have been not only impractical but also downright stupid to install such infrastructure with a material that was going to decompose within a few years. Thus chemistry staved off rot, but not without potential environmental hazards we are only beginning to understand.

In recent years chromated copper arsenate (CCA), applied under pressure to coniferous woods, replaced creosote, and this "pressure-treated" lumber indeed withstands exposure to and contact with water or damp soils and repels insect infestation, but the chemicals used to treat it are toxic. The extent to which these agents have transferred to hands or leached into soils—and perhaps threatened fruits and vegetables—is a subject of continuing study and debate, and manufacturers and critics of the stuff urge people to wear gloves and use an efficient mask when working with it. In some parts of the world discarded CCA-impregnated lumber is considered toxic waste. These concerns have in turn led to the advent of wood preserved with alkaline copper quaternary (ACQ), which contains no U.S. Environmental Protection Agency–listed carcinogens. Since January 2004 ACQ has been mandated in the United States in many applications.[18]

Poisons and irritants are not the only problems with wood. Among the greatest problems with using wood for structural applications are those that result from attack by insects and lower-order organisms that chew or eat it. Carpenter ants, termites, and certain bees and wasps ingest wood fibers, sometimes for food, sometimes just to produce their own building materials. Trees and houses are often infested with ants that tunnel throughout the tree stem or house beam as they establish their home in yours. Their enormous colonies can gnaw miles of tunnels in a house. Termites will eat you out of house and home—literally. Powder-post and other beetles lay their eggs in the ends of pores or other tiny crevices in porous woods such as oak and chestnut. Once hatched the larva eat their way through wood looking for the nutrients in the living or recently created cells, leaving a trail of fine wood dust in their wake on the basement or barn floor.

Fungi are the real culprits in most cases of wood decay. While some species are resistant to fungal attack (cedar, black cherry, black locust, and some of the oak varieties, for example), many others, both hardwoods and softwoods, are not. Hickory, as hard as it is, is not a match for these little, lower forms of life, as long as there is water around in some form. Pine is a little more resistant, but anyone who has left a pine board on the ground for long knows that is not saying much. With ample supplies of both water and air, and moderate temperatures (75°F–90°F) wood-attacking fungi will make a feast of the minimally resistant woods. If it is cold enough (below 40°F) or hot enough (more than 105°F), fungi have little effect. This, one might suppose, is one of the few comforts of working in wood in Antarctica, Spitzbergen, Greenland, or Death Valley. In temperate climates most fungi do not die when conditions change but merely become dormant. When wood dries out, the decay stops, a condition many mistakenly think of as dry rot, as if it were a special condition in which decay occurs in dry environments. In fact it is only suspended decay, stopped because of unavailable moisture. When the water returns or temperatures moderate, the decaying action resumes.

When Swedish explorers in Stockholm harbor finally found the sunken ship *Vasa* in the late twentieth century, many were surprised that the ship was intact, largely unchanged from its condition when it had sunk shortly after its launch in 1628. The reason for the preservation was that there was not nearly enough oxygen in the water at the bottom of the harbor to enable the various fungal attackers to destroy the substance of the ship. But bringing it to the surface—and hence to an oxygen-rich environment—would have led to almost immediate collapse of the waterlogged wooden fabric of the craft. Shrouding the ship in a tent and keeping the air supersaturated while workers chemically treated the wood allowed for stabilization of the structure and hence the preservation of the failed flagship of the Swedish king's fleet.

Some woods also have properties that make them problematical for certain uses, even as they are an advantage for others. Resinous trees, such as pitch pines and other conifers, were considered extremely valuable for the substances that could be tapped or otherwise extracted from them for naval uses—tars, pitches, turpentine, and other goods important for shipbuilding. But the gummy stuff so prized by the empires built on wooden ships also collects on power saw blades and drips on areas one might rather keep clear, quite literally "gumming up the works" of some machines. Lignum vitae, whose high resin content makes it ideal for some industrial uses, was less than ideal as the material for a shuttle in early looms. Its hardness would have been a plus, but its resin would not have been so useful if it stained the textiles. Beech, a somewhat less hard wood that could be smoothed to a waxy finish, served as well, and without the resin problems. A hardwood that did not take a smooth finish would also have been problematical in the loom business, since anything but a smooth surface might well have snagged fibers and fouled the finished product.

Wood on the Move

"Wood moves." Hang around a group of woodworkers for any length of time and you will hear this pithy statement. It is not meant existentially—it literally encapsulates one of the woodworker's persistent challenges: coping with the effects of water on wood. Even if it is relatively uniform, swelling and shrinking are conditions with which furniture makers, carpenters, and anyone else living with solid wood have to reckon. (Plywood moves to some extent, but not nearly as much, since its cross-banding structure—thin sheets glued together so that the grain is at a ninety-degree angle to adjoining layers—tends to balance the movement in all directions.)

Trees, like all plants, must have water to live. But water becomes a problem when using wood for human purposes. It takes up space in a tree, and when the tree ceases to be an active plant, it usually will evaporate, but unevenly. As a piece of wood or an entire tree stem dries, stresses and strains brought about by irregularities in the grain will change the shape of the wood.

Wood swells in humid atmospheric conditions and shrinks when the air dries out. Residential heating in winter exacerbates the problem since it generally lowers humidity. Furniture and architectural elements in older, less comfortable eras and places were not subjected to extremes of heating and cooling because they were for the most part sufficiently distant from the radiated heat of inefficient localized fires.[19] The rate of change in humidity, and therefore the swelling and shrinking of wood, was moderate and gradual in these conditions.

Most people know this as *warping* and use the term to encompass the various distortions of what was once a straight piece of wood. Shrinking and swelling, the results of drying or rehydrating, do not occur at the same rate in all parts of a board. Wood is not strictly uniform in density or in grain throughout; it is a natural substance whose physical characteristics vary with its growing conditions and environment. In addition, the manner in which a tree is cut in relation to its annual rings can affect movement caused by moisture-content changes. The term *warp* in fact encompasses a panoply of distortions from original dimensions. *Cupping* occurs across the short, end-grain side of a board, so that looking down its length reveals a curve much like a supine shallow parenthesis. A similar bend over the length of a board (and along the grain) is termed a *bow*. Often bowing and cupping are the result of differing rates of shrinkage across the radial or tangential planes of a board. A flat-sawn board that includes the pith will shrink in such a way that the outer edges cup away from the pith, which is drying tangentially while the outside edges dry radially. *Twist* occurs diagonally along the length and breadth of a board, so that one corner is higher than the other three. A *crook* might be described as cupping along the length of a board; if you stand the board on its narrow edge both ends are higher than the middle. A *kinked* board has a localized crook to it, usually the result of a knot.

Knots can be attractive elements in the design effect of a board, but for the most part there is a reason "clear" woods are favored for many wood uses. Technically a knot is the trace that a branch has left on the tree stem. Early branches in a tree's life often grow for a time, die off as they are overshadowed by branches in the tree crown, and are gradually grown over by the cambium layer. Such knots are termed *encased,* or *loose,* knots. When the tree is cut and milled into boards, these knots often drop out of the lumber. *Tight,* or *red,* knots were still alive when the tree was cut and were intergrown with the tree's annual rings. They tend not to fall out of the board after it is milled. Depending on the axis on which the tree was cut relative to the knot, its appearance can be anything from a nearly perfect circle to a flattened oval in appearance. In any case, knots are regarded as defects in the commercial lumber business, with boards graded with regard to the number and tightness of the knots within a set length of board. Pieces with loose and numerous knots are judged poorer grade lumber and often subject to greater irregularity in dimensional distortion. Knots are also a problem because they tend to be difficult to cover with paints and finishes and are often hard to smooth with ordinary tools.

In most contemporary residences wood is still a major building component, especially in interior support structures such as beams and wall framing and in such details as cabinet frames, drawers, and doors. Designers keep trying to get consumers to put stainless steel or enameled or plastic surfaces everywhere, but usually fail to do so because the modern stuff looks as fashionable or "homey" as a television series from the 1970s. So consumers put up with or only quietly grumble about doors and drawers that stick when it rains or gets humid, and they believe builders when they are told that the house will "settle," as if it were in some way animate, when it is often merely twisting and cupping that causes walls to crack and floors to creak. Smart carpenters and furniture

FIG. 17
Curly cherry plywood panel. On this one-quarter-inch-thick panel, a succession of thin veneer slices has been book-matched to produce a herringbone pattern.

makers know when to leave small spaces for expansion or to fit joints more closely if conditions are humid. Novices, if they are observant, learn when and where to make the joints tight and when to leave some slack. To contend and manage wood movement is to understand not merely how to measure and cut but also how to calculate the action. Builders who scoff at physics and engineering and boast of their "on-the-job" training are better left to the netherworld of anti-intellectualism that bedevils those cultures that erroneously—sometimes disastrously—scorn the book and the classroom, as if what takes place therein has no bearing on what takes place outside of it.

Veneering, or Covering the Plain with the Fancy

The first veneers—thin sheets of fancy grained or figured wood—were manufactured by carefully and slowly sawing rare logs with fine-toothed saws. It took skilled craftsmen many hours to cut uniform sheets as thin as one-eighth of an inch. Cabinetmakers often used them in the order in which they were sawn from the log, matching the edges to produce symmetrical patterns that closely approximated mirror images. [FIG. 17]

The prized figured veneer log is called a *flitch,* and the veneer slices a *book;* consecutive sheets positioned as mirror images are termed *book-matched.*

The Industrial Revolution and the plethora of woodworking machines that were developed in the early nineteenth century changed the economic calculus of veneering. Initially water and other power sources were employed to drive up-and-down saw blades set to cut thin pieces of fancy woods. Not long after these powered saws were developed, inventors and engineers came up with huge heavyweight machines that passed a large and extremely sharp blade straight down (or up) against a log that was tightly secured. Off would come even thinner sheets (as thin as a hundredth of an inch today) cut like microthin boards from the log. When slicing the veneer of a desirable formation such as that found in the crotch of a tree, one kept the log in one position and produced a "book" with gradually increasing and then decreasing widths of the figure, as the knife moved from the narrow outside edge of the formation through the wider middle and back to the other narrow edge. In other cases millers adjusted the orientation of the log periodically to reveal the most complex figure.

The other method developed for producing thin sheets of wood involved a radically different interpretation of the finished product. Enormous lathelike machines were developed in order to rotate the log against a spring-loaded knife blade, producing (ideally) one continuous sheet of thin wood of a uniform thickness. If there were trees with annual rings of uniform width running parallel to the center of the tree, then conceivably the sheets produced by such machines would have shown little or no figure, only evidence of medial rays, resin canals, or other distortions from absolute straightness. But trees are never so, and thus the rotation-sliced veneers had grain patterning reflecting the times in each revolution that the knife cut through minor fluctuations in thickness or straightness of rings. This type of veneer is most commonly used in most modern plywoods, which require large expanses of unbroken wood to be cost-effective.

This technology may well have come from parallel technological developments in food production. The United States Patent and Trademark Office records of the nineteenth century show a multitude of successful applications for labor-saving devices in the home. Among the most numerous are those for apple parers and corers. The basic technology for the hand-crank apple parer and corer remains the same in the early twenty-first century as it was in the 1850s. The apple is impaled on one end, the crank turned, and a spring-loaded knife blade slices off the skin in a uniform thickness, the spring action keeping the blade uniformly close to the surface being removed. In the case of the apple it is the skin that is discarded; in the veneering machine it is the skin that is saved. Indeed, the whole log—less the bark—is transformed into the desirable "skin." Technological innovation often works like that. In the papermaking industry, the expensive method of producing paper in individual sheets by pressing and drying a mash of boiled rags and eventually wood pulp in frames gave way in the early nineteenth century to the Fourdrinier machine, which produced a continuous roll of writing stock on a belt that ran through a vat of the mash. The veneering lathe cutter was the reverse of this idea, in large part. Similarly, the moving assembly line, which most people credit to Henry Ford's inventive mind, was in fact the mirror image of the *dis*assembly line pioneered by pork packing houses in the 1870s and 1880s, decades before Ford used the concept for assembling cars. In the case of the swine, a continuous chain moved the animals as butchers cut them into pieces; in Ford's rendition, the workers also did not move much: The car began as a few parts, and as it moved down the line, workers attached more and more parts until a finished automobile emerged.[20]

The technical difficulties involved with the use of veneers are considerable. Since complex figure is what makes the wood desirable in the first place, the swirls, "eyes," "flames," and all the rest, with their inherent stresses and strains, are not only present but also in very thin form, and therefore in most cases in very fragile condition. At least in a large piece of wood with interesting figure there is some element of strength in the substance and interconnected nature of the grain. Solid thick pieces of figured wood split, cup, twist, and crack, and veneers do as well, only more so. To defeat the cracking and peeling wrought by the expansion and contraction that changes in temperature and humidity bring, artisans first laid down a thin sheet of some other less impressive wood species with its grain running at a ninety-degree angle to the eventual finished surface. *Crossbanding,* as this process was called, alleviated some of the effects of wood motion and separated more expensive veneered furniture from cheaper goods, although it might have taken years to show the difference. This was, at least during the nineteenth and twentieth centuries, akin to the way careful preparation of one's personal qualities such as education and propriety were thought to separate the "better sort" of people from the "common sort."

The grain and figure of expensive woods could be faked. Paint was the most obvious and cheapest method for coating a secondary or "base" surface with a flashy top layer. Mineral- and vegetable-based pigments and later coal tar and other chemical colors were used on all sorts of plainer structural substrates to give the appearance of more expensive or more stylish materials. Swirling a darker (or occasionally lighter) color on a contrasting painted surface to look like fancy grain was a widely practiced technique, not merely on furniture but on doors and woodwork as well. The earliest practitioners used feathers and fronds to create the effect of wood or marble grain, the latter most often on columns or pilasters or other large architectural elements. Eventually, and certainly by the late eighteenth century, metal combs and small handled graining tools were available on the mass market. Dipping one in paint and brushing it across a contrasting colored wet base paint in the general pattern of real wood grain produced a passable imitation and in some cases a doppelgänger for the real thing.

The practice of layering inexpensive wood with a topmost "finish" surface was not limited merely to reproducing grain to fool the eye. Artists, carpenters, and cabinetmakers made fancy surfaces with paint and varnish, and modern collectors prize old furniture still in its original paint, and for good reason. The palette of much of this early furniture and sculpture is often bold, and the painted decoration of some eighteenth-century Swedish, Norwegian, Finnish, Austrian, and German furniture is remarkable to contemporary eyes.

The quest for the "genuine"—often confused with the search for the creations of those without formal or academic training—sometimes leads people to steer clear of the outwardly decorated, as if it were some sort of sleight of hand concocted to fool the innocent among us. Veneering often conjures up such suspicion. For some nineteenth-century critics and worriers, veneering, painting, and plating brought to the surface issues that previously had resided in the realms of philosophy and religion. Were things as they seemed? Were people as they seemed? Fancy fabrics mass-produced by industrial methods and fashion made accessible through mass-circulation magazines clothed the bourgeoisie with the surface of respectability and refinement, a textile equivalent to fancy wood glued to pine or poplar. But what was underneath the glistening or otherwise elaborate surface? Was the "secondary" wood under the veneer or the "base" metal of silver plate—all common, coarse, and unrefined—analogous to the *real* person under the outer layers of clothing and "manners"? Was a trick of the eye—trompe l'oeil—dangerous or merely entertaining? Eventually the idea of layers of civilization over a barbarian nature played a role in the development of two important theories, one of the nature of human behavior and the other of the origins of American politics. By the end of the nineteenth century Freud had based his theory of personality at least in part on the idea of civilization as a series of civilized layers atop a barbarian self, and Frederick Jackson Turner similarly proposed that the frontier experience in the United States had stripped off civilization's layers to reveal the savage, who then rebuilt and reconstituted civilization, albeit in a compressed time and in a manner that he thought promoted self-reliant democracy.[21]

In our own time such concerns have vanished when we encounter sheets of plywood in the lumberyard or at the construction site, whether we are in Delhi, Dubai, or Denmark. One might say that plywood has the advantage of democratizing beautiful wood by making available more product for the consumer at a price far less than for a solid piece of the same wood and in a form that resists warping and twisting better than the solid material. This very democratization is perhaps what drives up the status of solid woods in domestic life. It is an old pattern of consumer and class behavior. As the average citizen gains more access to the goods once available only to the wealthy, the latter develop new standards of exclusion to stay ahead of the "pursuing" bourgeoisie and working class.[22]

Veneer is not only on plywood but also in the wood-appearing plastics and metals applied to particle board and substrates that never were part of the forest. We are not suspicious of it; we don't worry so much about the "base," unless we think it is made of some substance that can hurt us, such as a carcinogenic adhesive or moisture protectorant. Whether this means we are less concerned with the "problem" of the surface of things or people or whether we are so numbed by the omnipresence of things that are not entirely as they seem remains a question for a small number of philosophers in our midst. Perhaps the surface is all that matters to us. Politics and popular culture certainly suggest that.

2

Shelter

Fly over just about any major city in the world and you will see a consistent pattern of building. The urban center will be densely packed with tall buildings, some commercial, some residential—and none made of wood. A little farther out, smaller and shorter buildings are more common, although a scattering of medium-height structures (five to ten stories) still dot the landscape. In older cities many of the more modestly sized buildings are built of wood. Even farther from the urban core—in the suburbs—the vast majority of buildings are both residential and for the most part wooden. In cities in which there is a sizable population of desperately poor people, enclaves of tiny wooden shelters are common, often made of salvaged materials. These days many of these are composed of industrially made materials. Eventually the city and suburbs disappear from the airplane window, replaced by the woods and agricultural fields of the countryside, where nearly all buildings are made of wood.

It is not surprising that wood has virtually disappeared from the center of cities, except for interior trim and furniture and the occasional tree in the park or along the street. Real estate is expensive in the city and stronger building materials have made it cheaper to build vertically rather than horizontally. Wood does not have the necessary strength to build skyscrapers. As soon as iron and steel technology and the elevator made it practical to build up rather than out, wood was obsolete.

It was not merely the economics of urban real estate that consigned wood to the dustbin of building materials in such places. Densely packed with people and buildings, cities filled with wooden structures were cataclysmic fires waiting to happen. As they periodically burned throughout history (more about that in chapter 9), cities gradually became more "fireproof" as they were rebuilt in stone and brick, although some of the interior structures of newer buildings continued to be made of wood until the latter twentieth century.

While steel studs and concrete have replaced wood in apartment buildings, skyscrapers, and other commercial buildings, houses in most of the industrialized world still consist of wooden beams, studs, plates, and rafters. Improvements in building and woodworking technology made it easier to build wooden houses, barns, and factories in the nineteenth and twentieth centuries, especially after people discovered that balloon framing allowed them to construct structures with large interior spaces and relatively lightweight supporting members.

In North America spruce is the wood of choice for framing; elsewhere pine or other softwoods are used. Sheathing buildings once required nailing a layer of one-by-six- or one-by-eight-inch boards to a skeleton of "two-bys." Now it is done with large (four by eight feet) sheets of plywood or chip board (made of glued-together chips of wood); with pneumatic nail guns, two strong-armed builders can close up a house in a day. Once the exterior was sheathed, builders protected the house with a layer of horizontally oriented boards attached so that they shed water. By the second half of the twentieth century, however, wooden clapboarding was being superseded in many areas of the United States by vinyl siding. The first incarnations of vinyl siding were easy to spot for what they were, but by the 1980s plastic with the appearance of wood grain texture was available. Whatever the protestations of the vinyl and other plastic goods manufacturers, they were still fake wood, easily spotted as such and lower in status than the real thing. This may have to do with the higher maintenance costs of wood (painting, together with replacing rotten or split boards), which some see as a marker of status and wealth, rather than merely a pain in the neck.

Origins, Log Buildings, "Cabins"

For most of human history, however, wood was the exclusive building material for nearly all structures. The earliest forms of shelter made of wood probably antedate the Neolithic era. Builders in wood in prehistoric times used natural formations they found in forests and scrub areas. These they interwove, filling the interstices with more pliable plant materials, mud, or some combination of the two to provide a temporary and effective protection from the elements. Eventually, as people developed tools for transforming wood into different shapes and forms, so too did housing for humans and domesticated animals evolve to more complex and efficient forms. Binding sharp stones to sticks with flexible plant fibers or strips of animal hide enabled people to cut and shape wood; binding heavy rounded stones to sticks enabled people to pound it. Weaving and binding grasses and other such materials made for effective roofs and walls. In other buildings, wattle and daub (mud and straw or other plants), stones and mud, or just plain mud (and later, bricks and mortar) were used to block out wind, rain, and snow by infilling the spaces between the rough outlines of the logs or timbers. With the advent and evolution of metallurgy, the quality and quantity of

FIG. 18
Detail of half-timbering. Castle Combe, England. Photograph by Yoshio Komatsu,
Built by Hand. *Courtesy of Yoshio Komatsu. The curved timbers appear to be a*
natural formation of the sort much in demand in the shipbuilding industry.

edged tools increased and more complex and "finished" wooden elements appeared in buildings.[1] [FIG. 18]

A limited spectrum of tools and the difficulty of moving heavy timbers meant that early builders in wood used medium-sized trees that
were near to their building sites. Those near rivers could have used them
to float logs downstream, as loggers continued to do for centuries, but
even with a river or stream nearby, big trees still were problematic.
Moreover, the difficulty working some common woods, such as oak,
meant that builders gravitated toward those that were both straight-
grained (easier to split) and soft, such as pine, which has not only these
qualities but also great strength and versatility as a building material.

In northern climate areas pine, spruce, and fir forests were abundant
and remain so, largely because of the replanting that became standard

practice in the twentieth century. Initially, pine building was for the most part accomplished by horizontally stacking logs on top of one another, with notching at the corners to lock them in place. Ancient builders used this formulation to construct permanent immobile structures for housing. The Native American hogan, a common building form in parts of North America, uses horizontally laid log walls and a roof of some combination of mud and grasses and whatever else local tradition established as an effective weather shield. The long house, a large structure in which many families lived, was employed in a variety of forms in North America, in both the eastern and Pacific Coast forest areas. Traces of evidence relating to building in wood in ancient Egypt reveal that as early as 2600 BCE there was considerable trade in timber for building. For furniture and buildings Egyptian craftsmen used cedar, ash, and other woods native to more distant parts of Africa and the Middle East and may have made use of the mangrove that grew along shorelines farther south and east.

Farmers were the majority of the population in most of medieval Scandinavia (and in the world around 1000 CE, for that matter). They lived in small, usually one-room dwellings with an open smoke hole in the roof. The oldest surviving such buildings, though enlarged from their initial floor plan, date to the thirteenth century, and written sources suggest that this form and structure go back several centuries before the survivors were built. These records indicate that Scandinavian housing was sometimes built using stave construction: hewn beams and planks set vertically with their lower ends in a sill (the bottom horizontal section of a building frame) and their tops set into a horizontal member at ceiling level, called the plate. Door posts were sometimes carved in elaborate designs, as were the horizontal support beams that extended beyond the outer walls to add space to the second floor storage and sleeping areas.

Storage and living quarters such as the Stave Loft (*staveloftet*) from Åi, now in the Hallingdal Museum in Nesbyen, Norway, and the Norwegian storage building now at the Skansen open-air museum in

FIG. 19
Staveloftet *(Stave Loft)*. *Åi, Norway, ca. 1250. Heavy timbering, closely cut joinery and elaborate carving demonstrate a high degree of mastery of the materials. The upper floor has built-in furnishings and the entrances rich carving. The bottom floor was for storage. (Photograph before 1908, courtesy of the Hallingdal Museum, Nesbyen.)*

Stockholm illustrate the sophistication of Norse log building technology. The former is a "building on a building," consisting of a vertical-boarded, gable-roofed storehouse resting on top of a very heavily timbered, horizontally laid structure of much smaller floor area (twenty by twenty feet) than the building above, which measures approximately twenty-seven by thirty-three feet. [FIG. 19] Generations of builders doubtless reworked and tinkered with the storage house, but the traces of early engineering competence and woodworking skills are still evident, and it is a building of startling, even breath-taking appearance, with the upper story cantilevered about seven feet over three sides of the lower structure.[2]

It is likely that the ninth-century expeditions and raids of the Norsemen exposed them to similar building traditions in Russia, central Europe, and the northern British Isles, although the wood was fast disappearing from the latter region even at this early date. While stone construction gradually superseded log and stave building for large institutions in some areas of Norse contact by the seventeenth century, the abundant forests and improved woodworking technology of the home region helped maintain these traditions in prosperous areas until the eighteenth century, and in more remote regions until the later nineteenth century. Stave construction persisted along the Norwegian coast while horizontal log construction was more prominent to the east, in what is now Sweden.[3] In part this may be a result of strong traditional connections to the Finnish and Russian roots of horizontal log construction and the westward migration of peoples who carried with them knowledge of building methods.

There seems to be a curious reversal of what we may normally think of as the evolutionary process in building from coarse (or at least less refined) to more worked and manipulated natural materials. Stave building used planks or at least split logs. Log buildings employed members left in the round and laid horizontally, chinked between courses of logs with mud, grasses, or ferns. Preferring the latter seems to be retrogression. But inhabitants of the interior of eastern Scandinavia favored horizontal log structures because they offered a stronger base on which to build, and behind which to hide, providing dwellers with a sense of security from marauding humans and animals that the stave building may not have supplied.

These houses were not merely structures to keep people relatively dry and somewhat warm; they were (and still are, if we think about it) spaces created and invested with cultural and social meaning. For centuries people all over the world have seen the house as a metaphorical representation of the human form, the door akin to the human mouth, the house's avenues for the outflow of human waste like similar parts of the body, the hearth a physical reference to the source of life, and the enclosure itself as a representation of the womb or some other primeval safe place.[4] The house protected inhabitants from both the tangible and the invisible (namely, spirits) that they believed threatened their safety and survival. Perhaps the more aggressive and better-armed peoples of the Scandinavian coastal areas felt less need for protection (though their stave houses were not exactly flimsy) than did inhabitants deeper inland.

The association of horizontal log buildings with a retrograde architectural style also means that most people have never really looked at them with care, or have assumed all such structures were built with the crude craftsmanship found on the North American frontier. The decorative work on both interior and exterior surfaces and the tight joinery found in Scandinavian examples undermines this notion and the related linear conceptions of architectural technology. Complex carvings, chamfering (the process of relieving the sharp edge of a squared timber), and other forms of cut-out decoration were common in log buildings of Scandinavian farms and villages. Logs were sometimes faceted on the inside, and supporting structural elements were often carved or cut to form graceful curves on the undersides of the cantilevered supports. Ventilation openings in storage barns and lofts took the form of hearts, circles, and elongated shapes, and highly carved heavy vertical posts are found on many of the surviving log buildings in Norway. While it is possible that these buildings survived because they were the most elaborately carved inside and out and the most substantial buildings on the landscape, they nonetheless indicate an advanced sense of decoration and design.[5]

European settlers in the New World brought with them traditions of building in wood, though many of these were quickly altered by contact with émigrés from other parts of the Continent. Anglo-Americans populated much of what is now New England, but they were joined by French settlers in Canada; by Dutch, German, and Scandinavians in the colonies of New York, New Jersey, and Pennsylvania; and more remotely, by Spanish settlers in Florida and southern Louisiana, though contacts with the last group were infrequent. Native Americans and their solutions to dwelling challenges had some influence on settlers, especially in the first years of settlement.

Americans like to think of the horizontally laid log building as their invention, the "log cabin" in the treasured lore and myth that surround and shroud the beginnings of the United States. The idea that no one would have thought of this solution to shelter before the seventeenth century in any part of the world is both bizarre and indicative of some Americans' historical myopia, or at least of their willingness to be seduced by an attractive myth.[6] Moreover, it ignores Native American traditions embodied in the long houses of the eastern tribes and the hogans of the Southwest.

The horizontally laid log building, then, is not in any real sense the product of some visitation of practical genius on the immigrants to a forested land from which the idea fortuitously sprang. It is instead a practical solution to the housing shortage that greeted the newcomers: One can accumulate enough logs for the job more quickly than enough stone. It is much quicker to put up log walls, though it requires more care than meets the eye to get them straight and tightly interlocked at the corners. The inspiration and the know-how was already present in certain parts of North America or was brought here by immigrants from places other than England, people who fortunately happened to be located in the middle colonies, in what is now southern New Jersey and Delaware, the colony of New Sweden.

To make the logs stay together, each one had to be notched at its ends, to accommodate the logs that intersected it and formed the corners. The simplest notching system employed the removal of a semicircular portion of wood that corresponded to the outline of the bottom half of the log intersecting it on each end. Called (for obvious reasons) *half-round* notching, it was a fairly simple conception that worked well enough. Lay the first log down, cut the notches, lay another log parallel to it on the opposite side of the planned structure, notch it on both ends, and lay the two remaining logs into the notches (on top of a course of mud, stones, or some other sealant, since the second logs will be one-half

FIG. 20
Partly constructed log cabin. Joker Mine, Medicine Bow National Forest, Albany County, Wyoming. Courtesy of the Library of Congress, Washington, DC.

the thickness of a log off the ground). The first course of logs, defining all four walls, is now in place. Remove semicircles of material from the two logs laid on the first two, put in the chinking (sealant) on the top edge of the first two logs, and lay down the first two logs of the second course, and so on. [FIG. 20]

The half-round (also called the *cradle* or *saddle*) notch functioned well enough and required less work on the corner joints and sometimes a bit less wood, since there was a considerable amount of mud between the logs. [FIG. 21] More complicated log corner joinery, such as the square notch, produced tighter joints and less space between the logs through which wind and creatures could enter. Common in Scandinavia was a joining technique in which each log was cut on top and bottom, removing a section that resembled a trough with outward-flaring sides. The outside edges of the log were faceted at the ends so that the resulting joint was tightly locked and the logs lay close on top of each other along their lengths, provided that they had been trimmed flat. In many cases Scandinavians also trimmed logs to a hexagonal shape that they then notched accordingly. [FIG. 22] Similar to this in complexity and effectiveness was the half-dovetail notch, which required the ends of the logs to be trimmed into rectangles. Logs running along parallel sides would

FIG. 21

Corner notching variations. Half-saddle and V-notch. Kristi Hager, photographer. William & Lucina Bow Ranch, Silver Bow County, Montana. c. 1870. Square-notch. Stanley Mixon, photographer. Friends Select School, 16th and Race Streets, Philadelphia, PA, [n. d.] Courtesy of the Library of Congress, Washington, DC.

FIG. 22

Corner notching. Sami Folk Museum, Inari, Finland.

be cut on their top edges at a compound angle (angled down and toward the inside) to receive intersecting logs cut to fit into the receiving slot. The double angle of each cut ensured that the logs could move outward in neither direction—the angles locked each member in and the weight of the logs made it impossible to pull against the joint.

Horizontally laid log buildings were also built in Japan and Korea, and more than likely in China as well. Often constructed of smaller diameter logs and used for, among other things, storage buildings in which ventilation was important, these buildings have logs joined with notched corners like their Western counterparts and date back at least to the seventh century CE. Their characteristic rooflines are rather different from those of the West, making use of the upswing curve of the joints in a four-facet hip roof, a form in which four sides rise from the eaves and meet in a point in the middle of the building.

The extra labor involved in more complicated joinery made it uncommon in the United States and Canada, possibly because there was a smaller local supply of skilled workers than in Europe. Or it may have been that Americans were less interested in the log house as a long-term residence, since they envisioned their future in a more technologically advanced—and certainly more fashionable and refined—building style. In spite of the hot air of political managers and hucksters in the nineteenth century, most Americans considered the log house a temporary abode and a sign of little refinement, maybe even a sign of barbarism. It was a stage in human development to be overcome, as was its relative, the sod house of the Great Plains. In Abraham Lincoln's case the log cabin was used not only to promote the candidate's homespun honesty and humble roots but also as an indication of how far he had come.[7] Stone and brick were—and still are—symbols of permanence and wealth.

European and European-American Framed Building

Leaving the log cabin for rectangular and relatively smooth beams required tricky and dangerous work with axes and adzes. After securing the log to a cradle on the ground, the worker removed the bark with a spud, snapped a line if chalk was available, and then climbed on top of the log, bent over and made a series of shallow cuts up and down the length of the log with a long-handled felling axe. Now it was time for the broadaxe, so named because its blade (which took different outlines in different cultures) had a longer and more rounded sharp edge than did the felling axe. Most people used the broadaxe while standing alongside or on top of the log, swinging the curved or angled handle of the axe at the side of the log, chipping and splitting away the small sections of the round edge delineated by the shallow cuts made with the felling axe. Repeating the process on all four sides of the log ideally produced a square and straight building timber.

At this point the timber was suitable for barn beams and in places in the house where the wood would not ordinarily be seen. Broadaxed lumber still had the telltale marks of the cross-grain cuts first made before "hewing to the line." Removing these cut lines and smoothing the beam or the faces of boards split from a beam required an adze, a cutting tool that resembles a modern gardener's hoe in blade-to-handle orientation. That is about as far as the similarity goes, however. The adze handle is shorter than the hoe's, since the tool is swung much like the broadaxe and often while standing atop the lumber. The blade is usually curved slightly inward and beveled on the inside edge. An adzed beam was meant to be seen and its surface is characterized by a slight rippling along the length of the piece. The cut marks many moderns refer to as adze marks are in fact the cut marks of the felling axe made in the first phase of dressing the log.

This form of "dressing" made the tree civilized, useful for human consumption, just as dressing a chicken or a side of beef or pork readied the meats for human cooking and consumption. For some people squared timbers were a sign of social elevation above the log building, which was a little too close to the natural and the bestial for comfort. Often North American pioneers first built a house of logs because it was quicker to get it completed, a consideration of no small import in the harsh climate of much of the continent. What happened afterward was often revealing. Some homesteaders eventually built around the log interior and covered it with more refined and manipulated wood or plaster.[8] Others simply built another—usually larger—house and consigned the log house to another use, often as a livestock shelter. These alterations are not merely a testament to the practicality of the homesteaders; they reveal the genuine conception of the log house as "coarse" and "rough" and as a temporary condition in a process that repeats the stages in human development, compressed into (hopefully) a few years rather than a few centuries or more. Whether dwellers of these houses actually thought of themselves as recapitulating the history of civilization is unimportant. Their actions speak of a sense of progression from log to beam and plank, from roughness to finish, from coarse to refined.

English settlers from East Anglia and the Midlands—the areas from which many of the seventeenth-century immigrants to North America came—carried over building practices considerably different from the experience of Scandinavians. By the beginning of the Atlantic migration in the early seventeenth century, the English Crown controlled much of the remaining wooded lands of the realm, and the rest was largely in the hands of well-to-do landowners. Inferences drawn from *Domesday Book* (1086) suggest that about 15 percent of Britain was forested, and estimates of the extent of the woods in about 1600 put the amount of forest cover at 7.7 percent.[9] The English rural landscape had been logged off years before the American adventure, transformed into tillage and pasture land. Wood was in short supply and vernacular building showed it. Earth, field stones, and cut stone were the available materials and composed the external fabric of many houses, barns, and other outbuildings. Where wood timbers were put to use, they were as skeletal structure, the intervening spaces filled with brick or stone and some combination of mud, limestone, and straw or other material to bind the infill. Called timber framing, this method, which dates as far back as the Roman Empire, still made use of a considerable amount of lumber, often of heavy dense woods such as oak, which was one of the more numerous species in the English woodlands. The great virtues of this method were in its limited use of wood and the overall strength of the structure, a result of the species used and the intricate joinery and bracing employed. (See figure 18.)

Early English housing frames may well have resembled a tent or A-frame arrangement, with beams sunk into the ground at equal intervals and leaned against a beam at the roof peak, or ridge. The internal triangular shape of the structure had the disadvantage of space that was less useful than that of a square-walled structure. One way to partly counteract the angled wall was to make use of naturally occurring bent logs, or *crucks,* that, split lengthwise, could form a gradual transition from vertical wall to pitched roof, a framework that once completed resembled the pointed arch of a church window. Set on stones at ground level, these beams were made solid by their attachment to the ridge beam at the roof apex and by tie beams that ran between them horizontally and across the span, from one side to the other, called collars or ties, depending on their placement. The weight of the roof, pushing down on the crucks, was countered by the horizontal strength of the ties and collars. Naturally occurring crucks were a relative rarity, however. The idea of the slanted roof atop a vertical wall frame eventually took the form of the post-and-beam building with which we are today familiar, with similar bracing and supports on the top floor.

English builders of the medieval era eventually developed methods for constructing strong buildings of multiple stories, using their knowledge of timber framing and enlarging upon the traditional forms. To the braced supports and joinery of the cruck and later buildings, they added extra bracing that connected horizontal members to vertical posts to produce a diamond or Saint Andrew's Cross pattern on the exterior of the building. Fifteenth-century German timber framers probably developed this art to its highest or most complicated form, constructing multiple-storied buildings that cantilevered over their first-floor area, often in steps that further jutted out as each floor was added.

It took no particular genius to see that timber framing of the English and German variety made no sense in North America. It was not merely that there were more trees in the New World; trees were so varied and densely packed that they were in some ways a menace. They not only got in the way of agriculture; they were a hiding place for wild animals (whose nature, size, and numbers were constantly exaggerated by reporters to the Old World) and people thought to be bloodthirsty savages, even though nearly all of the Native Americans were anything but that. But to Europeans convinced that they were the bringers of salvation to a corrupt world, the riches of the New World environment were there for them to tame, harvest, even despoil. Firmly believing that they had found the answer to many of humanity's vexing and fundamental questions, they saw no contradiction in using force to make the landscape more like what they had left and its original inhabitants into either their victims or their servants, if possible.

Chance encounters with other Europeans of a different building tradition or Native Americans with their own solutions to the problems of making a dwelling led to a rapid abandonment of the cruck or timber frame tradition in the colonies of the English New World. A substantial number of carpenters and joiners made their way to North America in the first generations of immigrants, implanting a set of building skills among people accustomed to complicated joinery.[10] The ubiquitous trees enabled settlers to adapt log, stave, and board fabric for the walls and skeleton of buildings. Combining the skeletal idea of English cruck and timber framing with the horizontal (and, less commonly, vertical) character of Scandinavian and Native American building eventually brought about the post-and-beam and board housing of early American settlements.

Joinery

The key—in more ways than one—to the success of frame and board architecture was not merely in understanding how to cope with weight and stress in a building but in how to join the elements in a fashion that kept the building standing. Without proper joinery the enterprise would not work. In the modern age this seems easy enough. Get out the hammer (or more likely the pneumatic nail gun) and attach the frame together by banging in the nails. The friction between the wood (which has been stressed into compression along the edges of the nail) and the metal nail holds the wooden pieces together. Screws are more effective, but more expensive, and take longer to insert. Metal plates and supports are also effective in linking wooden frame members, since they can eliminate contending with the weakness of an end-grain joint or the labor of cutting interlocking joints.

Nails, pins, and other metal parts were available by the seventeenth century, but few if any were used as we employ them today. Metal wares were expensive to make, wooden materials cheap or free for the taking. Getting ore out of the ground is much more taxing and slow than harvesting trees, especially in North America. Protestants may have seen some sort of divine calling or encouragement to labor, but there was plenty of work to keep the Divinity happy with them without messing about smelting metal ores, which in fact doubled their work since smelting required fuel—usually wood. Farming, hunting, spinning, weaving, cooking, tending the animals, and defending oneself or paying taxes did the labor-as-virtue trick.

For nearly all the joinery jobs of building construction, therefore, interlocking or fitted joints, often held together with wedges or wooden pins, were the order of the day until the nineteenth century. In houses in remote regions this is still the norm. Many are still framed with heavy interlocking timbers, the spaces between the frame filled with lightweight materials in tropical areas and more substantial infill in harsher climates. [FIG. 23] Today in the West we usually find these joints in furniture and not housing, with the exception of the revival of post-and-beam housing for the small market that wants it. Furniture joints unencumbered by metal are glued together, as they have been for centuries, though the modern aliphatic resins and other synthetic glues are superior in most ways to their organic predecessors, which were made from animal hides or bones. These old glues were of little use in house

FIG. 23
House. Kinnaur, India. Photograph by Yoshio Komatsu, Built by Hand, *Courtesy of Yoshio Komatsu. Heavy frame construction with wood infill of this partly finished house shows mortise-and-tenon joinery.*

construction. Their adhesive strength could be impressive, but not enough to counter the enormous stresses in a building. In any event, once one had gone through the labor of cutting and assembling the big joints of a building, drilling holes for the pins—even with a hand auger—was little enough work for the reward of a pin that everyone knew would be secure.

The complicated joint most modern-day people know is the tongue-and-groove, a joint most often used to connect boards on their long sides. Primarily used to conceal nailing, as on floor boards, it is also an effective joint in uses of minor stress, and it has the advantage of (in general) doubling the glue surface one would have by simply pressing together the edges of two boards. It works for joining the vertical sides of frames to the horizontal sides, as in a cabinet door. But it is less useful for connecting the leg of a table or chair to its top or seat. Downward pressure on the table top (a pile of books) or chair seat (the sitter) can cause separation because there is little to prevent the tongue from spreading outward. Thus the effective joint for these uses is the tongue-and-groove's more muscular (or at least more inherently rigid) cousin, the mortise-and-tenon. [FIG. 24]

The tenon part of the joint is a rectangle fashioned on the end of a piece of wood, but its size is further reduced from that of the rest of the piece. That is, once you cut out pieces on opposite sides to make the rectangle, you then cut a small bit off one or both of the remaining sides. Why narrow it? The answer lies in the strength of a mortise, the hole into which the tenon fits, as opposed to that of a groove. Once the tenon, if properly cut, is inserted, it cannot wiggle around much or dislodge as easily as the tongue in a tongue-and-groove joint. It is often further

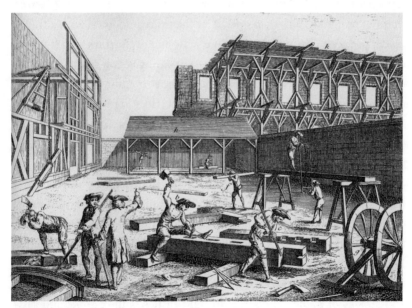

FIG. 24
"Charpente" (Construction), Plate I, volume 2 of Denis Diderot, L'Encyclopédie (1751). Chopping a mortise and paring a tenon in the center foreground; pit sawing in the middle ground; frame construction in the background.

strengthened by inserting a pin or treenail (pronounced "trunnell" or "trennel" by the British and their offshoots), driven through a hole in the stock in which the mortise has been made, intersecting the tenon, and continuing out the other side of the piece. Movement is possible, but failure unlikely, especially since the tree nail was generally made from very dense and hard wood. Varying levels of swelling and shrinkage in woods, different tensile and other strengths, and other physical characteristics (information that was gained through years of experience and passed down from master to apprentice) helped eliminate the guesswork in making secure joints, which in the end held up buildings as modest as a one-room house or as ambitious as a multiple-story church.[11]

The supernova of the joinery firmament for modern aficionados is the dovetail, so named because of the shape of the joint. Now used primarily to connect drawer sides to the drawer front, dovetails get their strength from its multiple connections and, more important, from its shape. You pull a drawer to open it. If the sides are joined with a trapezoidal connection situated so that the wide end of the trapezoid (the end of the "dove's tail") is facing the front, the pulling action cannot dislodge or pull away the front from the sides since it is quite literally pulling against itself. It is a clever and powerful interlocking joint, challenging and nearly impossible for the untrained and unpracticed woodworker to cut accurately,[12] though enough skill and practice make it seem matter-of-fact for the professional.[13] [FIG. 25]

Uncommon now, save in the work of fine furniture makers, the hand-cut dovetail was frequently used in building joinery, and especially in the frame construction of large buildings such as churches and formal public halls, but not for the country seats of the nobility. In pre-Norman (that is, before 1066) England and early medieval Europe, great halls built of wood frames for the wealthy did not long endure the sieges and attacks of medieval wars. Forts and castles built of stone and protected by water fared better. Whole cities were encircled by stone walls, some as much as twenty feet thick.[14] Inside the walls—until they burned—were wood-framed structures.

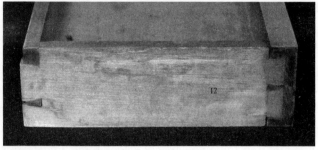

FIG. 25
Dovetailed drawer. Maple and pine, ca. 1840.

Churches may have survived better than baronial halls because they were consecrated space and less likely to suffer depredation, though zealots of the Protestant Reformation exacted a heavy toll on the interior decorations of many of the old structures. But enough survivors in whole or in part exist in Great Britain and Norway to demonstrate a stunningly complex system of joinery employing combinations of dovetails, pegs, mortises, and tenons that anchored the huge buildings and secured their interior spaces.

Sacred Buildings

The engineering marvels of these churches arose out of necessity. To support the huge weights of their superstructures and the collective mass of their congregations, church builders placed requirements on both the wood and the joinery that extended the previously known or conceived limits of the material and its building technology. It did not always work. Cecil A. Hewett, whose *English Historic Carpentry* may be the last word on the subject, quotes the *Anglo-Saxon Chronicle* on the collapse of a floor in 978: "some there were sorely maimed and some did not escape with their life."[15] Scattered references exist detailing similar, though perhaps smaller, disasters that befell churchgoers in buildings whose craftsmen fell short of complete competence for the magnitude of their task or the architect's (or cleric's) ambition, so the disaster of 978 was not a unique occurrence. Many more such incidents probably escaped the chroniclers' attention. So the stakes were high and the pressure to perform competently intense, and failure catastrophic.

Medieval European churches, educational institutions, and public buildings expanded and established both the engineering capabilities of their builders (save those unfortunates noted above) and the possibilities of wood as a construction element. Further to the east and north, however, was another tradition in church building that continues to attract the attention of those who know where to look for it. The stave churches of Norway and the onion-domed churches of Karelia (the borderland between Finland and Russia) represent a different approach to the use of wood in the enclosure of sacred space.

The essential form of the stave church is that of free-standing massive supporting posts—often sections of the entire stem of very large trees—set directly on or into the ground, linked by horizontal members and the interstices filled with vertical planks or half-round sections of tree stems. In most cases the corner posts do all or nearly all of the work of supporting the immense weight of the superstructure, although some of the larger and later-built churches have a central post to help support the weight of the church spire, which sometimes rose several stories and required considerable understructure.

The first such churches were probably built sometime before 1000 CE, though archaeological evidence is scarce. No churches survive from this era. Norwegian researchers have established that approximately seven hundred stave churches were built and all of the twenty-seven that survive in one form or another date from the twelfth and thirteenth centuries. The architectural roots of the stave church are a source of considerable debate among scholars. One theory traces the form of the floor plan to similar ancient churches of Anglo-Saxon England. Given that the Vikings made their way into England, Scotland, the Hebrides, and Ireland, it seems comfortably inside the realm of possibility that the invaders returned with architectural knowledge that they passed on. But it is just as possible that the knowledge went from Norway to the British Isles. Moreover, early theories of the origin of the form and construction method suggested a link to the East, perhaps because linguistic theory and migration research traces the origins of many of the European peoples to Central Asia. Finally, as the architectural historians Christian Norberg-Schulz and Gunnar Bugge note, pre-Christian sacred buildings almost certainly utilized the vertical tree-stem system of support, with stave infill.[16]

Nearly all of the surviving churches are strikingly vertical. They rise several stories in height; the angles of the shingle-clad roofs are steep, often set at more than sixty degrees from the horizontal. The exteriors are dominated visually by the first floor vertical posts and fill and the roof shingles, rather than any exterior wall surfaces, except on the gable ends of the buildings. The largest churches also have complicated systems of intersecting and stacked roofs of decreasing size as they rise to the spire. [FIG. 26]

FIG. 26
Stave church. Borgund, Norway. Photograph by Robin Strand.

The interior of a stave church conveys a complicated first impression. At once the immense rise of vertical space—much like that of the more well-known Gothic cathedrals of European areas to the south and west—indicates engineering unexpected of wood as a building material. Competing with the lightness of the space is the heaviness of the thick and densely situated wooden structure, nearly all of which is visible. Colonnades of treelike columns, braced with X-shaped interlocked beams in the second-story galleries, frame the central open space. The similarity to the Gothic cathedral—minus the flying buttress—seems unmistakable, yet to assume that stave church construction was "borrowed" privileges the engineering technology of others at the expense of the vernacular builder, and assumes both a relatively linear history of ideas and, more troubling, that knowledge originates in those societies that latter-day analysts desire as their roots, usually for some political and presentist reasons. "Native genius," as the architectural historian Sibyl Moholy-Nagy phrased it in 1957, takes its cues from the environment and the needs of the locality and is not always subsumed by the academic culture of outsiders, even if subjected to it by military or other conquest.[17]

The other striking visual quality of the stave churches is the intricate carving on both interior and exterior surfaces. Most noticeable on the outside are the carved figures at the ends of the roof peaks on the upper stories of some of them. With the appearance of dragon's heads (complete with long tongue), they remind onlookers of the pre-Christian heritage of the Norse peoples and are an unlikely relative of the gargoyles of churches in the rest of Europe. Not all stave churches have these accoutrements, but even those lacking them often have some sort of vertical decoration at the ends of the roof peak. The visual complexity of the roof surfaces is further carried by the diamond pattern of the shingles, and the corner posts and door frames are often intricately carved as well.

It is on the inside that the indigenous carvers have left their most conspicuous marks. Columns, posts, braces, and especially door frames appear to lay waste the notion that these churchgoers and their leaders were somehow joined in an opposition to visual delight and sophistication. Complex intertwining relief carving suggests possible linkages to Celtic designs—and hence to the Celtic peoples of the British Isles and coastal France. One door frame in the Borgund church (ca. 1150) appears to have elephant heads, birds, and other stylized beasts integrated in a swirling design consistent with the worldview reflected in the dragons on the exterior.

It is only a distance of a few hundred miles from Norway to Karelia, and thus many recent travelers have compared the stave churches with the Russian log churches to the east. There are superficial similarities: both types are made of wood, and both are complicated structures of great height. But the Karelian and Russian churches emerged from the tradition of horizontal log building and utilize octagonal floor plans in many instances to maximize interior space and to provide a strong interlocking system where the logs meet. There is similarity in the use of shingles installed in a diamond pattern, but there is little further elaboration akin to the use of carving in the stave churches, and the onionshaped domes belie a relationship to Asia that is absent from the Norse churches. The most famous of these is probably the Church of the Transfiguration, located in Kizhi, Karelia, and built in 1714.

Stave churches in some ways seem to resemble houses of worship in East Asia, particularly the pagoda. The multiple roof lines and decorative sculpture on roof peaks suggest this, but it seems a stretch of the imagination, part of the seemingly unending search for *the* origin of the human race and its accomplishments. The angle of pagoda roofs is almost always much less steep than that of the stave churches, and the support systems differ as well. Older structures in China, Korea, and Japan and in much of Southeast Asia employ a series of closely spaced vertical posts and interlocking rafters and braces to carry the weight of relatively flat and often tiled roofs to the base of the structure. [FIG. 27]

The native forests of East Asia—filled with large cedar, red pine, cypress, mulberry, and varieties of the aromatic sandalwood—allowed for the construction of extensive religious complexes and extremely

FIG. 27

Pagoda. Mount Hagaro-san, Yamagata Prefecture, Japan. Fourteenth century. Photograph by Yoshio Komatsu, Built by Hand. *Courtesy of Yoshio Komatsu.*

large sacred structures within them. Close grained and of great strength, these woods were both workable and able to withstand great weight. Chinese, Japanese, and Korean woodworking technology was at least the equal of that in the West and Southern Asia. Metallurgists and smiths had developed laminated edges for cutting tools and finely crafted hand axes, hatchets, and adzes to shape trees and tree parts into uniform shapes and smooth surfaces. Joinery traditions were considerably different from those of the West, although the mortise-and-tenon and dovetail were certainly in use in the Orient as they were in the Occident. Pinned joints in which one element slips into or is set in another coexisted with joints involving keys that once fitted together could not be pulled apart without breaking the wood itself.

Making metals in East Asia was no less expensive in time and labor than it was in the West, but the response was somewhat different. In Europe the narrow-bladed bow saw was developed, whereas the Asian

FIG. 28
Shinto shrine. Toyama Prefecture, Japan. Photograph by Yoshio Komatsu, Built by Hand. Courtesy of Yoshio Komatsu.

saw consisted of a long handle and short blade that cut on the pull stroke instead of the push. Asian chisel makers adopted the same metal-saving strategy as did those who made saws—short blade and long handle, the better examples with laminated steel blades.

Sophisticated building techniques and architectural styles can be tracked several thousand years back in East Asian history. At the moment the Romans were building the classical structures of their empire, Shinto shrines of a different but no less erudite architecture were constructed in Japan. [FIG. 28] While ornament and color are common (paint especially so), many shrines exhibit elegant simplicity in design and construction, without the use of color.

Stilt Buildings, Floors, and Roofs

In areas of Asia in which rainfall was heavy and inundation frequent, the practice of building on stilts was common. But this architectural tradition is more complicated than simply staying above the wet ground. To understand the nature and evolution of the wooden house we must consider two important elements in the conceptual structure of enclosing space—the roof and the floor.

We can learn a great deal by asking what seem to be simple questions about what we take for granted; that is, we assume, as do most people, that we know what our, and others', ordinary lives mean. That assumption helps us miss the obvious and in part explains why we tend to dismiss new information about daily life as if we already knew it, when in fact we don't. Most people in the industrialized world do not think historically or cross-culturally when they encounter the material world. They see what they consider "primitive" as the end of a linear progression that has been in motion—merely at a slower pace—just as their own has been. What we moderns do not see is that "other" people today may be living in or on structures and forms virtually identical to their forebears' of centuries or even millennia ago, the products of architectural and technological thinking appropriate to climate, place, and cultural values. When initially developed, these older forms in some cases may have been well in advance of those of our own vaunted ancestors, who were for the most part scrambling around on dirt and in huts of no particular sophistication while the denizens of stilt houses in Southeast Asia were living in complex structures elevated from the earth.

Stilt or elevated housing is not just housing intelligently set on poles in water or in places periodically flooded. Housing in parts of Indonesia, Africa, and Southeast Asia is often built off the ground not only to avoid floods but also to exert some control against rodents and other invaders of living quarters. The elevation of the floor is also a cultural statement of separation from the earth and from the dirt. The dynamics of how we view the relationship between the earth and where we "keep house" lie at the roots of how we conceive of ourselves and our society. In western European traditions of building, we can find ample evidence of this clearly defined separation in early Minoan, Egyptian, Greek, and Roman architecture, and most certainly among the classical culture of Rome. The ruins of Pompeii, Ostia, and Herculaneum reveal that people in these cities paved over the earth with stone and, in some cases, tile.

The idea that the earth was somehow "base" continued for centuries after the classical era. In much of Europe and North America, those not of middling or great wealth trod on earthen floors in their one- or two-room houses, as did nearly all slaves. Flooring elevated one not merely physically but also on the ladder of civilization, figuratively tending to the divine. This stratification of civility ultimately was taken to extremes. In the late-eighteenth and nineteenth centuries European ladies of wealth and status demanded—and sometimes got—textile coverings on the befouled grounds or streets of cities, but not merely to avoid soiling their shoes and clothes. Even inside parlors of distinction, little footstools were a common accoutrement for the delicate lady, who daren't "dirty" herself on the floor.[18]

Vernacular housing from the coastal areas of the Indian subcontinent to Oceania is in many instances based on traditions of building technology that embrace strong structural members supporting a wooden floor. With axe and splitting technology, the trees of the tropics could be utilized for the complicated architecture of stilt housing.[19] While some of the forest in these areas could and does grow to gargantuan height and girth, some species grow thinner but nonetheless long and straight stems, making them ideal candidates for substantial posts on which to set houses, and beams to carry the loads of large structures.

In these houses, the floor—whether of longitudinally split logs placed to expose their relatively flat sides, planks shaped by axes and adzes, or woven and rigid mats—was a symbolic as well as practical statement of civility and civilization (in the Western conception of cultural development). The wood used to provide this platform was *worked* with tools, transformed from its native and natural state into another form, although obviously still of the original substance. The floor, then, is a statement of *artifice,* of the use of the human hand to transform the natural into the useful and the *artificial.*

The raised floor was also characteristic of Japanese architecture from at least the sixth century onward, when Buddhism arrived in Japan from China. Buddhist tenets of simplicity reached into architecture, reducing the ornament that was characteristic of high-culture architecture and focusing design efforts on achieving an environment that promoted the peaceful contemplation and harmony Buddhism promised. Japanese architecture made use of a lightweight but strong native plant, bamboo, especially for interior walls and other partitions.

Native in many forms throughout most of the temperate and tropical areas of the world, bamboo is technically a woody-stemmed perennial grass, but one that can grow to heights of thirty feet and to a diameter of well over eight inches. Hollow on the inside, it is nevertheless a strong, stringy, and lightweight material with which one can build some structures or parts thereof. In Japanese housing, bamboo has been used for roofing, and bamboo lattice has been used as the base for the plastered area between wall panel frames. For centuries furniture has been made of bamboo, as have been a variety of other instruments for daily life.

Mundane or sacred, roofs of necessity and expectation embody more artifice than floors. Open to the heavens a building is merely a corral or cage. Even wind protection is minimal. A roof makes the fenced-in area a structure that seemingly defies gravity. At its most fundamental, the roof appears to be magical, sitting above the ground. We think that we intuitively know why a roof does not collapse, but that is actually learned information, based on our daily experience of seeing buildings. In fact, very few people actually understand how building loads are carried to the ground. Thus the structure that supports a roof and transfers weight to the posts and infill of walls is not merely physical; it is metaphysical. Wood in this and other visible structural situations is a sinewy material once alive but now dead and put to the use of support, of gravity defiance, or at least (for those of an engineering bent) gravity transference. But in fact we cannot *see* the forces that the architects and engineers tell us about when we look at a Chinese temple or a stave church or a neighbor's garage.

Solutions to the challenge of spanning and covering a space for the most part involve a skeleton over which are laid overlapping layers of material that will shed rain, snow, and ice. Vertically aligned poles (in small structures) or beams are laid between the eave and the peak of the roof, radiating outward if a circular roof, running parallel if a rectangular roof. In framed architecture commonly found in the West, the eave-to-ridge members are called *rafters.* When properly covered with thatching material (usually tied grass), or horizontal weather boarding and layered tiles, slates, or shingles, the roof is rainproof. Framed houses built in the first centuries of European settlement in North America tended to use rafters set widely apart with horizontal bracing members, or *purlins,* placed between them. But the abundance of lumber made practical a method of closely set rafters joined by thin slats nailed to them, and to the slats were nailed the shingles. More wood was used, but less joinery.

Testament to the importance attached to the framing and joinery of the roof is the way in which it is left visible on the interior in many cultures and across many centuries. The cynical among us might suggest that this is merely the result of sloth on the part of those early or ethnically different builders. But that explanation loses whatever potency it might have when we look with any care at that framing. Whether the building is an ornate European cathedral, a house in tropical South America, Africa, Oceania, or Indonesia, or a Maori *marai,* much of the framework is elaborately decorated with carving or paint. [FIG. 29] Decoration directs onlookers to notice the roof, to be drawn to the qualities of protection and mastery of the material—wood, grass, bamboo—that constitutes it.

The shape of the roof as seen from the outside is also meant to draw the onlooker's gaze, to consider it as having both a form and a function unknown in nature and in fact *meta*natural. The upswept eave lines of pagodas and various secular Asian buildings are not intended merely to shunt rain away from the sides and posts supporting the structure, though they may indeed do that. Vernacular housing in many parts of Indonesia displays what to foreign eyes is an ebullient, even flamboyant roofing tradition. Built of bamboo or native hardwoods, the rooflines arc dramatically upward at each end, the ridgeline resembling the hull of a boat. Practically, these roofs are not necessarily what one might expect in a region where the rains fall heavily and often. But that is not

FIG. 29

Marai. Rotorua, New Zealand. A holy meeting place of the Maori peoples of New Zealand, marais are highly decorated, with carving on nearly every surface that retells their history and pays respect to the powerful forces of their faith.

the point. The Toraja peoples of South Sulawesi, Indonesia, believe that their ancestors migrated from the China-Cambodia borderlands and build roofs that pay homage to that journey, and they decorate the framing members and gable end supports with color and carved decoration. [FIG. 30]

We have come to accept the unseen reasons for the roof's stability and permanence, much as most people accept that a huge multiton passenger jet will actually fly, although in some part of the human brain it seems impossible, even if we understand the theory of lift. But when a roof collapses, almost no one blames the material, and for the most part rightly so, since the failure is usually that of the engineer who did not properly calculate the stress and strain or the potential weight load to be carried, or the architect who pushed the envelope a bit too far, or the builder who was incompetent or sloppy. The physical properties of the

FIG. 30

Tongkonan house. Sulawesi, Indonesia. Photograph by Yoshio Komatsu, Built by Hand. *Courtesy of Yoshio Komatsu. A masterpiece of functional and aesthetic construction, these houses are both sturdy in their frame construction and imposing in their shiplike form.*

material are fixed and knowable; it is we who fail to honor them or to see that metaphysics does not trump the physics of wood.

Balloon Framing and the Disappearance of Wood

By 1840, advances in technology, industrial organization, and engineering changed the way people in industrialized societies built in wood. Harnessing water and steam power and improvements in metallurgy (for saw blades and nails) when coupled with the factory system, brought about a mass-production system for timber processing and nail making.

At about the same time, balloon framing similarly revolutionized architecture in wood. This method relies on a series of closely spaced vertical planks (commonly called *studs* today) that run from the sill plate (the horizontal plank laid on the foundation) to the roof, or head, plate (the horizontal plank to which the rafters are connected). There are no immense heavy corner posts (though they may be twice as thick as studs) and no load-bearing large beams between the corners or in the interior of the building to carry the weight. Gone are the heavy diagonal braces between the frame members, as in European timber frame structures. In their place are short criss-cross spacing braces between studs or diagonally oriented exterior boards applied over the studs. Walls can be nailed together flat on the ground, stood up, and nailed together at the corners without fretting about whether a mortise will accept a tenon. The same lightweight construction technique is used on the roof, with closely spaced rafters nailed to the head plate and ridge pole, which in fact is likely to be a plank running the length of the roof, sufficiently wide to accept the angled edges of rafters.

Balloon framing caught on rapidly in the United States and Canada because lumber was cheap and long studs and planks were available, since there was enough virgin timber to seduce people into thinking the supply was endless. Nailing was easier than joinery, and the timbers were lighter than the hulking posts and beams of earlier framed buildings. Americans used pine rather than the heavier and harder-to-work oak, though in some instances and for some specific uses (the ridgepole, for example) builders still chose hemlock, oak, or maple. Balloon framing also meant that placing windows and doors was easier, as long as the carpenters knew enough to install weight-distributing headers (thicker beams, often nailed together) above and sills below the window opening. Interior spaces could be more flexibly arranged, since the load was borne through the outer walls or tied to chimney stacks.[20] Interior finishes were usually plaster over laths, which are thin wood strips nailed horizontally to the vertical wall studs and across the ceiling beams.

These houses eliminated wood as a visual element in the interior. Gone were decorated beams and visible means of support in favor of smoother, and hence more refined and polished, plaster walls. Stonelike surfaces surrounded inhabitants of the new houses, that is, until the inevitable cracks appeared. Wood became the material of smaller goods—furniture, food preparation utensils, musical instruments, toys, gunstocks. The house became less organic, more artificial: less a sure protective enclosure, more a technological tour de force, a trick of civilization and a witty and smart construction with its strengths—and weaknesses—hidden from view.

It was not so in the barn or the stable, however. In these large buildings, with great spaces for the storage of fodder and animals, the old big timbers, mortises, tenons, and treenails were still important, although they too were finally done in by the balloon frame, at least by the 1870s and 1880s. Mass production not only penetrated the manufacturing processes for lumber and nails; it exerted a powerful influence on the manufacture of entire buildings. As the web of railroad lines grew in the nineteenth century, the factory idea and the concept of interchangeable parts that had been applied to the mass production of clothes, shoes, and simple furnishings was grafted onto the building industry. Precut buildings of all sorts could be had by mail and delivered by train, and were a boon to isolated farmers in the Great Plains. But even if farmers obtained the wood in plank form and cut it all themselves, balloon framing beat big-timber barns for ease of construction, as the fears of collapse subsided the further one got from the old timber-frame barns in time and space—until the first tornado, that is.

Just as the clothing industry had established standard numerical sizing for bodies by the outbreak of the Civil War, so too boards for building became available in standard sizing. Standard two-inch-thick planks in various widths were common by the end of the nineteenth century, although present-day owners and rehabilitators of nineteenth-century houses will swear that the standard is a cruel joke. The complainers forget that the original builders did not have to worry about exact thickness in wall studs since the plaster and lath coat would amply smooth the differences between studs five and seven-eighths and six inches wide, while differences in thickness mattered not at all. Today we use wallboard that is machined to a constant thickness and that shows every wavy imperfection in wall stud widths. That is what often happens when we apply new technology over old. The older technology has within its entire job process moments when it is possible to make corrections and achieve exactitude; these are different from those of the modern building process. Interrupting or altering the older process of correction invites havoc.

Manufacturing the Materials

Saws do a fine job of cutting wood. But the resulting product is rough to the touch and slightly uneven. This is of no concern for framing lumber, even in the attic, since rafters in many societies have ceased to be a part of the decorated or finished interior. But rough-sawn floorboards are unacceptable to most consumers, and not merely because of splinters: They also fail the social test. They are inappropriately rough and coarse for civilized spaces. Furthermore, they fail the test of cleanliness, one that became increasingly important in the nineteenth century. Acceptable in the barn, they were unthinkable in the house, unless one was too poor to afford finished lumber.

It is certainly possible to smooth floorboards by hand, with sharp tools (adzes, planes, and drawknives). Mechanizing the sharp edge to smooth a board required a conceptual shift, or at least a shift in how the edge would be applied to the wood. Water power could most easily generate circular motion—the wheel turned by the water transferring the motion to the spindle or axle. Attach several wide sharp blades to a cylinder (all at exactly the same distance from the center of it), attach the cylinder to the power source either directly or by means of belts or gears, feed the rough board into the spinning blades (making sure that they are spinning towards you), and the board will be shaved smooth if you don't attempt to shave too thick a layer at one pass. Planing machines were in use before the American Civil War in the United States and in Europe. In big mills they were behemoths with long beds on which to rest the planks. The machine was for the most part built of wood and strapped together with iron bands and bolts to withstand the tremendous force generated by the waterwheel and the chatter of the plank being smoothed

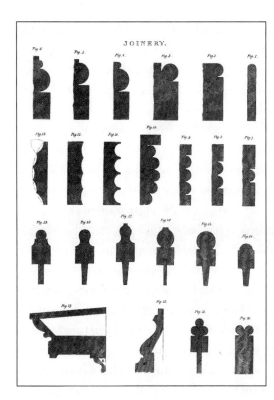

FIG. 31
"Joinery." Encyclopedia of Architecture *(London: Peter Jackson, 1852). This plate presents just one of the multitude of decorative possibilities in moldings and trim catalogued in the many building guides published in the nineteenth century.*

by what was essentially a succession of violent whacks with a sharp blade. In truth the finished product was not exactly smooth, but it was free of splinters and fine enough for footfall.

Aside from cabinetry and furniture, the other use for wood in the industrialized house was for trim; decorative wood shapes applied to places where walls met floors and ceilings and around other areas of visual interest or problematic junction. Producing moldings was a time-consuming and difficult process. Curved profiles were usually the product of the skilled woodworker wielding a series of planes. [FIG. 31] Plan books, magazines, and carpenter's and builder's guides were sources of information about different shapes and profiles, and by the latter half of the nineteenth century these were readily accessible. With more information available, demand for more variation in decoration emerged as well.[21]

Windows with glass panes were especially challenging for carpenters and builders. Included in the basic gear of the woodworking shop were tools for making the decorative thin pieces of wood (*muntins*) that held the panes in the window frame. Called sash planes, they were a vital part of the carpenter's repertoire because of the numerous windows included in houses from the eighteenth century onward. Pine was the most common choice because when devoid of knots (as it had to be for the thin bars to remain intact) it was relatively easy to shape with a plane, and it had enough tensile strength to withstand the pressures on windows from the exterior and the weight of the glass. Moreover, pine was a serviceable wood for this purpose because it took well the oil- and lead-based paints of the sixteenth through the nineteenth centuries, protecting the sash against the weather.

Even as window glass became a mass-produced industrial product in the nineteenth century, the old form of closing off window openings endured. Shutters (the name does tell you a lot) are panels that are attached to the frame of a building with hinges so that they can be closed in inclement weather. They have an ancient history and have been used in one form or another wherever the weather is foul enough to need shutting out, a condition which pretty much includes everyone for about as long as people have been putting windows in structures. Wooden shutters familiar to most people at the outset of the twenty-first century are of basically two forms—frame-and-panel with no openings for air flow, and the more complicated frame with a series of movable slats that can either let air circulate easily or be turned to shed rain.

Construction of the frame-and-panel variety entails using the mortise-and-tenon joint to connect the frame parts, and setting the panels into grooves on the inside edge of the frame, all the way around. The smart shutter maker does not glue or secure the panel, but lets it "float" in the groove so that the inevitable expansion and contraction of wood in the presence of moisture does not crack the frame or the panel. White pine was the wood of choice in the United States, but any wood with similar physical qualities—strength, lightness, and easily worked into panels or small slats—could be used. In Japan, for example, where much more attention was paid to constructing lightweight components for room partitions as well as for window treatments, cedar and red pine were probably the most commonly used woods. These and similar woods were also used in Korea and China, where paint was widely employed as a protectant and a form of decoration. These coniferous trees take oils and lacquers effectively, a result of the absorptive qualities of the wood fibers and perhaps also because the coatings were often derived from other plants.

Wood, Stick, and Brick: The "Morality" of Materials

In the 1930s Walt Disney Productions created a series of cartoon shorts under the general title *Silly Symphonies*. The most famous of these was *The Three Little Pigs*. While many analysts have interpreted the story as a paean to the steadfast and stolid values of hard work and diligence, and a comforting moral tale to calm economic fears during the Great Depression, there is more to see in this porcine parable for the purposes of this study. Recall that the Big Bad Wolf blew away the houses of straw and sticks built by the two ditzy pigs, who had made sport of the working stiff who built his house of brick while they played. In trouble as the voracious lobo came after them, they found refuge in the stodgy pig's brick house.

Brick and stone are the strong and safe building materials, at least in this cartoon and for the most part in real life. It is certainly true that they do not burn as wood will and that they are probably a lot more resistant to high winds. But even in those stolid and solid homes there is the problem of the roof. Even slate is still attached to a wood frame, and probably to a complete skin of boards nailed to the rafters. Fire ravages stone and brick houses too, but it leaves the walls standing, though weakened by the intense heat. If abandoned, wooden parts of brick and stone houses rot away, leaving the shell. Travel through the remote highlands of Scotland or the Hebrides and you will see stone shells scattered about the countryside, the legacy of the Highland Clearances that took place in the early nineteenth century, when the lairds and other holders of great tracts of land removed the crofters who had previously paid annual rent to farm the big man's land. For all its weaknesses as a building material when compared to stone, and for all of its strengths (easier to work, lighter, able to be built as a lightweight frame through which wires and pipes can be passed), wood at least has the decency to disappear when it is no longer needed. Abandoned buildings first seem to buckle along the roofline, lean one way or the other, and then collapse. Then they finish the rotting process that began their fall to the ground and eventually become compost, leaving their traces in the metals and other inorganic materials with which we once held them together or embellished them. Maybe we want to see those relics and ruins standing as a silent witness, or as a memento of those who went before. But it is difficult to separate the urge for history and memory from that of reverie.

3

The Rub of the Grain

Consider the following piece of furniture. Made of straight-grained, bird's-eye, and blister maple, its function is to hold tablewares. Probably made around 1840, it is essentially a two-door cabinet with three drawers on curved, heavy feet. It measures forty inches wide by forty-seven inches tall by twenty inches deep. On the top surface are set two smaller, shallower drawers in a small frame. The piece is elaborately decorated, not in paint but in wood. The front feet are surmounted by full columns with turned capitals. The cabinet doors are framed with four pieces of cherry, attached at the corners with mortises and tenons. Inset in the frame is a rectangle formed by cherry strips with squares in each corner. Inside the rectangle on each door is a central panel of blister maple that has been hand planed to a four-faceted shape that comes to a blunt peak in the center. The sides are framed as well with an elongated panel of plain maple. The front of the large drawer has been covered with a mosaic of maple and cherry veneer squares to resemble a checkerboard, with a larger central decorative device that resembles panels of a quilt. On the back are shipping instructions indicating the piece was made in northeastern or north-central Pennsylvania, and shipped via Binghamton, New York, to Whitney's Point, New York. The only metal fastenings are in the brass hinges of the cabinet doors, the brass cabinet door lock, the bolts that attach the pressed glass knobs to the drawers and one door, and four screws in the back to hold a now-vanished decorative backboard. Later in its life someone used wire nails to reattach once-glued molding pieces that had fallen off. [FIG. 32]

The screws were a premium fastening in their time, effectively attaching the vertical backboard to the chest from the back, where they could not be seen by onlookers. The back of the chest is crude and unfinished, and the shipping instructions are painted in black paint on the

FIG. 32
Sideboard. Maple, bird's-eye maple, blister maple, pine, cherry veneer. Northeastern Pennsylvania, ca. 1840.

lower part of that side, indicating that this side was never to be exposed. From the front it is a complex and visually arresting work of the cabinetmaker's craft. The mortise-and-tenon joinery of the frames of the cabinet doors is obvious (and meant to be so) when the doors are opened. The drawers are dovetailed and when opened the end grain of the pins contrasts clearly with the face grain of the tails. The backs of the drawers are also dovetailed, but there are fewer of them, from five in the front to three in the back of the large drawer and from three in the front to two in the back of the small drawers. (See Figure 25.) It is far more beautiful and elaborate than it needs to be to hold tablewares.

What do we learn from this piece of furniture? First, we can see that the cabinetmaker made allowances in the amount of work to be invested in the parts of the piece that no one would normally see. The back is rough, even messy. The drawers have less work in the parts where the back of the drawer meets the sides. But how did its makers (it is possible, but unlikely, that a single person made the piece) get from boards (sawmills probably cut up the trees) to sideboard?[1] This piece of furniture—common yet several steps above the ordinary in its degree of refinement—provides an opportunity for understanding how and why we shape wood to fit our needs.

Tools for Refinement

The tools required to make this sideboard were numerous, complicated, and highly specialized. The raw materials are woods that had been singled out from the run-of-the-mill lumber because of their special physical and visual characteristics. But to make a useful cabinet that would endure normal human use, planks had to be squared and flattened, the parts had to be accurately cut and engineered to give the piece the strength necessary to enclose storage space, and the article assembled and decorated to please a client. For the wood to reveal its inner secrets of art and wonder, the finished surfaces had to be smooth so that the shellac or varnish applied would shine and enhance the figure and grain.[2]

The enormous number of discrete tools dedicated to woodworking establishes several important points about the significance of wood in the lives of nearly everyone. First, the practices involved with woodworking stretched across most of the day-to-day lives of people, because some form of woodworking was part of their everyday experience, whether they were making a box to carry dung or constructing fine furniture. Second, wood was essential to nearly every other trade, especially as a material for making patterns and the hand tools and machines of other trades, from waterwheels to tobacco presses. Third, wood was the major component of nearly all forms of conveyance, save animals on which people rode (although many forms of saddles for camels and elephants were at least in part made of wood). Fourth, nearly all containers were made of wood until the twentieth century. Fifth, virtually all agricultural implements produced before the twentieth century also were wooden, except for their cutting edges. Since agriculture has been for most of human existence the occupation of nearly everyone, this is no small point.

Transitory Certainty: Planes, Planing, and Jointing

Woodworking novices usually find out the hard way. Newcomers to the trade inevitably take a look at the plans and drawings, get dimensions, and start cutting the boards to size. Trusting in both the machines and the expertise of the sawmill to produce straight lines and in the stability of the wood, they are trapped in the multiplication of error that each succeeding step of the project inevitably brings. Perfectly measured and carefully cut joints somehow do not fit cleanly and snugly. The labor of virtue thus results only in the craft of the hopeful child. Sometimes the project is beyond rescue; the next stop is the kindling pile.

What has happened in this sad scenario is not necessarily a betrayal by the sawmill—moisture and the stresses and strains of grain conspire to change the dimensions of the wood. The first job that the makers of our sideboard had to get done was to plane and joint their wood, making it straight, flat, and square. For most of the history of making furniture in which the joinery mattered, cabinetmakers got their straight edges and surfaces using planes.[3] Archaeological remains and pictorial sources suggest that the plane as known in the West probably originated in ancient Rome, or perhaps the hinterland of the Roman Empire. These early planes were most often a combination of a wooden rectangular solid (the stock), or metal sides and sole with wooden infill, with a cutting iron or blade wedged against a bar across an opening cut in the stock through which shavings escaped.

Our cabinetmakers probably had more than one hundred wooden-bodied planes. They more than likely made some themselves for specific jobs and bought others for general use from toolmakers and, later, tool companies. Making a wooden plane appears to be a relatively simple task. Take a block of wood and work it over with other edged tools until it is as straight and smooth as possible. Then chop a rectangular hole (the bed) through the block, wide at the top and narrow at the bottom. Lay the sharp iron on the bed, and (in some cases) on top of that place a *cap* iron, a flat piece of steel with a slight curve at the bottom. Drive a wooden wedge down into the bed area and you have locked the iron in place, ideally with the sharp edge barely protruding from the bottom.[4] Pushing (or in some cultures, pulling) this tool, if the blade edge and bottom are straight, slices off extremely thin ribbons of wood, leaving no splinters or gnarled edges.

This is the essential form of the plane, unchanged from culture to culture. Traditional Japanese planes differ from their Western counterparts only slightly. In general the wooden body, or stock, is not as tall. The Japanese iron is generally thicker and made of laminated steel, whereas Western irons are of a single piece. Japanese plane makers seat an additional wedge iron in the blade opening, but there is usually no additional wooden wedge. The bodies of some wooden Chinese planes have a characteristic wavelike shape, and there is a long tradition of decorative carving of the bodies of European planes. Slight regional variations, primarily in the wedging mechanism or decoration of the stock, can be found throughout the rest of the world.

The stock allows the user more force and control than a chisel offers, allowing the worker to use leg strength and, if the plane's flat bottom is long enough, to automatically cut off high spots in the work piece, eventually yielding a flat surface. Adjusting the depth and straightness of the cutting edge often requires a delicate touch with a hammer or mallet. Driving the iron down is easy—simply tap the top of the iron and wedge. In reality, however, only the surest hands and dumb luck get the iron wedged evenly and at the perfect depth. To get the iron loose and reset requires tapping the stock, either on top, near the front (some planes had a wooden button inserted for this), or on the back edge of the plane, termed the *heel*.[5] [FIG. 33]

Our cabinetmakers used jointing planes—some as long as two feet—to get dead flat and square surfaces and shorter smoothing planes to put a shiny, glasslike finish on boards. They formed the flattened pyramids

FIG. 33
Wooden planes. United States, England, Scotland, 1840–90. Only a tiny fraction of the wooden planes available to carpenters, joiners, and cabinetmakers, these were made in tool factories. Many woodworkers made their own planes from slabs of hard woods such as beech and irons they obtained from smiths. Brass plates in the right foreground are for detailed work.

of the door and side panels using special planes with their irons set at a low angle so they could shear off unwanted wood without tearing out pieces of the blister maple. They made rounded moldings and trim with planes that had curved cutting irons and soles. The sideboard is testament to the early preparation of its materials. Its joints are still tight.

Sometime after this sideboard was made—around 1840—wooden tool manufacturers began to drive individual toolmakers out of the market (at least in the United States and Europe). The same forces that were transforming the Industrial Revolution into the steel age also altered the making of planes. The wooden-body plane was gradually superseded by tools made entirely of cast iron and steel. While there had been metal-edged and metal-soled planes since the Roman Empire, the all-metal plane did not take over the market completely until the latter decades of the nineteenth century.[6]

Initially, and perhaps for reasons of consumers' suspicion of change, many planes were a combination of wooden bodies and a metal blade-adjusting mechanism. But soon these became dated, favored by the old-timers and the out-of-step. Exactly when innovation trumps the emotions and practices of tradition is unclear. It is not an irreversible linear process; sometimes the innovation turns out to be a net loss in some way. For much of the history of technology and its cultural components, though, there is a tipping point, when the new replaces the old and has either compromised with it or somehow become associated with what is safer, easier, and better. In the case of the plane, the process seems to have moved along fairly quickly, though the "golden age" of the metal hand plane was a relatively short one. [FIG. 34] Big factories had been using water- and steam-driven planers since the middle of the nineteenth century, but small shops were still the province of the hand plane until the electric motor, developed in the later nineteenth century and miniaturized not long after that, gradually took over smoothing and shaping. Carpenters and cabinetmakers still had hand planes for years after power tools had replaced them, but they became the occasional tool for the quick hand touch-up rather than the center of the la-

FIG. 34
Modern metal planes.

boring activity in the trades. Some specialized planes remained in the kits of a few tradesmen, such as coopers, whose craft remained relatively untouched by mechanization until metal and plastic barrels replaced those of wood. Not long after World War II had ended, small powered planes and other tools became part of the amateur woodworker's equipment as well.

In their heyday metal planes offered several advantages to manufacturers and workers. For toolmakers, it was easier to mass-produce iron and steel planes once the molds for pouring the molten metal had been made. It was also easier, with new powered machines, to produce a flat surface on the tool that could be guaranteed to remain more or less so, after any initial "fine-tuning" (called "fettling") of the metal body surfaces. Artisans liked metal planes because they offered an easier way to adjust the iron by means of a "frog," a screw mechanism that could raise or lower the cutting and cap irons without having to resort to tapping the blade and wedge in place, then back out, then back in, ad infinitum until the desired depth was achieved.[7] The frog also had an arm that protruded from its top that allowed for lateral adjustment of the edge to keep it parallel to the sole. Manufacturers also replaced the wooden wedge with a cam that locked the cutting element into place in the throat. Some planes had adjusting mechanisms for the mouth opening as well, to keep the opening of the throat as close to the thickness of the shaving as possible. [FIG. 35] Applying a smooth downward pressure while planing with this setup would tend to force the shaving to curl up from the wood, rather than splintering from it, producing the gossamer ribbons of wood that are the stuff of woodworkers' dreams.

Modern powered jointers have replaced long jointing planes, making this process much easier and faster than the job was in the hand-tool era. It is difficult to understand why some woodworkers still get bamboozled by the appearance of the board when it is relatively painless to square and level the piece. The answer may lie in the urge to get on with the more interesting parts of the job, since jointing is admittedly not a particularly sexy part of woodworking. In addition, it is tempting to think that the more dramatic work of the thickness planer, which shaves and smooths a board, but does little to remove humps and hollows along the board's length, will do the jointer's work for us. Jointing has none of the immediate gratification of the waves of tiny curled shavings produced by a thickness planer doing its work on the wide surface of the board. Jointing also requires patience. The first pass over the spinning blades set between two long flat steel or iron tables distressingly points

FIG. 35
Plane. Record Manufacturing Co., Sheffield, England, ca. 1980.

out that only the ends or the middle of the board are reaching the knives. A huge number of passes, it seems, will be needed to straighten the edge. Wiser (and not necessarily older) woodworkers will advise the newcomer that a jointer is one of the two power tools they should first acquire, but most of us still wait until we have acquired four or five other power tools first. Jointing seems too picayune and even a little precious, until one figures out how and why what was a masterpiece in the plans ends up looking like part of the set from *The Cabinet of Dr. Caligari*.

Most boards make the woodworker really toil to get them ready to use, and the harder the board and the more "interesting" the grain, the more difficult the job is. The very qualities that brought the consumer and the maker to the board may undermine the shaping and dimensioning that are essential to making items that look as people expect them. Plus there are also the trials of temperature and humidity. All conspire to make certainty in this procedure transitory and even illusory. In 2003 I built thirty-one new kitchen cabinets of kiln-dried northern black cherry. I religiously jointed each piece of wood surface and planed them carefully to within a tolerance of one-sixty-fourth of an inch. I was careful to seat the panels in grooves with space to expand and contract. The doors were as close to perfect as I could get them, with a few of the mortises and tenons a little dodgy but nonetheless effective. Within a few months of installation in late winter, the carcasses of the tallest cabinets were spreading apart in the dry heated winter air of the house. One of the doors took longer to make its individuality known, but eventually it developed a kink on the reverse side of a tiny knot that in the end bent the top of the door nearly a full inch from the cabinet front, while it butted flush against the bottom of the cabinet. Wood movement made a mockery of the accuracy of my machines and my certainty of success. In summer the recalcitrant cabinet door retreated to near straightness, only estivating while awaiting winter's opportunity to thumb its nose at my attempt at mastering nature with the toolmaker's and engineer's metallic certainty.

Sawing

Our cabinetmakers were more skilled and more respectful of the limits of their work. Once they had gotten their stock flat and true, they set about cutting the boards to length and width. Given the probable origin of the sideboard—northeastern Pennsylvania—they may have used Anglo- or German-American bow saws, since that region was home to many immigrants from those areas of Europe.

Their saws were for the most part updated versions of ancient tools. The earliest saws were likely to have been combinations of animal bones or teeth bound to a wooden handle. In regions of the world in which soft and straight-grained woods were abundant, such a tool might have made modest inroads in cutting. Saws in ancient Egypt (2000–3000 BCE) resembled modern-day bread knives, their form limiting the amount of force that could be applied to the work piece. By the era of the Roman Empire, however, nearly all varieties of hand-powered saws wielded in the twenty-first century were known and in use.

Like saws for felling trees in the forest, workshop saws employ a succession of sharpened teeth, alternately set (slightly bent) to cut across the grain and form a kerf somewhat wider than the thickness of the saw blade (so it won't bind). Cutting with the grain (*ripping*) requires a different set and fewer teeth per inch, since rip saws perform best when the teeth in essence shave off tiny strips or chips of the wood with each stroke, rather than slicing minute pieces across the grain.

The varied saw types are keys to both the complexity of wood management for the woodworker and the finish requirements of the final product. If you are cutting fence rails to length, you probably need only the sophistication of the buck saw, with its big teeth and wide and deep gullets between them. But making a smooth and precise cut in a one-inch-thick board for finer work (for a floor or the sheathing on a building) requires more teeth per inch. Hence the panel saw most Westerners recognize, a wooden-handled trapezoidal piece of flat steel with small teeth. It is still too coarse for delicate work, though fine enough for cutting the shoulders of a tenon, for example.

When people decided to spend serious money at the cabinetmaker's shop or, in some cities, the furniture wareroom, they expected tight and in some cases decorative joinery. Hence a coterie of small saws designed for precision rather than cutting speed have evolved over at least the past two millennia. Anyone who has cut wood with a panel saw, frame saw, or bow saw knows that the blade flexes, making straight cuts an accomplishment (especially if you try to cut the piece too quickly). A shorter saw blade with a stiffening piece of steel or brass applied to the top edge ("backed") will cut more slowly (because of its short stroke and the finer teeth it usually has), but its form also makes it easier to get a straight and smooth cut with the narrower kerf that fine joinery demands. It is almost certain that the sideboard's tight tenons in the door frames and carcass were cut with back saws.

Curved cuts, however, need thin, narrow, and flexible blades. They are made with coping, fret, and bow saws, whose blades are held in tension by a winding or screw mechanism. Coping saws get their name not because they helped artisans contend with life's vicissitudes but because they were often used to cut the intricate profile of a piece of molding into the end of another piece of wood to make a tight fit where the two pieces met. (This is called *coping* the joint.) The thin blade and tensioning mechanism allow you to turn the blade while the saw is in the cut. The modern mechanized versions of the coping saw are the band saw, a large machine with a long (around eight feet) continuous blade, and the jigsaw, a small machine equipped with a reciprocating narrow saw blade that can be worked to cut tighter curves.

Most saws used in Europe and the Americas do their work on the push stroke; the exceptions are some miter, dovetail, and narrow-blade saws. Pulling provides control rather than power and allows for intricate work, since a narrow blade would buckle, bind, and perhaps snap if pushed. Nearly all Japanese and other Asian saws cut on the pull stroke, and woodworkers regardless of their heritage or ethnicity for the most part agree that the pull stroke is not only more accurate but more efficient as a cutting stroke. The difference in orientation has to do with the cutting and posture positions of Eastern and Western cultures. Westerners elevate the work piece and push down on it, using the force of their braced legs, shoulders, and triceps for power. This may be a result of Euro-American traditions of seating oneself well off the floor in day-to-day activities, rather than close to it, as is more common in Asia, and especially Japan. In the Asian workshop this seating tradition may have translated into placement of the workpiece near the floor, accessible to the sawyer with a short-bladed saw (as most Japanese saws are), who can then stand on the piece to steady it. Westerners either lean on the elevated piece or clamp it to a work surface, also elevated thirty to thirty-six inches off the floor. Put another way, the culture (the direction) of the saw teeth may be a part of the culture of the seat.

Simple Shaping and Shaving: The Windsor Chair

In chapter 2 we saw how carpenters developed a spectrum of interlocking and pinned joints to connect the framing members of various buildings, from the modest one-room log house to the monumental sacred structure. Whether the joinery made a long timber out of shorter ones (by means of a *scarf* joint) or linked two (or more) pieces of wood meeting at an angle, the shapes of the mating pieces were the result of working the wood both with and against the grain, removing the waste with a cutting action.

Different trades demanded different joinery. What worked for a church steeple was not necessarily effective for the boatbuilder or cabinetmaker. Thus the shipbuilder's hatchet may have had a blade curved differently from that of the carpenter, who was not much interested in creating the curved surface of a boat hull. But some sophisticated furniture forms could be made with relatively rudimentary tools. The hand adze excelled, for example, in those projects in which large flat surfaces were to be shaped, as in chair seats made of single wood planks. Plank-seat chairs and benches were effective carriers of sitters' weight and longer lasting than woven seats of rushes or other flexible plant material. But as anyone who has sat in a church pew can attest, a flat surface and the fat of the human gluteal area are not a comfortable fit. We do not know exactly when someone figured out that hollowing the seat to more closely reflect the actual shape of the human bottom was a good idea for the purposes of comfort, but the idea certainly has remained important in most cultures in which sitting on furniture is common.[8] In the United States the most common example of this transition is the so-called Windsor chair, a form common during the time our sideboard was made. Derived from English chairs made since the early eighteenth century, the form continues to be produced in the twenty-first century. A lightweight chair, the Windsor generally is composed of a shaped plank seat and turned legs, arms, and back, the latter tied together with a crest rail into which the back (and sometimes arm) spindles are inserted, or *let.*[9] [FIG. 36]

The virtues of these chairs are many, from the standpoint of both the makers and the users. Windsors emerge from a long tradition of what some furniture historians have called *stick furniture,* or furniture that at one time was made from small branches or pieces of *riven* wood (that is, wood split from the tree stem in wedge shapes). With minimal shaping, boring, and cutting, inexpensive serviceable seating can be had. The simplest form is the stool, a plank or round sawed from a tree stem into which three or four legs were inserted. There are pictorial references to stools and more elaborate stick furniture that date to about 1500 BCE, and there is ample evidence of this form in all cultures in which people sat on furniture.

For the maker, the greatest virtue of stick furniture was the relatively small amount of processing necessary to produce a seat. Accurately drilling round holes in the seat bottom to receive the legs and in its top for the spindles of the chair back and sides, drilling the legs to receive stretchers (the pieces between the chair legs that provide stability), and cutting the legs to the right length were the chief tasks, in addition to cutting the seat plank and smoothing it. The Windsor most likely emerged not directly from stool-type stick furniture but from a blending of that tradition with the English country wooden crafts associated with agriculture—fences, gates, and pens—that made use of pollarded and coppiced trees and the small-dimension trees unclaimed by the Crown or landowners.

Windsors were mass-produced by country carpenters and joiners who had easy access to small lumber, especially in Anglo-America and England. Foot-powered lathes enabled chair makers to turn leg stock into decorative and repeatable patterns to support the seat, which was the only large-diameter piece of wood in the chair. The Windsor's splayed legs, braced with turned stretchers, distributed the weight of the chair and sitter more efficiently than do the traditional square-leg designs of heavier chairs made of rectangular or square frames and panels.

FIG. 36
Windsor chair. Ash and pine, ca. 1780-1810.

Inexpensive and for the most part easily harvestable raw materials were boons to rural and eventually urban manufacturers. A Windsor could also be made with a small number of tools, another virtue in economies of constant scarcity. The minimum requirements included a cutting tool (an axe or saw) to remove branches or harvest saplings, a drilling tool (an auger), an edged tool with a curved blade to shape the seat (an *inshave*), a knife to remove bark, a froe to rive pieces from the log, a foot-powered lathe to turn legs and other spindles, a mallet or hammer, and a hatchet or adze to shape and smooth the few flat surfaces of the piece (seat and crest rail). Simple planes to smooth wood and a curved-iron *spokeshave* to smooth the back and side spindles were useful but not essential.[10]

The design and construction methods of the Windsor chair enabled makers, even in small shops, to stockpile parts and mass-produce them,[11] which was especially handy during times of low demand for a carpenter's or joiner's work. Windsor forms—chairs, benches, and stools for the most part—were made, consciously or not, with the least technology and labor needed to produce a product that lasted long enough to satisfy a customer, and maybe long enough to outlast the maker. They were, in the end, accoutrements serving the goal of comfort and convenience rather than the essential functions of shelter in all its manifestations and meanings. Buildings were of longer term and sturdier stuff, or were supposed to be so.

Consumers liked Windsor furniture because it was inexpensive, comfortable, and lightweight. Its heft was an advantage because of the way most people used furniture. Until the nineteenth century, ordinary people moved chairs, benches, stools, and even tables around, setting them at the perimeter of a room except for meal times. Since much of the work in the pre-1800 household demanded that women stand or walk (in the case of spinning on the large wool wheel), chairs and tables easily rearranged were a necessity.[12]

For Windsor chair makers, its construction and design was an advantage because they only had to fashion round holes and round tenons, thereby avoiding the labor-intensive square mortises and tenons of furniture such as our Pennsylvania sideboard. The round joinery of the chair might come loose, but the multiple spindles meant that this probably would not happen all at once, though occasionally the splayed legs were forced apart by too much weight or poor craftsmanship in the making of the tenon or the hole. But splattering a chair, however embarrassing and even painful that may be, is not of the same level of calamity as the roof falling in or the floor giving way.

Joinery: Pounding, Chopping, and Paring

Nomenclature often is a telling clue in cultural history. In the case of the woodworking professions, contemporary language is less revealing than that of the past, at least in western European and American traditions. Open the telephone book and you can easily find "carpenters" and "cabinetmakers" and "woodworkers." Carpenters do the coarser tasks of building: they frame houses, put down floors, and install kitchens. Cabinetmakers are generally associated with the craft and aesthetics of the furniture-making trade. Woodworkers are a more generalized lot, encompassing the above two groups and anything else related to the material. But confusion arises because the term *cabinetmaker* might be applied to those who make (some might scoff, "knock together") the "boxes" that become kitchen or bathroom cabinets. This matters most to fine furniture makers, who recall the traditional, guild- and union-based titles conferred on craftsmen according to their skill level.

But who hears the term *joiner* now? The term was once a common part of the language and joiners part of the hierarchy of woodworkers, ranked above the carpenter and below the cabinetmaker. *The New Shorter Oxford English Dictionary* defines joiner as "A person who (as a profession) does light wood work and esp[ecially] constructs furniture, fittings, etc. by joining pieces of wood."[13] Cabinetmakers are similarly described as "skilled joiners," linking them to the original Anglo-European meaning of the word "cabinet," which was derived from the Old French word that meant "a secret receptacle or repository . . . a case or cupboard with drawers, shelves, etc. for storing or displaying objects."[14] For the wealthy collector in late medieval and early modern Europe, the most elaborate product of the cabinetmaker's art was the "cabinet of curiosities," a rectangular box (usually on a stand) whose doors opened to reveal a multitude of tiny drawers and shelves in which precious "curiosities" were stored. The cabinetmaker's honored position originated in the demand for decorative storage and display appropriate for the contents and status of the collector, whose assemblage was testament to his or her surplus riches, superior education, and allegedly greater abilities of appreciation.

Complex joinery such as that we saw in the last chapter's treatment of ecclesiastical buildings is also common in East Asian architecture and furniture. Chinese, Japanese, and Korean woodworkers made use of the mortise-and-tenon, to be sure, but the aesthetics of Asian furnishings often demanded a more complex joint. An emphasis on slender and graceful legs and concealed joinery led cabinetmakers to joints whose complexity was great, yet often hidden from the viewer. Finishing differences between end grain and face grain impelled craftsmen to the miter joint, in which two or more pieces of wood are joined at forty-five degree angles so that only face grain shows. (End grain not only looks different, it absorbs more finish and thus often becomes darker in color than does face grain.)

The miter joint, if merely glued or nailed together, is weak, and almost useless if any stress is applied to it because the entire joint surface consists of end grain. The way to strengthen the miter joint's weak character is not through exhortation or encouragement or deprivation, but by introducing face-grain glue surfaces. Cutting a mortise-and-tenon on the inner faces of the miter provides an interlocking but invisible joint that provides face-grain glue surfaces. A simpler method is to cut a groove across the corner of the joint, into which a "key" or spline is inserted and glued. But this is objectionable to some people because it is visible.

Joining three pieces of wood (a leg and two horizontal pieces, as in a table or chair) requires more complicated joints, at best interlocking and with a minimum of mortising. Eastern and Western furniture makers and their clients came to favor thinner supports and reinforcing parts in refined and highly finished pieces. In part, this prejudice for the dainty over the heavy and blocky is part of the artisan's and the client's desire for a showy tour de force, a form of competition to produce effective functional pieces of furniture that appeared to flaunt traditional assumptions about the limitations of wood. One way to challenge long-held views of the appropriate size of furniture parts was to use denser woods of fine grain that were less likely to split or otherwise fail. Another method was to use more bracing, such as stretchers between legs, while paring down the diameter of the support or brace. This raised problems when joining the pieces, however. Thin legs and aprons (the frame under a table top or seat) left little room for tenons and mortises. Furthermore, in a complex joint of three (or more) pieces of wood, the risk of hollowing out most of the leg (into which the apron was most likely tenoned) was acute. Chinese furniture masters developed joints so complicated that the various-sized tenons they used resemble a city skyline in miniature, all of which were let into correspondingly cut mortises.[15]

Without glue or pins it is nearly impossible to construct a joint that cannot be separated in at least one direction, since the pieces have to be fitted or slid together. The most knowledgeable woodworker narrows the number of directions in which a joint can be separated to one, ideally the direction in which the object is least likely to fail or slip—that is, the direction in which normal use exerts the least pressure. Since a roof exerts most of its force outward and downward to vertical wall supports, the cleverest designers and builders dovetail rafters, floor joists, and ceiling plates into vertical supports with the wide part of the dovetail toward the exterior. Thus the forces of gravity and stress are working against the strength of the strongest part of the joint. On a smaller scale, the weight of a tea-table top and its supporting central column will tend to force its low, splayed legs apart, so cabinetmakers dovetailed the legs into the column base, counteracting the outward thrust of the weight above. The best drawers are made with the side walls dovetailed to the front part of the drawer, on which you pull to open it. Theoretically the joint might slide outward, in the direction in which it was assembled, but glue is usually sufficient to hold that joint because outward is not the direction in which stress is commonly applied. A tenon that sits tightly in a *through mortise* will have a more difficult time wiggling than it will in a shallower joint. But through mortises allow no room for mistakes in making the joint and are hence the mark of the confident and skilled cabinetmaker. [FIG. 37]

For the house framer and the maker of more sophisticated furnishings and goods (such as vehicles), the major disadvantages of angled interlocking joints were that they required additional tools and more time to do the job. Chief among the tools were chisels. Generally made of a metal edge and a wooden handle, chisels are driven into wood to sever fibers and then remove stock. Ancient chisels had cutting ends of bone or stone, and the technology of making the cutting edge and the methods of sharpening it have changed dramatically in the industrial age. Western chisels are usually drop-forged, flattened on one of the dimensional sides (as in one-half inch, three-quarter inch, and so forth), and

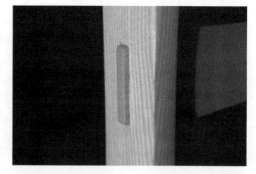

FIG. 37
Ted Blatchly, table with through mortise and tenon. Ash and walnut, 2005.

ground to sharpness on the other side. That edge is then honed or stropped (usually on a leather band) to remove the tiny burr formed on the leading edge by the grinding or sharpening motion. Japanese chisels for the most part are hand-made by smiths, with a softer base layer forged to a harder layer, a process that yields a sharper edge.

The heavy work of chopping out a mortise requires a chisel with a thick blade and flat sides so that you can pound it into the wood and lever the chips out. It's a crude process. For more delicate work there are paring chisels, which have beveled top edges on the long sides as well as on the cutting edge. They are pushed with the hand rather than pounded with a mallet. Curved edge gouges are used for carving details. Turners press long-handled gouges and scrapers against a spinning piece of wood to produce (one might say "turn out") spindles, bowls, plates, and other pieces that are round in cross-section. A complete woodworker's shop might well have two or three dozen chisels and gouges, while the timber framer might need but a few.

Understanding the methods of shaping wood with saws, augers, gouges, and chopping and scraping tools encompasses only a small part of the intricacies of shaping and finishing. Yet learning that much is no small achievement for the worker with hand tools only. What sits in the workshop or on the ground is a great distance from the standing tree, but it is still coarse, unfinished, unrefined, and to many, uncivilized— still too close to the wilderness, the beasts, and the "country matters" that were subject to ridicule and scorn by the well-to-do. In the case of the framework of the house, coarseness did not matter because of the finished layers applied later in the process of building. But visible coarseness was for many people an embarrassment; it meant identification with those seen as "lower class."

Shaping and Curving

Framing the side panels and the door fronts of our sideboard are curved pieces of molding. Putting those curves on wood, both over the length of a board or across its edges, introduces a new series of problems. Moldings and trims for houses were first made by carefully chopping out the unwanted material with adzes. Just as the straight-bladed adze cannot leave a smooth flat surface on a beam, neither can the curved adze head produce a smooth curved surface, whether hollowing a vessel or chair seat or a length of molding. The action of the adze is essentially that of a big chisel, and like the chisel on a flat surface, both force and control are difficult to achieve. It takes no great imagination to figure out that if a flat straight plane accomplished much of this for the flat surface, introducing curved irons and soles to the plane form could bring more control and about as much force to the job as one could with an adze.

A workshop equipped for making curved surfaces would include at least fifty to one hundred planes to meet the demands of an informed clientele. As the Industrial Revolution democratized housing in many parts of the world in the late nineteenth century, the demand for complex (and hence higher status) molding and curvilinear decoration for houses increased dramatically. The diversity of molding possibilities and the relatively low volume of such materials used in housing and furniture left room for the small molding shop and the individual carpenter and joiner as the industrial production of coarser construction materials drove the small manufacturer of those goods out of business. Short runs were unprofitable for big factories and milling operations, and long runs were likely to be money losers because taste and fashion were fickle, changing quickly because of the wider dissemination of visual information that began in this era.

For the smaller molding and shaping businesses, however, the down side of this opportunity was the expense of completely equipping a workshop with all the necessary planes to accommodate the demand. One way around this was to use planes with interchangeable irons and adjustable "fences" to guide the cut. Combination planes date to the sixteenth century, but these were rare and designed for cutting different widths of straight-sided grooves. The heyday of the combination plane in the United States was the later nineteenth and early twentieth centuries, when the Stanley Rule and Level Company of New Britain, Connecticut, introduced the "45" combination plane, advertised as "seven planes in one."[16] It was supplemented in 1897 by the more expansive and complicated "55," which offered 42 cutters as standard equipment and 41 additional special cutters, as opposed to the 21 standard and 30 additional cutters of the "45." [FIG. 38] Called "a planing mill within itself," it was in many ways an engineering marvel, but it may have been too complex for most woodworkers. The long setup time for each operation may have disappointed those who had figured the "45" or "55" would relieve them of carrying a bunch of wooden planes to a job, but who then discovered that they did not have the patience to fiddle with the apparatus every time they needed to cut a new profile. A few manufacturers produced (and one or two still offer) combination planes, though they are for the most part the domain of the dedicated hand-tool user or the collector.

FIG. 38
Stanley "55" combination plane. Stanley Manufacturing Co., New Britain, Connecticut, ca. 1900. Wooden molding planes ca.1875-1900. In the background are the fifty-five cutters that came in with the tool.

Turning

Straight lines are about exactness and precision, about forcing a set of values unknown in nature on wood. Perfect circles, cylinders, and symmetrical curves are about *improving* the curves nature produces, of transforming the wild into the graceful aesthetics of the salon and of fine art. The turned columns and capitals on the front of our sideboard might have been made by carefully shaving and smoothing rectangular stock with a curved drawknife or spokeshave, but the process is time-consuming and the results often disappointing in their irregularity. A lathe brings the regularity of the machine to the workpiece, and with considerably less effort than it would take to do the job with other tools. As the lathe moves the wood, the turner presses a gouge against it, maneuvering the edge into the workpiece to get the desired shape. Until other sources of power were discovered, the power for the lathe still came from the user (or in some cases, a helper). But even in early forms it was a more efficient way to shape wood—within its limitations—because it multiplied the force brought to it.

Material survivals and visual evidence demonstrate that turning is at least as old as the Egyptian dynasties of five to six thousand years ago. Traditional furniture from sub-Saharan Africa shows lathe work as well. The simplest of these machines, and the form most likely used for thousands of years, is the bow lathe. The turner simply pushes a bow back and forth, the bow's string supplying the rotation to the work-piece. With the other hand the woodworker holds the cutting tool against the spinning wood. People usually used this lathe to produce small articles such as the legs of stools. Portable and easy to set up, they continue to be used in parts of North Africa and Asia, among people with little capital to invest in machinery.

Another simple form, the pole lathe, is still occasionally used, primarily in central and northern Europe. Pole lathes operate by means of a large pole braced and secured on the ground. A string is attached to the top of the pole, wound around the wood to be turned, and attached to a treadle secured to the ground or a brace. Pressing down on the treadle spins the workpiece and pulls the pole down. Releasing the treadle allows the pole to spring back for the next pass. Both forms only allow work on alternating spins of the workpiece, but the pole lathe generates far greater force than the bow lathe. Easily fabricated from materials in the field, pole lathes were common among the rural English chair "bodgers" who made rough chairs and other traditional "country" furnishings (such as Windsor chairs), fencing, and other accoutrements of farm life well into the twentieth century.[17]

Impressive as the work turned out by these lathes was, the great advance in turning came with the introduction of power sources that spun the stock continuously in one direction. One of the new machines was the wheelwright's lathe, a two-wheeled apparatus powered by an assistant who turned a crank on a large wheel, transferring power to the smaller lathe wheel by means of a leather belt or rope. Used by wheelwrights to turn wheel hubs, these lathes were also useful for turning larger pieces, such as staircase newel posts and even columns and mill shafts.

The mechanical advantage (the power) of the wheelwright's lathe over less sophisticated machines was partly counterbalanced by increased manpower requirements. Thus the lathe was in limited use until the power came from running water, a treadmill, or (eventually) electricity. Then the lathe became at once a fast, efficient, potentially dangerous, and almost magical tool. As it turns at high speed, the margin for error with a sharp tool is far less than that of the older forms, yet the speed with which material can be removed is breathtaking.

The history of turning appears to have had two distinct phases. The ancient Egyptians developed highly sophisticated techniques that do not seem to have had much of an impact on other cultures or for thousands of years afterward. In the rest of the world, squared mortise-and-tenon joinery and carving, rather than rounded forms, were common from about 3000 BCE onward. Chinese, Japanese, Korean, and Southeast Asian furniture tended to be composed of wooden members shaped with drawknives or other cutting and scraping tools. Superior metallurgy, especially in Japan, may have had the effect of not "pushing" other technology, since the efficiency of a sharp drawknife made short work of shaping. In tropical areas bamboo provided a sturdy round material for objects when needed or desired, which also might explain why turning seems to have generated relatively little interest in Asia when it was becoming popular in Europe.

Late sixteenth-century European wooden articles such as chairs and spinning wheels indicate a dramatic increase in enthusiasm for turned legs, spokes, and smaller items. The dearth of turned elements in furniture and architectural woodwork until that time may be related to the amount of work it required to turn items, relative to the labor involved in producing square or rectangular profiles. A moderately skilled woodworker could also make serviceable, undecorated spokes for wheels with a spokeshave, a two-handled tool with a small curved blade in the middle. Pulled toward the worker over a piece clamped into a shaving bench, the spokeshave quickly removed the corners of a piece. What, then, was the point of turning, at least until people harnessed exterior power sources, especially those that did not require other people or animals?

A technological explanation for the surge in the manufacture of turned legs, spindles, and applied turned decoration is, in the end, only a partial one. If there was what some economists call a latent demand, or a desire on the consumer side of the market for goods known but not yet available or financially accessible, whence came it? We cannot assume (though many analysts of the past have done so) that wanting circular things is "natural." Nor could we call it just common sense. When you investigate instances in which common sense is used as an explanation or justification for some action or other, it turns out not to be so "common" after all, but in most cases merely the tactic by which people assert that what *they* believe is what all should.[18]

Why make rounded furniture and tools? Rounded ends of spindles and spokes may have been easier to insert in a chair seat or a rim, especially if the artisan had a good auger to drill the holes. The tool is easier to use than the chisels and mallets required to make a mortise. Lathe work also allowed craftsmen to expand the availability of "fancy" design, often used in what seems to modern eyes odd or strange places. Turning opened up a new horizon (and a new cross-section) for the finished surface. Woodworkers could more easily imitate certain forms in nature, such as fruits and seeds. This was a second big step away from nature—first straightening, then imitating.

Turning also offered the chance for a new type of tour de force, a new opportunity for further artifice, and a new opportunity to try to create sensational effects with wood. Turners carefully pared away wood that old-timers thought essential for support in a chair or in a heavy chest or cupboard. With bigger and faster lathes the legs of a piece were made lighter and a source of decoration, with the deepest incisions a small wonder to the eyes of those accustomed to older rectangular forms. Cabinetmakers set heavy articles of furniture on bulbous turned "feet," some of which were cut so far back at the place where they joined the carcass of the piece that there seemed hardly any support at all. In some cases the deep turning did not work; the piece failed, collapsing the chair or chest. The artisan had made the mistake of pushing the decoration too far into the heart of the wood, perhaps not knowing that the grain or the species could not support what had worked elsewhere. But that was the game of the enterprise.

Spheres and ovoid shapes are also much easier to produce on a lathe. In this case turners could make multiple pieces from a single wood spindle by turning them next to one another, paring away as much wood as possible between the shapes. Once removed from the lathe, the pieces could then be separated with minimal cutting and finishing. The enthusiasm for these forms is evident in the ornament applied to some of the otherwise heavy case furniture of the era, such as the English court cupboard. Halves of spheres or of egg shapes were routinely applied to the exteriors of pieces and finished in a color different from that of the main body, or were made of a different wood altogether. Craftsmen glued

FIG. 39
Chest of drawers with doors. White oak, red oak, chestnut, white pine, soft maple, American black walnut, cedar, maple, cherry, beech, cedrela, snakewood, rosewood, lignum vitae. Boston, Massachusetts, 1650–70. Courtesy Yale University Art Gallery, Mable Brady Garvan Collection. The broad palette of woods fine cabinetmakers employed served both practical and decorative purposes and illustrates the international trade in woods as early as the mid-seventeeth century.

half-spindles to the legs of these pieces as well, producing a sort of visual joke or pun—the delicate spindly leg pasted onto the real rectilinear leg. (These were usually made by turning a piece and cutting it longitudinally or, less commonly, by gluing two boards together, turning them, and splitting the piece at the glue joint.) [FIG. 39]

Turning also made possible the introduction of fancy work into the structure of tools and machines. The efficiency of a spinning wheel is not enhanced by having decoratively turned spindles in the supports or as spokes in the wheel. But turning may have been seen as an appropriate— even necessary—cultural statement for a job that was for the most part the domain of women, done in or near the house. The tools for spinning and winding yarn and flax all operate with a circular motion, but only a few must themselves be round in cross-section; yet nearly all are so by the early seventeenth century. Perhaps the popularity of rounded shapes was a manifestation of high-culture ideas that women were (allegedly or desirably) further from nature than men, and thus had an elevated status (if only in words or superficial material forms). Ordinary people may have applied this to their own situation in small ways—even in the mundane realm of tools—whatever the realities of their economic and social lives may have been.

Turning also made it much easier to make bowls and other hollowed vessels. Faceplate turning, in which what is ultimately to be the bottom of the bowl is attached to the powered end (the *headstock*) of the lathe, eliminated the need to laboriously hollow a block of wood with an adze or inshave. It opened new worlds of working wood for decorative and practical uses. Gouges shave away the core of a block of wood and similarly shape the exterior of the piece. One result of the rise of the lathe is that the bulbous and intricately grain-patterned burls that grow on many species were more commonly used by artists and artisans than they had been when only hand tools were available.

Turning, then, is as much about humor, fancy, and distance from nature as it is about more formal considerations of design and function. While it is possible that greater access to better technology enabled more people to own turned items, the demand for turned forms seems more likely to have arisen from the long effect of the international outreach of European nations that began in earnest in the early sixteenth century. The possibilities of trade and the centralization of political power in England around Elizabeth I, for example, drew the wealthy, powerful, influential, and aspiring elites to the court in London, as they had for monarchs and the aristocracy in other European capitals. As economies became less local and more regional—even national and international— the countryside became more "protoindustrialized" than it had been before, when technology revolved about the central landholder's estate and economy. Ultimately the early trades and activities associated with the production of goods for a wider market became the soil from which the Industrial Revolution began. Richard Arkwright and James Hargreaves, for example, are considered two of the fathers of that great change. They made their marks by multiplying the machine capacities and power that the people of the hinterland had already begun to use.

This does not diminish their inventive genius; it merely pays some attention and credit to those who came before.

Nails and Screws

Nails and screws changed almost everything about joinery. Nails cleaved the carpenter from the cabinetmaker and sent the joiner to oblivion, a relic found in old census and tax lists, identified in the dictionaries and encyclopedias of bygone trades. Worse than that, once wire nails were available, the joints they held together were even weaker than those of the rectangular cut nail. The latter held the pieces by shearing the wood fibers and establishing more resistance between nail and wood. Round nails slid in between layers, paradoxically promoting cracking while making for a bum joint.

Screws, however, make for a very strong joint. They go back at least as far as Archimedes in ancient Greece. For the most part early engineers didn't bother with wood joinery when they thought of using screws. They thought in bigger—much bigger—terms, employing the screw as a mechanism to move large quantities of material, such as grain in grist mills or water in canals. The screw was also an important part of the Industrial Revolution, since the moving screw needed only a circular power source, and that was readily available once water power had been harnessed. The idea that one might draw two pieces of wood tightly together with a spiral was also not new, but it was impractical. Smiths could make cut nails by pounding a piece of stock flat and then cutting it, a process easily repeated (in concept) in a rolling and cutting mill. Round wire nails came later, as metallurgical technology evolved and mills could produce more flexible wire stock.

Screws and their relatives, bolts, were a technological challenge of a different order. First, they were round in cross-section, which was considerably more difficult to accomplish at the forge than was the flat bar that could be pounded against an anvil. Even more challenging was the spiral, which had to be cut in a continuous motion in the metal blank. The knowledge to do this was not arcane, but the process was not an easy one, requiring at least a modicum of regularity in the thread (essential in bolts, important in screws). One could use a lathe with metal-cutting tools for the task, especially if the metal to be cut was relatively soft, like brass. The job was correspondingly more difficult for steel and iron. Perhaps this explains why there were only four wood screws in our sideboard, to hold up an otherwise unsupported top.

Nails and screws, in concert with prebent and predrilled pieces of sheet metal, for the most part signaled the end of joinery in housing, with the occasional exception of the joints between rafters and horizontal wooden beams, called plates. Rafters are still sometimes notched to fit over the plate, but even this joint has been increasingly abandoned in favor of a pierced metal connector. These are an effective way to join framing members because they allow for driving nails only into face-grain surfaces that hold the nail far better than does end grain. Floor joists, for example, are more effectively held to a perpendicular carrying beam by a metal hanger shaped to cradle the joist with "ears" through which nails are driven into the face of the beam. The alternative had been notching the beam and the joist so that the joists literally sat on the beam. Less experienced or less ambitious carpenters simply nailed the beam into the end grain of the joists. This was the lazy, ignorant, or crooked builder's tactic, since the weight of people and furniture and the house above could push the support beam down and the nails out of the end grain.[19] Metal hangers, plates, and other devices for joining wood have made house framing easier and faster, but whether they have improved the job in other ways is open to question. It is not clear that these technologies have made housing cheaper for the consumer, though they may have done so for the builder, who can charge retail prices for the goods, having bought them at wholesale or "contractor's prices." Labor costs are what contractors try to avoid since there is no profit in them; materials costs bring a profit. Mass-produced furniture and furnishings of the past 100–150 years similarly show an evolution away from wood-to-wood joinery to the use of metal fittings—first nails, then screws and plates, and in some cases pneumatically driven staples and nails.

Smoothing

Our cabinetmakers passed some of the tests of skill in manipulating wood (and hence nature)—the ability to impose straight lines and make pieces fit together. Now they had to find ways to impose a smoothness of surface so precise that it reflects an image and glistens like a mirror. In nature the most common (maybe the only) instance of this condition is the surface of a body of calm water. Other than that rare condition (no current and no wind), such reflectivity occurs only in manmade materials like glass and polished metals. In essence what woodworkers do when smoothing a surface is to manipulate wood into a condition that is both artificial and a testament to their skill with tools. The surface that is smooth to the touch and "finished" to reflect an image also lets us see the beauty once buried in the grain or concealed by roughness.

Grain and figure are the distinguishing marks of different tree species, and in their diversity they illustrate the infinite variation of nature. However much they may look alike, no two boards or trees are exactly the same. Painting the surface makes them disappear. Whether we like it or not, the surface matters in architecture, art, and other aspects of our immediate surroundings. Humans may choose rough surfaces for aesthetic and cultural reasons, but those conscious and unconscious choices are in a context in which the artifice of the smooth surface is the norm, or at least the standard. Some may prefer the adzed beam, exposed because it reminds them of an imagined earlier era, or favor walls and furniture made with bark-skinned logs or branches in the camp or cabin in the woods, but those are reactions as well as assertions. There is nothing wrong or odd about these predilections, but they in no way contest the refined standard of the smooth and reflective surface finish in objects.

Smoothness is important even if you don't care about grain and figure. Paint a rough surface and it will still be rough. Hidden wood surfaces, surfaces so far from view people cannot see them well (such as the high ceilings in a church), and those in rough areas (barns and factories) do not need smoothness; the quality they must have is strength. In the end, getting smoothness out of splintery and fuzzy surfaces was another way to separate craftsmen from ordinary woodworkers.

There are two methods to tame the rough—abrasion and cutting. Abrasion is the way most of us go about it, whether by hand or by machine. We sand things, usually with sandpaper, a product made by gluing sharp pieces of some mineral to a sheet of paper or stiffened textile. But sandpaper is a fairly recent phenomenon. The concept has probably been around for centuries, but not the product. It was too expensive to produce and harder to manipulate before sophisticated systems for measuring the size of the grit were available. Sanding did not get the attention other technological and practical problems received because there was another method that got the job done—cutting and burnishing with planes, scrapers, and other metal implements. Once people had figured out how to make steel and how to sharpen it on a stone, making smoothing planes was a fairly straightforward operation, at least on the surface. (Planes for shaping wood into curves are another matter, as we have seen.)

Since no trees and no boards are alike in the configuration of the grain and the presence of figure, no one plane or one angle of the iron would work all woods. Planing against the grain is usually an invitation to disaster. The iron splinters the wood instead of slicing it as the metal edge digs into the face of the grain. Figure, whether swirls, bird's-eyes, crotches, curls, blisters, or quilts, will tear out in chunks, leaving ragged depressions in the surface that sometimes are as deep as a sixteenth of an inch. Taming the very qualities that make certain woods aesthetically pleasing and desirable often requires as much time and labor as planning a project, processing the raw materials, and measuring, cutting, chopping, gluing, and assembling it combined.

You might get somewhere close to that glasslike smoothness with a razor-sharp plane with which to remove microscopically thin layers of wood. This method takes what seems like forever, and it does not work in every case. About the only tool that can achieve this surface on wood is the scraper. It is the simplest of tools—a flat, thin piece of metal with a tiny burr curled over its edge.

FIG. 40
Scraper and shavings.

The scraper finishes the smoothing job of planes by slicing off ultra-thin layers of wood as it presses down the remaining ends of the sliced fibers, thereby leaving a lustrous surface. [FIG. 40] Sometimes veneered surfaces can be worked with other smoothing tools without damage, but highly figured woods and thin modern veneers are easily ruined with coarser implements. There is little doubt that someone spent many hours scraping the panels and drawer fronts of our sideboard, especially the geometric marquetry of the large drawer. Scrapers are the essential finishing tools for inlaid and marquetry surfaces, as well as for just about anything that is veneered. Because the other materials used in many in-laid designs, such as ivory, brass, mother of pearl, and silver, are so different in their hardness, to smooth the wood parts requires small tools that can be worked up to the edge of the inlay. Scrapers are the best—some would say the only—tools that can do this. Marquetry—a picture or pattern made of wood pieces—requires a tool that can safely smooth woods with different physical characteristics as it encounters a plethora of grain direction changes in a very small space.

For all their simplicity, diverse geographic origins, and age as tools, scrapers are to this day among the most mysterious instruments in the woodworker's arsenal. Getting the edges of the tool flat and straight before creating the burr is a task easily bungled, even with modern clamps and jigs to aid the unsteady or uncertain hand. Running a hard metal tool, called a burnisher, consistently across the edge at the proper angle is even more challenging, probably because the veterans of this activity know that a confident stroke is the key to success, as it is in painting or just about any sport in which there is a stroke of one kind or another. Believing that the burr you cannot see (but can feel) will really do what it is supposed to do is a similar test of the heart and hand. Finally, actually scraping the piece on which you have labored for hours to get to this point in the project is an act of faith.

No wonder, then, that almost all woodworkers opt for sandpaper when confronting the smoothness problem, especially with figured woods. You can force sandpaper, rubbing it harder to remove material, whereas trying to force a plane or a scraper simply means a crooked surface or an injury. Better yet for workers of the past fifty years or so, an electric motor can do the forcing. Just follow the progression of sandpaper grades from coarse to smooth grits, skipping none of the increments, and you are guaranteed a smooth surface, at least to the touch. It works extremely well, in the end, especially if the woodworker proceeds all the way to the extra-fine grades.[20] But however fine the grit, sanding is still abrasion, and the surface does not have the sliced and slippery gleam of the planed and scraped.[21]

Mechanizing the Hand

Throughout the nineteenth century furniture companies strove to build pieces that economized on labor-intensive joinery. Large-scale planers and jointers were some of the first machines developed after mechanized saws were in use. These were huge machines with long beds on which to rest planks. The wood was moved forward at slow speed and fed into a head in which knives were spinning, until the plank was relatively smooth and of uniform thickness. Jointed, planed, and cut to the proper size by machine, the factories and mills had more or less eliminated the work of generations of apprentices and workers armed with jointing and smoothing planes.

Joining technologies, however, were a somewhat more difficult challenge, even if dovetails could be jettisoned in favor of nails and screws. First, some consumers objected to nailing, since it was hardly decorative and was easily distinguished from older and higher-end work. Put another way, it was carpenter made, rather than cabinetmaker made. Second, the process by which the interlocked dovetail joint was produced seemed out of the realm of duplication by machine. One strategy required breaking away from the dovetail shape while maintaining the sort of wood-on-wood joinery that status furniture connoted. The Knapp dovetailing machine actually did not produce dovetails, but a row of circular dowel-like protuberances on the side of the drawer front that were a result of pressing a series of spinning bits with a void in the center of each. The drawer side was drilled and shaped to accommodate the small cylinders and the whole thing glued together. It was a functional success and for a time a commercial success as well. [FIG. 41] Dovetailing jigs and machines were subsequently developed in the twentieth century, in both large format for the factory and smaller sizes for the home workshop.

Elsewhere in the construction of furniture, complicated joints were replaced or altered by machines. Except for products of the Arts and Crafts Movement of the latter nineteenth and early twentieth centuries, mortise-and-tenon joinery was largely replaced by the tongue-and-groove, which was far easier to automate, or screws and other fasteners. While one could certainly cut tenons on powered saws, mortises presented more difficulties, since about the only way to mechanize chopping them out of a workpiece was to drill a series of holes. The continuous tongue-and-groove joint got rid of the pesky mortise problem, since the groove also could be made by running the workpiece into a spinning blade or bit. It is an effective joint when properly glued, but more likely to fail than the mortise-and-tenon, which can be pinned, and, in the case of the through mortise-and-tenon, has less opportunity to wiggle loose.

The Arts and Crafts Movement celebrated the mortise-and-tenon joint in the face of the factory revolution that was making it obsolete or at least less profitable for big business. It was a conscious effort to resurrect hand production or at least the making of an entire piece from start to finish, rather than as a result of assembly-line production. In England, Europe, the United States, and Canada colonies of artisans turned out work that celebrated visible joinery, and heavier, plain-style furniture, often made of quarter-sawn oak that showed the wood's medial rays in dramatic fashion.

Mortise-and-tenon joinery did not disappear in the twentieth century, but new methods of cutting the mortise were developed for the small shop and solitary woodworker. The drill press could be used to make mortises, albeit with the rounded ends of the drill bit. With the

FIG. 41
Drawer side joined by a Knapp dovetailing machine. Pine. United States, ca. 1880.

right jig, the tenon could be similarly rounded, and for the really adventurous a round pin could be inserted through the joint, with a drill and a length of commercially produced dowel. Eventually drill attachments that made holes with right-angle corners were perfected. The most common machine in use today—the hollow-chisel mortiser—employs a round bit surrounded by a hollow square chisel blade that shears off the wood remaining as the spiral drill bit cuts and extricates the wood from the hole.

Bending

It does not take much experience with wood to discover that it bends. Boards and beams once straight often take on the graceful, if unwanted, curves of winding, cupping, twisting, and kinking. While this may be the source of angst and anger among those who have to redo or fix warp after the fact, the flexibility of wood has been an opportunity for other woodworkers. All sorts of artifacts of ordinary life make use of this quality: seats, splint baskets, boxes, and furniture frames. Bending can relieve the maker of time-consuming joinery, carving, and hollowing, as well as making it easier to produce articles such as sieves or winnowers, in which holes are desirable. Rather than massive amounts of drilling, an open weave will do the trick.

Wood can be bent permanently in its green state—that is, freshly cut from the tree—and even when dried if it is thin enough and the bending demands are not too rigorous. It can also be bent by subjecting it to heat and humidity—that is, by steam bending. The ancient Greeks and Romans knew about bending, as did woodworkers in ancient Asia and North America. Once a piece of wood has been steamed (about one hour per each inch of thickness for dried wood and about half that for green wood), it can be bent around a form, clamped, and then dried in that shape. Nails and screws (or bolts) are the most common fasteners used to hold the wood to the form, although some pliable woods in thin sheets or strips can be held together by weaving (as in baskets) or stitching with leather, sinew, strong plant materials, or wire.

When wood is bent by whatever means, the inside of the curve is compressed and the outer surface is stretched. At some point, if enough force is applied to the bend, a piece of wood will splinter or break on the outer surface or crack and shear on the inner surface.[22] Some woods are more desirable than others for some bentwood functions. Of the thirty hardwoods and twenty-four softwoods whose strength properties are analyzed in R. Bruce Hoadley's *Understanding Wood,* shagbark hickory has the highest modulus of elasticity (the measurement of stress and strain when wood is bent) and the highest modulus of rupture (the measurement of the force needed to break the species). If you want a wood that is stiff, and therefore harder to flex, but that would return to its original shape when stress and strain were removed, hickory is the best choice, especially since you can put great stress and strain on it without breaking it. Hence archers might want to use hickory for the bow. Early golf clubs were fitted with hickory shafts for the same reason. A golfer's movement puts stress on the shaft, since the hands move through the swing rapidly and the weighted head even faster.

For those purposes for which returning to an original shape was undesirable, woods with low moduli of elasticity and relatively high moduli of rupture are better choices. These woods, such as willow or hazel, remain for the most part outside of the timber and lumber trades because they are not particularly useful in building or for most furniture forms. There is a long tradition in rural areas of bending wood to make fences and gates, barrel hoops, furniture, baskets, and boxes. Waste branches woven between stakes driven in the ground form a "dead hedge," a lightweight fence that is a common form throughout most of the world.

Hurdles, or wattles, are short lengths of woven fencing that can be moved from place to place and quickly assembled for controlling or penning livestock or for protecting gardens from invasion by hungry creatures looking for an easy meal. Hurdle makers weave long strips of flexible wood between closely spaced vertical poles (*zales*), wrap the strip around end poles with a twist, and weave in the reverse direction, interlocking the end into the weave below. Long a part of rural life, hurdles became an essential part of medieval and early modern sheep farming economies, in part because they best met the need to temporarily pen sheep for shearing and dipping. In parts of the world where free ranging still takes place (such as Lapland, with its reindeer), temporary metal fencing has for the most part replaced the wooden hurdle.

Wooden barrel hoops can be fashioned by quartering long poles (with a froe or billhook), trimming the corners, shaving off the inside edges (with a drawknife), and then bending them while green.[23] These were used for dry-goods (slack) barrels, since their ability to hold the staves together was variable with atmospheric conditions. Barrels meant to hold liquids were most effective if metal hoops were used.

Bending wood to make furniture has considerable advantages. Craftsmen using pliable green woods could make rudimentary seating and other furniture without needing a sawmill or the labor and skill of cutting and chopping mortises, tenons, and other joinery. Windsor chair makers could fashion curved backs. Once inexpensive nails were available, even the tasks of weaving and securing the wood were relieved to some extent. The Appalachian Mountain area of the United States was particularly known for willow furniture; today the region's bentwood products tend to be manufactured for a market based on fashion rather than necessity. In the late nineteenth and early twentieth centuries, an era of Romantic nationalism and mythic historical confusion in many industrialized parts of the world, "rustic" furniture became a minor fashion phenomenon,[24] to the extent that fake or faux rusticity could be found in articles made in factories. [FIG. 42]

The fad for rustic furnishings was an outgrowth of that era's passing interest in the "outdoors" as a counterpoint to the perils of urban life and the cultural interest in the "roots" of national groups. The former concern was linked to a combination of ancient fears of the effects of "crowding" in cities, combined with a dim popular and professional knowledge about the origins and spread of disease. The latter enthusiasm, shrouded in the fog of the ancient past, had less-than-glamorous ties to the racial nationalism, eugenics, and theories of racial and national "purity" that were to infect politics in the twentieth century. Such work and such cultural imperatives are a far cry from the enduring sculptural qualities of contemporary artists working with natural wood formations, their work for the most part rooted in both formal design aesthetics and a spiritual connection to both the forest and to the importance of preserving it.[25]

Craftsmen uninterested in using bark-on limbs and "twigs" for their creations still employed bending in their work. The traditional ladder-back chair is essentially four round legs (the back legs extending upward to form the back) with two or three sets of four round stretchers let into mortises or round holes in the legs to provide strength and rigidity. The top set of stretchers also functions as the frame for weaving or otherwise attaching the seat—of bent and woven ash splint in the settee shown below—and the back is finished by inserting mildly bent slats into mortises in the extensions of the back legs. Antique ladder-back chairs

FIG. 42
*Settee. Hickory, ash, splint seat. United States, ca. 1900–
1930. Better-quality rustic furniture, such as this settee, was
held together by round tenons and mortises; less expensive work
employed nails.*

FIG. 43
Ladder-back chair. Maple, ca. 1850. The turned chair parts are two steps removed from the barked branches of the rustic settee. Trees cut into planks are then turned to the rounds, but a machined round, not the vagaries of tree growth. While still somewhat crude, this chair shows a high degree of artifice.

indicate that most chair makers did not bend these extensions, or posts, to enhance the ergonomics of the chair, but some did shape the slats to accommodate the curve of the human back. [FIG. 43] The round posts were more difficult to bend because of their cross-section and mass than the relatively thin back slats; moreover, the extra bending would have cut into profits, since it would have increased production time for each chair.

Bending thin sheets of sawn wood, made flexible by steam or cold-water soaking, enabled mid-nineteenth-century craftsmen and inventors such as the New York cabinetmaker John Henry Belter to develop laminated rosewood furniture of great strength and thinness, and physical comfort. Belter developed a system whereby he bent his laminated rosewood chair backs in two different curving directions, thereby accommodating the body more ergonomically when he added upholstered seats and backs. In addition, the strength of bent lamination allowed him to remove large amounts of wood to produce elaborate pierced and carved surfaces on furniture that seemed to magically support its occupants on delicate legs. A master of both engineering and fashion consciousness in the United States, Belter's lightweight designs must have also been popular because they allowed sitters to convey the notion that they too were dainty, both in manners and in physical bulk.[26]

Bentwood furniture—in Euro-America usually associated with the late-nineteenth-century Austrian furniture maker Michael Thonet—is the product of a highly mechanized and industrialized process. Joined in the main with screws or bolts and bound with metal bands to minimize breakage on the stressed outer edge of the curve, these bentwood chairs had the greater strength of the bent grain and the portability of their slimmed down framework. With virtually no mortising and tenoning to add expense, they were a boon to the cost conscious while they incorporated the comfort of the curved surface or frame. Industrial chemists soon discovered that certain chemicals—anhydrous ammonia, for example—extended the pliability of wood even further, making tight curves even easier to produce.

Splints and Curves

Elsewhere in the United States at about the same time, another important use of thin bentwood was developed by people with an utterly different worldview from that of the urban Belter and his fashionable and wealthy clients. The Shakers, a utopian group of Protestants who traced their lineage to the Quakers and to the Protestant Reformation, began to make covered oval boxes by bending wood. Used for storage in the sect's highly regulated communities, Shaker boxes were light and strong and had many imitators in the American marketplace. They were made by soaking wood and bending it around a wooden form, where it

FIG. 44
Shaker-made box, inscribed "Lucy Davis" in pencil on lid. Pine. Possibly Canterbury, New Hampshire. Late nineteenth century. Bent and stitched box. Spruce. Finland, ca. 2000. Both boxes are bent around forms, although the modern Finnish artisan chose to bind the sides together with a natural fiber, rather than nailing them.

was clamped until the wood dried in its new shape. The ends of the bent side piece were usually nailed together. In the oval box on the right in figure 44, the bent side is stitched together with flexible and strong plant stems. [FIG. 44]

The combination of strength and flexibility found in certain wood species, as well as their availability throughout most of the world, explains why they were used for so many of the goods of everyday life. Unlike metals, wood could be shaped without the high technology and high heat required to forge and cast, and the transformation from the raw material to the useful product required far less work, and in most cases no fuel. The latter characteristic was especially important, since fuel was essential for other purposes and acquiring it consumed so much human energy.

In our time, consumers take bending wood to be mundane. Modern architecture in some places makes use of vast expanses of bent laminated wood for carrying beams or supports. The Finnish architect Alvar Aalto explored the possibilities of curved and bentwood surfaces in architecture in the Finnish pavilion of the 1939 New York World's Fair. Wavelike surfaces defined the front facade of the pavilion, and Aalto repeated variants of that curve in a variety of materials and forms, such as his glass vases. Bent plywood is now a fixture in higher-end mass-produced seating furniture. From the 1930s onward, it was extensively used by industrial designers who sought to include suggestions of speed and streamlining in their works.

Nature, of course, bends wood all the time. Trees blocked from the sun will bend in ways that maximize their exposure to light, just as houseplants will turn toward windows. The compression and tension created by this movement distorts the pattern of new and old wood in the annual rings of the tree and has provided ready-made raw material for human use in ships and housing. Herding peoples of the Asian steppes, Native Americans of the Great Plains, and other peoples who lived in lightly framed and temporary vernacular housing have made use of flexible poles for the skeletons of their shelters. But natural bending does not always go in the direction humans want or need. Moreover, when we shape wood by bending rather than cutting, our aesthetic is more often than not toward the regular and the artificial, rather than the natural, whatever we say about it. Aalto and other industrial designers chose curvilinear forms that they thought expressed an "organic" style or a "machine aesthetic," rather than imitating nature. Bentwood furniture, whether in the form of actual branches bent for human use or Thonet's goods, is also hardly natural. Thonet's designs and those of his followers were certainly graceful and embodied an aesthetic defined over many decades of art historical performance and debate, but few could claim anything more than a passing relationship to the natural world. Rustic furniture made of bark-on branches or saplings was, on the surface, at least, closer to the woodlands, but it was only a superficial reference. That form of green wood furniture and furnishings is and was for the most part formalist, and the appearance a Romantic natural fantasy of the park.

The Sideboard in the Garden

Quantum leaps in the efficiency and possibilities of tools arrived as energy from sources other than the hand was harnessed, and the direction of machine movement changed from replicating hand actions. The longer view of the history of tools—from bone and stone to steel—raises one of those philosophical questions that often bedevil not only students of the history of technology but also virtually all investigations of human history: What comes first? It is true that to cut or smooth something well you need sharp bone, stone, or metal edges, and that the edge is for the most part held in a wooden body—as in a wooden plane—to make it more efficient and less dangerous to use. Making the blade holder, especially if the wood was a hard one such as beech, box, or lignum vitae, was a challenge. You had to use a crude tool to produce a more sophisticated and efficient one, after which you handed the old tool to someone junior or less wealthy, or more likely, stashed it somewhere in case the new one failed or because it offered the security of the past. Business and industry tend not to hang on to the old tools, arguing that they take up valuable space and that, ultimately, history is of little consequence. Space costs money, but casting oneself adrift from history is a dangerous practice, especially as world natural resources dwindle and policy makers and political heavyweights fail to connect their short-term goals with the long sweep of history. Operating from ignorance is never a good idea, though it may increase profits for a while.

The history of tools and wood isn't a smooth and linear progression, much as some historians and most amateurs would like us to believe. None of history is like that. People go up blind alleys and turn in circles all the time, and they have for as long as the species has been thinking about how to make life easier or safer. When we look only for the one stream that got us to the present, we miss all the whirlpools, ponds, and placid (if isolated) parts of the waterway and thereby miss the lessons of those areas. Even worse, we sometimes characterize those who ended up in those parts as odd, incompetent, or mildly amusing, a dismissive attitude that wastes the lessons of much of the history of human thought and labor.

Why bother transforming wood in the first place? After all, if it is such a trial to do so, why not choose log or stick buildings and furnishings made of branches, saplings, and materials woven from more cooperative flora, such as reeds, rushes, and vines? The answer, I suspect, lies in the desire to *get away* from the woods, to get *above* the swamps and the inundation, and to stay warmer and dryer than is possible in the trees or beneath them. Setting apart one's habitation from the materials that made it connotes status and power, important components in the history of human society. The other creatures of the world probably do not care about what matters to us, although some of the ancients' belief systems may well have thought that they did, and that a peeled and squared wall exterior would discourage the beasts, serpents, and spirits from breaching the barrier between the habitation and the howling wilderness.

Working wood, then, is important culturally and practically, though it is likely that the first woodworkers did not think much or at all about the notion that they were using less of it to build than they had when logs were the only construction material. In chapter 1 we saw that harvesting trees from the woods involved a constellation of generally large cutting tools, as well as operations and equipment designed to transport logs to mills, which then sawed them into planks, boards, and beams. In chapter 2 we encountered other specialized tools to split shingles and clapboards from blocks and straight logs. Framing, sheathing, and exterior "skin" woods were not meant to be shown on the interior; they were to be concealed by plaster or paint or to remain out of sight in the attic or cellar. They were transitional woods, located in the cultural space between the lumber mill and the civilized world of the parlor. They were the roughnecks and coarse types, visible in the (usually) male areas of work and covered in the (generally) female areas of the house's interior. People could trust them to support and protect but not to treat a (metaphorically) soft hand with respect and proper decorum. Molding and trim, on the other (soft) hand, were not present for their strength but to play their part in refining the boys and men allowed in the house and on the carpets. In this chapter we have seen how diverse the challenges were to "refining" wood and how extensive were the solutions that engineers, artisans, and inventors developed.

Shaping wood for human purposes enabled the construction of "tools for living," to paraphrase the architect Le Corbusier. In turn the demand for more complex and decorative surfaces, further and further from the natural forms of the woodlands, begot an expansion of the spectrum of tools for cutting and shaping the wood. Function, failure, and invention proceed apace, reinforcing one another. People also believe that working wood into efficient and sophisticated joinery, aesthetically pleasing designs, and lustrous surfaces—as in our sideboard—also promotes a measure of polish and refinement among the people who live with the work of the artists and artisans in wood.

4

The Empire of Wood

Empires control space and people. The great powers of world history have generally managed to subdue other peoples by means of force, using their ability to move military might and goods over great expanses of water and land. When most people think about this, they envision metal—guns and cannon—but until the twentieth century, wood was the critical element in the imperial history of the world. On land, soldiers made their way to foreign parts on foot, but ordnance was dragged from place to place in wagons and carts or on its own wheels, all of which were for the most part made of wood. Some weaponry, such as the catapult, was also made primarily of wood, as were the soldiers' crossbows, longbows, and pikes. Overland conquest was limited by several factors: the nature of the terrain to be covered, the physical limitations of an army on foot, and the continuity of dry land. Seaborne navigation on wooden vessels expanded the distances armies could move and the extent of imperial reach from the contiguous to the global. Navies made empires and wood made navies.

Coastal and, by the fifteenth century, open-sea navigation were the keys to imperial dreams and reality. The ancient empires of Egypt, Greece, Rome, China, Mongolia, Persia, and Turkey were wrought primarily by military conquest over land, although the Romans, Greeks, Chinese, and Turks made some use of water transportation. The Romans in particular used triremes (some as large as one hundred feet long and sixteen feet abeam) to expand the reach of their empire to the extent that the Romans termed the Mediterranean Mare Nostrum, or Our Sea. It wasn't "theirs" for long, however. By the ninth century there were major shipyards across Muslim North Africa, the eastern shores of the Adriatic Sea, and the far eastern Mediterranean, importing timber from almost the entire Mediterranean basin. For about four centuries Arab powers more than held their own against the forces of Byzantium and trading cities such as Venice. But by the thirteenth century the Byzantine Empire had successfully countered Muslim domination in much of the eastern Mediterranean and by the end of the fifteenth century the Moors had been pushed out of Spain.[1]

Coastal sailing and oared boats and ships played some part in expanding the lands under the control of great powers before the sixteenth century, but by 1500 the path to empire seemed to lie in the mastery of navigation and warfare on the open seas. The ability to navigate by means of the heavens enabled mariners to leave the comfort and confines of sailing close to shore,[2] and inspired imperial dreamers to design and build larger and larger craft, both to handle the swells and storms of the open ocean and to withstand and deliver the heavy ordnance of warships.

Those nations with skilled navigators and shipbuilders and ample forests of oak, pine, and fir were well-positioned to become major forces in world politics. Once enmeshed in the global race for control of foreign areas and their natural resources, the demand for timber to build ships ultimately outpaced domestic supply. The imperial powers in general preferred native species over foreign woods, but the mathematics of timber supplies worked against them. It was really quite simple. A prime shipbuilding oak tree took anywhere from 150 to 200 years to mature; fir and pine took less time (100 years) but were less desirable for many naval uses. A large merchant vessel or warship required a huge amount of wood—a 1,000-ton vessel could take between 1,500 and 2,000 trees and sometimes as many as 4,000.[3] Even though most seagoing craft were smaller and less heavily constructed than these behemoths, the demand on the forests was immense. Naval and mercantile expansion eventually led to international trade in wood, both for shipbuilding and for other purposes, such as housing and fuel.

Dugouts, Rafts, and Canoes

The earliest known water-born vessels were probably single logs on which people rode downstream. Straddling a log has its disadvantages, among them being constantly wet and, in some parts of the world, subject to attack from snakes, crocodiles, and other hungry creatures. Rafts offered some amelioration of the latter problem and were certainly a dryer form of transportation. Papyrus rafts were used in Egypt and the Horn of Africa at least five thousand years ago. Bound sheaves of the paperlike material were joined together to provide an easily maneuvered small craft in calm water, but one that could not manage easily in rougher conditions. Most rafts are not intended for seagoing use. But given the right raw materials, they can be made for safe traveling outside of protected inland waterways. Giant bamboo (*Dendrocalamus gigantus*), when bound together and outfitted with oars, has been used in Asia for large three-man rafts that ply coastal waters.[4] Balsa (*Ochroma pyramidale*), an uncommonly lightweight wood of great buoyancy, has been used for thousands of years in Peru, both on the seas and on inland waterways. Ancient peoples more than likely managed to migrate great distances on it. [FIG. 45]

The dugout, a boat made of a single tree stem that has been hollowed out to accommodate passengers and goods, is for the most part similarly unseaworthy, unless equipped with outriggers that provide stability.[5] Used for thousands of years, this ancient form of watercraft is still in use in scattered parts of the world, where large, easily worked trees are plentiful and tools and money are not. [FIG. 46] In parts of western Africa, for example, people routinely fell iroko trees (*Chlorophora excelsa*) for this purpose, since the wood is durable and grows to great length and girth. Vinhatico (*Plathymenia reticulata*) is still the favored wood for dugouts in Brazil. In New Zealand, the Maori use kauri (*Agathis australis*) for dugouts and outrigger craft.[6] [FIG. 47]

A skeleton of lightweight wood members sheathed in a thin membrane, the canoe is only a remote relative of the raft and the dugout and has been in use for almost as long. The bark canoe in one form or an-

FIG. 45
Pahie and balsa raft. Plate 7 from Abraham Rees, Cyclopedia, *or,* Universal Dictionary of the Arts, Sciences and Literature *(London: Longman, Hurst, Rees, Orme and Brown, 1820). Pahies are the ancient and traditional watercraft of the South Pacific islands. A combination of dugout keel and plank sides, they were often equipped with outriggers.*

FIG. 46
*"Indian 'dug-outs' on the Chagres River, . . . Panama." Underwood and
Underwood, ca. 1890.*

FIG. 47
*"Maori War-Canoe at Tauranga, New Zealand." London, 1864. Maori war canoes,
or waka, were often of great length. The waka built in 1940 to commemorate the
Treaty of Waitangi was 117 feet long, fashioned from three kauri trees.*

other is an international watercraft. The most famous variety is that made of cedar and the bark of the birch tree. The paper birch (*Betula papyrifera*), so called because its bark consists of several thin layers resembling paper, has been the most commonly used "skin" material in North America, in large part because it resists water damage and can be cut from the tree in long sheets. In the traditional canoe as made by North American native peoples, the bark is then bound and attached to the skeleton and the bark joints sealed with pitch. [FIG. 48]

Other canoe-making traditions, such as those of the Inuit and other polar peoples, involve using animal skins stretched and attached to the wooden skeleton of a craft. The most familiar form to southern peoples is the kayak, considered one of the most effective designs of all sea- and freshwater small crafts. Some canoe makers use coated canvas to provide the skin of the vessel, and a smaller number of makers use wood strips for the exterior fabric. Canoes and kayaks have maintained a measure of popularity among outdoor enthusiasts, but the material of construction has changed to plastics or aluminum. While the advantages in durability and lightness of these materials are substantial, their popular-

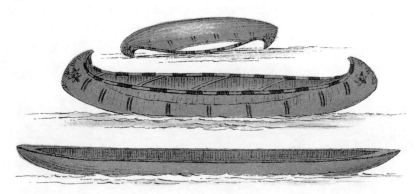

FIG. 48
Canoes. Plate 130 from George Catlin, The Manners, Customs, and Conditions of the North American Indians. *2 vols. (London: By the author, 1841). The craft at the bottom is a dugout.*

ity (along with that of synthetics in outdoor clothing) seems contradictory to the environmental ethos many outdoor enthusiasts espouse. Plastic clothing and watercraft will, for the most part, never decompose, thus adding to the nondegradable detritus the human race is depositing on this planet. Moreover, some of it is made from petroleum in one form or another, so it contributes in a small way to the depletion of that resource.

Plank Boats and Barges

People have sailed in wooden plank watercraft—boats and ships—for approximately six thousand years. The oldest such surviving craft is Cheops's Barge, built over 4,500 years ago and entombed in the Great Pyramid in Egypt. Discovered disassembled in 1954, its more than 1,000 pieces included cedar planks seventy-five feet long and several inches thick.[7] Smaller Egyptian boats made use of local woods such as acacia, which was available in shorter planks and pieces. Many of these boats show an irregular pattern of pieced construction, rather than long planks of relatively equal length.

Cheops's barge neatly links wood, watercraft, and empire. The cedar for the barge more than likely came from Lebanon, where it is native and relatively common. Historically, cedar is relatively scarce in Egypt and has been imported for millennia. An excellent wood for watercraft because it resists water-induced decay and degradation, it was an attraction for expansion-minded Egyptian leaders as early as 2500 BCE. Here is perhaps the progenitor of the pattern of expansion and empire that was to dominate much of world politics from 1500 CE through the end of the nineteenth century: ships for trade expanded the demand for wood for more ships.

Ancient Egyptian boats large and small did not have a keel, or longitudinal timber extending along the bottom, and they were joined and bound together without metal fastenings. In the former characteristic, they were akin to northern European plank watercraft of the Bronze Age, about 2000–1500 BCE, but in the latter they differed substantially from the plank boats of Scandinavia, Ireland, Britain, and Germany.[8]

These made use of metal spikes and other fasteners to hold planks together, as well as wooden joinery that locked pieces in place. Flat-bottomed boats such as those unearthed in Irish and Scandinavian bogs indicate a thriving tradition of river and lake craft in use throughout Europe, quite apart from the influences of Roman adventuring in the European hinterland. Danish war "canoes" outfitted with large "beaks" for ramming other boats enabled the Danes to establish hegemony over surrounding coastal areas, and Danish open-sea plank-wood vessels of approximately 300 BCE may have been precursors of the great Norse sailing ships of several hundred years later.

Archeological survivals and contemporaneous pictorial renditions reveal a high level of technological know-how and woodworking skill in ancient Greek and Roman ships. These ships were remarkable because they were extremely heavy (as much as one thousand tons), and because they were so competently constructed. Massive oak, fir, and cedar timbers were most commonly used for the hulls and other large parts, and small amounts of acacia and olive were fashioned into pegs, pins, and other specialized parts.

Boatbuilders laid the keel first and then worked their way up the hull, thereby creating an interior storage space (for merchant craft), areas for rowers (in warships), and anchoring for masts. Oak was the choice for structural elements, because of its rigidity, while fir often was used for planking. The pins, or treenails, holding fast the joints of the oak, fir, and cedar were usually of the more flexible local Mediterranean woods, into which a bronze spike was pounded to expand the end grain and tighten the joint before flotation. Exterior surfaces were smeared with pitched textiles and sheathed in lead to protect against marine borers, which could in a relatively short time wreak havoc on a seaborne craft. Attached to the hull with copper spikes, this protective layer was overlaid with a beech outer hull. The heavy planking and support structures for internal spaces were joined to one another by pinned mortises and tenons that were cut to tolerances so exacting that the finished ships needed little if any packing or caulking; once floated, the joints swelled in contact with the water to produce a watertight vessel.[9]

Ancient and early medieval merchant ships in the Mediterranean and on the western coast of Europe were for the most part sail powered, and many were single-masted vessels that stayed close to shore. Warships, however, were of a different form. Equipped for long journeys with many sails, which were deployed only in the absence of the enemy, they were powered by oarsmen when in contested and dangerous waters or in combat, situations in which more control of the craft was essential for survival. These vessels were designed to carry as many rowers as possible and to ram other ships to disable them. While oak was still used for the keel and a few other support structural elements, fir had the advantage of lightness, which enhanced the ships' maneuverability, and was employed whenever possible. Lebanese and Syrian cedar and pine were also used for vessels constructed in the eastern Mediterranean.

The ancient plank boat most familiar to people in the twenty-first century is the Norse, or Viking, ship. It represented a quantum leap in boat building technology in the eighth and ninth centuries (CE). Although there had been previous advances in keel design and construction in northern Europe, as evidenced in the fourth-century Nydam ship of northern Germany, Viking vessels employed more massive keels and complex joinery to produce seaworthy ships of graceful and more effective design. The Gokstad ship has a keel made of a single eighty-foot-long oak tree stem. With these larger and more dependable craft, the Norsemen were able to settle Iceland, Greenland, and (briefly) North America, reach deep into central Europe, and command much of northern Britain for many years. [FIG. 49]

Viking shipbuilders used big trees to build their craft, but they developed ways of making hulls that were not limited by the size of the trees they had at their disposal. They joined planks end-on-end with scarfed joints—angled cuts at the end of planks with mortises and tenons or tongues and grooves that fit together and were then held in place with iron spikes or treenails—to make the hull. In contrast, other ship- and

FIG. 49
*Wooden Viking ship. Excavated in 1904 from a burial mound in Vestfold, Norway.
ca. 816–20. Photograph by Robin Strand.*

boatbuilders of the time horizontally overlapped single planks (called *strakes*) that stretched from bow to stern. Viking boatbuilders were thus able to build longer boats with more streamlined curved hulls. Like the Romans before them, the Vikings used oar ports within the hulls of their ships, rather than the locks on the top edge of the hull, as other northern Europeans had for centuries. This arrangement allowed for a higher hull to be built, a design that offered greater protection from the high seas and a modicum of protection in battle.

Viking boats were as much sailing craft as they were oar driven. By the ninth century Norse boatbuilders had learned to set a large mast in the center of the vessel, attaching it to a thick plank often carved to resemble a large fish. (The mast was attached to the hump in the fish's back.) With the greater stability their improved designs provided, they could use larger sails to harness more wind power, thus enabling them to move more quickly, to make longer journeys, and to outmaneuver enemies.

Dhows, Junks, and Sampans

Along the eastern coast of Africa, throughout the Arab world, and well along the coast of the Indian subcontinent, the primary watercraft was, and for many people still is, the dhow. Usually two-masted vessels with a larger forward-angled foremast and a smaller vertical mast to the stern, dhows carried trade goods and travelers throughout the Indian Ocean. Arab navigators developed their own methods of celestial navigation by means of the *kamal,* which made it possible to determine latitude. The dhow's triangular sail, known as the *lateen,* was an effective form, easily managed by one or two seamen.[10]

Dhows are shell-built vessels, craft in which the planks were lashed together and sewn to the keel or, in its earliest form, to a dugout in order to increase the height of the hull. Most dhows, such as the shallow-draft *badan,* were small craft, designed for short-distance trade, but large ocean-going versions such as the *baghlah* and the *battil* were used by traders who voyaged deep down the African coast and far to the east. Oceangoing dhows laden with as much as five hundred tons of cargo from the Arabian peninsula, Persia, and the Indian subcontinent reached Chinese ports by the eighth century. Throughout the Indian Ocean basin the craft of dhow-building is still practiced with adzes and other hand tools in, for example, Zanzibar, which has been a commercial and cultural crossroads for hundreds of years.[11]

Sailors were taking to the waters throughout China, Japan, Korea, and Southeast Asia for about two thousand years before Westerners knew much about such travel, and the vessels they used for most of this time were sampans and junks, in many ways more sophisticated than the ships of the Mediterranean empires of the Hellenistic era. In much of Asia, bamboo was both an important building material and a technological inspiration for raft builders. It was a first-rate choice for rafts, both for the actual "planking" and for the ties that held the pieces together. Bamboo can be fashioned into strong flexible cord and the internal structure of the bamboo stem—with long strands running parallel along the length and periodic cross-growth that maintains the circular outer form—appears very much like the internal structure and exterior construction pattern of a hulled watercraft. Thus the plant itself may have been the inspiration for the design of larger vessels, though that is merely conjecture. [FIG. 50]

FIG. 50
"Sampans on the Water Front,"
Keystone View Co., ca. 1920.

Sampans are small river-worthy vessels built with plank sides and a small shelter amidship. They are most often poled, allowed to float with a river current, or moved along and piloted by means of a rear sweep, a long oar used to propel and steer the craft. Junks are in general larger vessels that range in size from the smallish reed carrier (often so loaded with light reeds that the bulk of the pile acts as a sail of sorts) to the four- and five-masted seagoing junk that can take on the rough waters that surround many of the islands and shorelines of the area. Sha-mu (*Cunninghamia sinensis*), the pine grown locally in China, was the primary wood of choice for ship construction, though builders of some large junks used teak as well. Laurel and camphorwood, dense and strong woods both, were employed for bulkhead structural members.

The irregular shape of a junk is testament to both the wisdom of the builders and the process by which it was conceived and built. One- and two-man craft working the rivers of the interior of China often had to navigate swiftly flowing waters. The irregular trapezoidal shape of the hull allowed for faster maneuvering, as did the use of two stern sweeps at different levels on the craft. Placed in front of the mainmast of some oceangoing junks was a centerboard (a retractable keel), a device in use in Asia long before it was known in the West. The structural strength of these craft is considerable, with heavy timbers used throughout, both vertically and across the beam. The masts in traditional junks were most often made out of teak logs that were treated to resist insect infestation by burying them in wet earth for extended periods of time.

Information about alien vessels, technologies, and foreign materials, especially woods, altered global material culture, for both the expansionists and those who one day encountered aliens on their shores. The boats and ships—all made from wood—were the essential carriers of those explorations and impositions. Mariners carried systems of information and beliefs to the people to whom they made their way and then back to their homelands. Potatoes and maize (corn) came from the Americas to Europe; horses made the reverse journey, where Native Americans welcomed them. Indigenous peoples hitherto untouched by

Europeans often suffered greatly after contact, from diseases for which they carried no antibodies and from the explorers' and imperialists' cultural arrogance, technological superiority, and military might. For the most part the impact on traditional cultures was negative. Hybridization and cultural exchange can be a good thing, but certainly has come with a steep price, at least for one side in the exchange.

Empires and "The Age of Sail"

Riches and the power to command and keep them were the driving forces in the imperial enterprises that reached across oceans in the fifteenth century. Coastal navigators from Portugal and Spain made their way around the southern tip of Africa by the latter decades of the fifteenth century; the Chinese had sailed through much of East Asia at least fifty years before that.[12] The success of the navigators in finding their way and returning, often with treasure and tales of even greater riches in the hitherto unknown places they allegedly visited, prompted European potentates, Arab leaders, and Chinese rulers to support further exploration and conquest. We in the twenty-first century like to think of these leaders as gullible clowns who ate up every word of these reports as if they were revealed truth. Perhaps they did so. But it is as likely that even leaders of a suspicious or questioning cast of mind still thought that expending large amounts of their riches was an investment that could—and did—reap huge returns, at least for some.

For the peoples of the late fifteenth century, news of exploration and encounters with alien peoples and landscapes was both a revelation and a revolution. Our new information tends to be of the minuscule or the mammoth, either in the realm of microscopy and electronics that are too small to see or in astrophysical form in which scientists tell us about what is so far away and so huge that it is in fact beyond our comprehension. But voyagers in the late 1400s brought back foreign flora and fauna, and sometimes even people who looked so odd that they caused a sensation. That the explorers *returned at all* was miraculous, given that they were traveling in small ships on unknown and often treacherous waters and were away for long periods of time—years in some cases.

These voyagers made real the fantastic tales that had been passed down in oral tradition for generations, or demonstrated that what had been once accepted as true was in fact fable and fairy tale. We wonder at the gullibility and greed that fueled the murderous quest for the Seven Cities of Gold by Spanish conquistadores, but those armor-clad imperialists believed that the evidence of great riches they found among the native peoples of the New World was the tip of a golden and jeweled iceberg, as did their counterparts in other European nations. Similar tales of wealth mesmerized the powerful and wealthy in the Indian Ocean basin and in East Asia. Oceangoing trade with Korea and Japan was an important part of the Chinese economy and foreign policy. By the twelfth century, commerce was ongoing between China and Angor (Cambodia) as well as with Korea and Japan. By the early thirteenth century China's Sung dynasty had established itself as a naval power in the region, but that hegemony was not to last. In 1279 Kublai Khan's Yuan (Mongol) navy defeated the Sung forces, and subsequent Mongol forces maintained control (of sorts) over much of the China trade into the fourteenth century. Yuan traders eventually reached Ceylon (Sri Lanka) and parts of the Indonesian archipelago, gradually displacing many of the Arab traders in the port towns and cities of these regions. The Japanese successfully resisted the Yuan navy, especially the expeditions of 1274 and 1281.

In the latter half of the fourteenth century the Chinese effectively ended Yuan hegemony, and the succeeding Ming dynasty ultimately took over the seagoing empire that the Mongols had established. Early in the fifteenth century the Ming emperor Zhu Di commissioned the expansion of the Chinese navy and seven expeditions westward to the Indian Ocean. These excursions, some of which consisted of hundreds of ships and thousands of mariners and soldiers, reached as far as Zanzibar on the eastern coast of Africa.[13] Chinese oceangoing vessels were among the largest in the world in their day. Some were in excess of five hundred feet long, or almost five times longer than the largest European ships. Many included watertight storage areas on board, a feature largely unknown in Western craft until the nineteenth century.[14]

By the mid-fifteenth century, the great age of Chinese naval exploration and adventuring was over, as Chinese leaders chose to withdraw from imperial expansion and sought to distance the empire from the rest of the world. An imperial edict of 1436 banned the building of oceangoing vessels and soon after that leaders ordered that all sailing boats be limited to two masts. Historians have argued that these were primarily cultural decisions, but some, such as Paul Kennedy, have argued that these decisions were strategic as well as cultural. He concluded that land-based threats from foreign powers to the north and failures at expansion to the south, as well as a conservative political bureaucracy, put an end to the Chinese navy as a competitive fighting force. Leaders feared that imperial designs and efforts overextended Chinese resources and weakened the fabric of Chinese culture.[15]

The other great naval force of the fifteenth and early sixteenth century was that of the Ottoman Turks.[16] From the Crimea and the Aegean Sea in the west and the Levant in the east, Turkish land and sea forces emerged as the greatest threat to the dreams of expansion that many European powers harbored. By the first quarter of the sixteenth century the Turks had occupied Damascus and much of the Nile basin in Egypt and had reached the Indian Ocean. By 1530 they had pushed into Hungary and were at the gates of Vienna. By the middle of the sixteenth century the Ottoman navy had gained the empire a major foothold in North Africa and the eastern Mediterranean and had raided ports as far west as Italy and Spain.[17]

Western and Turkish voyagers caught up with and surpassed Chinese navigation technology by the fifteenth century. While all three great regions of oceangoing navigation had the mariner's magnetic compass, Europeans, Turks, and other peoples of the Arabian regions had mastered celestial navigation, while the Chinese had not. The Chinese had advanced boatbuilding technology, producing vessels of great size with advanced sail technology, but had pulled back from global expansion. Nonetheless, the international imperial scene in the fifteenth century was quite different from the one most westerners have learned in their childhood schoolbooks. When European explorers set out for foreign parts to the east and south, they found considerable competition. Turkish and Arabian merchants and military powers were a strong presence in the Indian Ocean. Chinese, Mongol, and other traders were already involved in the "spice islands" trade in Indonesia, Ceylon, and India when Europeans first arrived. Their presence meant that access to the "East Indies" was not simply a question of getting there and getting rich, but an enterprise loaded with risk from pirates and the militaries of other empires intent on keeping what they had and what they dreamed of getting. The path to the riches of alien lands was potentially one of conflict, not only with the Turkish, Arab, and Chinese forces already there but with the other European national states. Finding a way westward, therefore, was not merely an adventure; it was an imperative. Sea power was—nearly—everything. With the navigational technology in place in the late fifteenth century and a faith in its continuing improvement, enterprising and grasping European rulers demanded big ships and the big timber to build them.

Wooden Politics

The demand for wood and wood products—shipbuilding materials, clapboards, barrel staves, boards, ashes (for potash), "naval stores" (tar, pitch, and turpentine), and charcoal for metallurgy—impelled the Dutch to seek these products from the natural riches of northern Europe and the Baltic Sea region. By carefully and cleverly working their way through the narrows of the sea and the politics of western Europe, they were able to build a formidable merchant empire and amass great national wealth by the fifteenth century, although the land space and the population of the Netherlands were relatively small.[18]

The sylvan landscape of the fifteenth century's imperial powers and imperial aspirants was one of the determinants in the imperial contests that would play out on the globe over the next four hundred years. China had seemingly infinite reserves of wood, but chose to abandon a navy and decimated its forests as its population expanded, increasing the demand for fuel and farmland. The Ottoman Turks were in control of wood species traditionally used for sailing vessels, especially in what are now Lebanon and Syria. Those European nations blessed with abundant resources of oak and pine were at an advantage if they possessed the technological know-how and the administrative systems of government to organize naval resources. Scandinavia had vast coniferous forests. These were essential raw materials for masts and for naval stores. The cold climate of the northern tip of western Europe was beneficial for the growth of more dense pine woods, since the short growing season tended to produce trees with tight annual rings. The relatively modest oak forests of southeastern Sweden were an important resource for the navy, and the Swedish kings commandeered the best of the wood for their use.

At the beginning of the sixteenth century Great Britain had an ample supply of oak, and many of the larger landowners began to replant the forests as they were felled. But this was a long-term solution to a problem that became pressing almost from the outset of England's great shipbuilding effort. Oaks took about 150 to 200 years to mature, and they were disappearing at a rapid rate. The Royal Navy seems to have paid little more than lip service to late-seventeenth-century warnings of the future depletion of those forests; shipbuilders continued to use home-grown oak for more than a century after John Evelyn published *Sylva: Or, a Discourse of Forest Trees* in 1664.[19] France, on the other hand, possessed vast forests of hardwoods throughout its northern and central areas and abundant coniferous forests in the southeast, and major rivers—the Loire, Dordogne, Seine, Rhone, and Garonne—that reach far into the interior of the country. These facilitated the movement of logs from the woodlands to the shipbuilding ports, but the distances between the woods and the Atlantic shipyards were great, and this compromised French shipbuilding efforts somewhat. Still, these timber riches and the security they offered in comparison to international trade may explain why French shipbuilders and imperialists paid relatively little attention to the oak and mahogany forests of the New World.[20]

Denmark, like the Netherlands, was a small nation undersupplied with wood for shipbuilding, but both nations used other means— trade and importation of Scandinavian and tropical woods suitable for shipbuilding—to develop impressive, if relatively short-lived, empires. Control of the Baltic Sea region was an important element in the imperial aspirations and success of Denmark, Sweden, and the Netherlands, and to a lesser extent, of Russia, England, and France. Sweden directly controlled access to the great forests of Finland, which was under its political control until 1809, when Russia claimed it after a protracted military conflict. European nations farther to the west and south had to contend with navigating the narrow waters of The Sound, the stretch of open water between Jylland and southern Sweden. To some extent, their geography empowered Denmark and Sweden, though neither country was able to shut out the other powers from trade in the Baltic region entirely.[21]

The Russian tsar's imperial designs and the ambitions of Russian traders, who sought warmer-water trading outposts and ports for their goods, were hampered by the Swedish, Dutch, and Danish presence in the Baltic, as well as by the difficulties of transporting timber, flax, iron, naval stores, and other goods from the southern hinterlands through Poland to the Baltic. Archangel, on the White Sea, was an important port, but so far north as to make it useful for only about one-half of the year. Even the establishment of Saint Petersburg as a Baltic port and capital of the country bore limited fruit for Russia's western aspirations.[22]

Trade in wood in northern and western Europe included another essential component for shipbuilding. As highly skilled and as well-crafted as these great oak, elm, and fir ships were, most nonetheless required some sort of caulking in the joints of the hull to remain watertight.

Although the expansion of wood in a marine environment was a positive physical characteristic, making joints tighter than when assembled, wooden dowels and pins could also work loose, in part because different wood species expanded and contracted unevenly. The sealant favored was oakum, a combination of hemp rope strands and pine pitch. Stuffed in the joints, the tarry mess provided the flexibility needed to keep ships relatively watertight. The source of the pitch most commonly used in making oakum was pine, which grew sparsely in Britain and the Netherlands, and insufficiently in Spain, Portugal, and Denmark. France possessed large tracts of pine, and Norway, Sweden, and Finland were rich in pine forests. So too was North America, but the logistics of long-range transport across the Atlantic initially made these resources less attractive than those of the Baltic region. Even the British, who had access to supplies from their North American colonies, particularly the masts brought from the hinterland to ports such as Portsmouth, New Hampshire, eyed the riches of the European Far North with envy.[23] British and Dutch timber traders pursued Norwegian and Baltic woods with vigor, even though the near continual state of war between at least two of the powers hoping to control routes made trade an increasingly dubious venture.[24]

The British managed to develop an extensive trade in the Baltic in the early and mid-eighteenth century, particularly for masts and naval stores. Stockholm, Riga, and Saint Petersburg were major points of access to the forest products of Sweden, Finland, Poland, Livonia, southwestern Russia, and the vast northern Russian hinterland.[25] By the last quarter of the eighteenth century, the British were importing large quantities of softwoods from Sweden, fir beams and boards from Prussian Memel and the Russian city of Vyborg, construction beams, battens, and small masts from Norway, and Russian hardwoods. Native British forests could not meet the needs of the country's transportation expansion (canals, bridges, and plank roads), industrial revolution (factories for the booming textile, ceramic, and other industries), and urban housing growth, as the countryside began to empty for the meager but nonetheless important monetary rewards of industrial work. With the defeat of the French in the Seven Years' War (1756–63) and the decline of both Sweden and Denmark as great naval powers in the North Sea, the vast coniferous forests of North America and Norway were opened to British trade and exploitation for masts and naval stores, thereby lessening some of the pressure to acquire these resources from the Baltic.[26] The Napoleonic Wars at the turn of the nineteenth century taxed the resources of the major European powers and restructured Baltic trading relationships between the British and the eastern and northern European powers, as the British began to tap the resources of Canada and the Indian subcontinent more expeditiously.[27] With the relative peace that characterized Europe after the defeat of Napoleon, British timber trade with Canada subsided from its high point in the mid-1820s, and English merchants reasserted themselves in the Baltic timber trade.[28]

They were not without competition, however. By the 1840s Norwegian timber traders expanded their markets in Britain, France, and the Netherlands, especially after the British altered their duty structure in a way that favored Norway over Canada.[29] By the 1890s, Scandinavian fleets outdistanced them.[30] By the late nineteenth century Norwegian timber traders were also a conspicuous presence in East Africa, especially in Zanzibar and Mombasa.[31] In the Gold Coast region of West Africa, late-nineteenth- and twentieth-century demand for mahogany for furniture and interior decoration opened up a trade that competed with the older established one for Caribbean and South American mahogany.[32]

France was one of the great losers in the imperial struggles of the latter eighteenth and early nineteenth centuries. Despite the nation's rich resources in the hinterland, its contractors were unable to meet the needs of the navy prior to the Seven Years' War. Dependent in part on the Dutch for naval stores and for some of the transshipment of shipbuilding materials from Baltic ports, France finally contracted directly with Dutch merchants by 1762, further weakening the Bourbon monarchy's power in the northern regions.[33] Close French economic relations with

Scandinavian suppliers expanded in the early nineteenth century, as they had for England and the Netherlands, and Norwegian and Swedish timber merchants found a ready market for their products among these nations in the middle decades of the century.[34]

We have seen that the Chinese had chosen to forsake a commanding naval presence in East Asia by the late fifteenth century. But while the 400- and 500-foot oceangoing Chinese vessels of the fourteenth century were impressive, it is not clear that they could have successfully countered the type of military ship being developed in western Europe at that time. The imperial domination that European nations were to enjoy from about 1550 onward relied upon large ships—built to carry heavy ordnance and to withstand the attacks of other armed craft—and the ability to maneuver them in fleets.[35] Large sail-powered vessels proved superior to local craft as platforms for guns and cannon, as the Portuguese, for example, demonstrated in the Indian Ocean in the fifteenth century.

A careful look at even a present-day map shows some of the traces of these imperial adventures. Goa is a small state of the nation of India, located on the western coast of the subcontinent. It was one of several trading outposts set up by the Portuguese in their heyday on the high seas. Greenland, which is for the most part a frozen outpost in the North Atlantic, until recently was under direct governmental control of Denmark.[36] Part of the state of Delaware was once New Sweden, as town names such as Christiana (named for the seventeenth-century Swedish queen) indicate. Most residents of New York City know that the Dutch were somehow once connected to Manhattan, and Dutch place-names are still prominent in the Hudson River valley. British and French imperial penetration in North America, Asia, Africa, Oceania, and the rest of the world is probably more obvious. Imperial power after 1500 was no longer solely a factor of geographic size (compare China with England or Denmark) or of land-based armies (though these were still important); rather, ordnance technology and seafaring had enabled even tiny countries to become great powers.

Cathedrals of the Sea

The critical importance of wood in international politics is evident in the magnitude and complexity of the task of building powerful warships and reliable merchant craft. A European warship of the sixteenth through eighteenth centuries required the labor of hundreds of skilled craftsmen and acres of forest trees. The numbers are stunning: A warship equipped with seventy-four cannon claimed approximately fifty to sixty acres of forest, or about three thousand loads of felled mature trees, each load constituting fifty cubic feet of timber.[37] The dimensions of the skeleton and cladding are what drove the volume of lumber into the timber stratosphere. At least three layers of planking several inches thick composed the hull. It was, moreover, a wasteful process; much of the timber cut for shipbuilding was cast aside, usually because shipwrights found its grain orientation unsuitable or because it was otherwise flawed. Some of it became very high-grade firewood. [FIG. 51] and [FIG. 52]

The *Blenheim,* for example, a 90-gun ship the British began on January 1, 1756, measured 142 feet 7 inches along the keel. Its gundeck was 176 feet 1 inch in length. At its widest beam it measured 49 feet 1 inch, its sternpost was 29 feet 8 inches tall, and it weighed 1,827 tons. Altogether, building this ship consumed 3,773 loads of timber, of which 957 were straight oak, 1,605 were compass, or curved, oak, 64 were elm, 281 fir, 102 of "square knees," 94 of "raking knees" (naturally curved stems and limbs, ideal for bracing), and the rest "thick stuff" of at least 5-inch-thick material, as well as an additional 206 loads of 2½-to-4-inch "plank." The craft required a total of 188,688 cubic feet of high-grade lumber. The ship was launched on July 5, 1761, five and a half *years* after it was begun. The *Royal George,* a 100-gun British ship of almost identical measurements, took ten years to build (1746–56) and consumed 5,750 loads, or 288,025 cubic feet of the same materials, with one important addition—216 of the planking loads were recorded as "Dantzic [*sic*] oak plank."[38] The British were certain that only English oak would do for the best and most seaworthy ships. But supplying heavy timbers (called

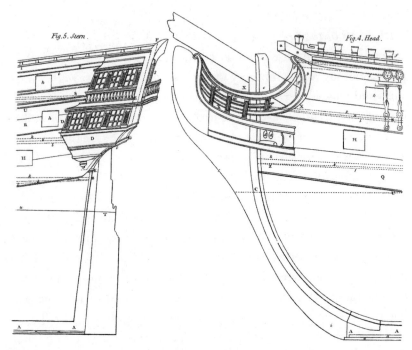

FIG. 51
"Construction of Ships." Plate 3 from the Royal Encyclopedia *(London: C. Cook, 1791). Some notion of the great size and complexity of these vessels is evident here.*

baulks) was a difficult task because most of these had to be hauled overland to English shipyards. As sources in southern and eastern England became depleted, there was an advantage to procuring some elements of the planking from the oak-rich forests of northern Germany, Poland, and western Russia.

Regulation and Resistance

Regulations concerning the use of timber in western Europe can be traced as far back as 670 in the Netherlands. Forests there were scarce, and agriculture relatively intensive, thus exerting even greater pressure to cut down the small number of trees. Confronted with the realities of their landscape and their imperial aspirations, the monarchs of the

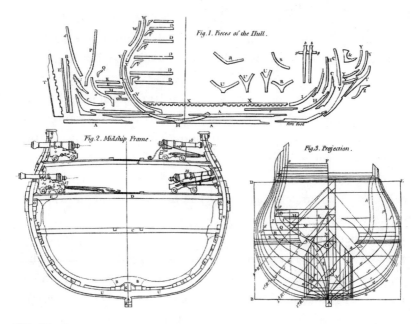

FIG. 52

"Construction of Ships." Plate 3 from the Royal Encyclopedia *(London: C. Cook, 1791). It is easy to see why naturally curved lumber was in such demand. Large ships were composed of complex curves and enormously thick planking.*

Netherlands initiated regulations to control the use of wood resources earlier and more generally than did other countries. These rules still allowed peasants to take wood for their heating and cooking needs but forbade selling it to outsiders. Live trees were not to be cut—only deadfall could be claimed.[39] Grazing rights, firewood rights, fence wood rights (usually branches), and rights to wood for repairing buildings were codified in a myriad of laws throughout much of Europe by the thirteenth century. These laws often set the peasantry in conflict with the aristocracy, which had a social and sporting interest in maintaining the woods in a wild state for their formalized hunting activities.

While there were abundant forests and hedgerow trees (the best source for naturally curved lumber, since they grew in the open and spread their branches) when the European powers began building their navies in the late fifteenth and early sixteenth centuries, the enormous amount of wood necessary to build a fleet brought about restrictions on local inhabitants' access to woods that they had once regarded as either fit for harvest for their own uses or for other traditional purposes, such as providing cover for game or separating fields. Since the monarchs of Europe claimed sovereignty over all of their domains, technically they were within their rights (so they thought) to claim all such prime wood for use in shipbuilding enterprises. Trees destined for the fleet were marked as royal property, and the penalties for being caught with such wood were severe.

These deterrents were not completely successful, and various forms of resistance and pilferage continued throughout the age of building great wooden warships. In Sweden peasant practices on lands tradition- ally used to grow fodder were challenged by the monarchy's demand for compass wood. Accustomed to cutting branches of large trees to allow sunlight to reach the ground, peasants were confronted by the Crown, which in the end allowed for what it thought was a compromise—the trimming of some branches. But farmers would have none of that solu- tion and used the new plan as an excuse to remove entire trees and clear more land for their herds, creating a conflict that was resolved only when the king set aside royal forest plantations.[40]

In England the "customary rights" of gathering fallen wood and branches upon which the country poor relied for their fuel needs were eroded in the eighteenth and nineteenth centuries, as farmers and other landowners began to crack down on gleaning and gathering.[41] Like game poaching and crop theft, wood theft was a common crime in the countryside. Landowners complained that wood gatherers pillaged fences, gates, hop yards (filching the long poles used to support the plants), hedges, coppices, and the forests for fuel and building supplies. Supervision and surveillance were at best hit-or-miss propositions, and daring and desperate "harvesters" occasionally stole entire trees of con- siderable size. Miscreants were usually fined, but the possibility of imprisonment or even transportation loomed over those unlucky enough to be caught. In Herefordshire between 1822 and 1844 convictions for wood theft outnumbered those for crop theft, save for 1840, when farmers and magistrates cracked down on the latter crime, in part as a response to the great pea-pilfering plague of the time.[42] Court records show that most cases of illegal collecting occurred in winter, when the poor were desperate to keep warm. Women constituted nearly half of those convicted of wood theft, and many children were involved as well. Although the value of much of the wood swiped was virtually nil, farmers chased down the gatherers with grim persistence, both for the principle involved and because there had been a substantial history of rural resistance to the wealthier farmers that included incendiarism and maiming domestic animals.[43]

The demands of commercial forestry in the Scottish Highlands in the eighteenth and nineteenth centuries also conflicted with tenants' traditional forest practices and access to wood, and there is some indication that ordinary people resisted commercial and governmental policies with success.[44] In southern France local resisters to the National Forest Code of 1827 intimidated charcoal workers and the innkeepers who housed government inspectors; they also harassed the latter "disguised" as women. The "Demoiselles" of the Saint-Lary forests fought a guerilla war for several years against the Forest Code's restriction of traditional peasant wood harvesting. Often as much dramatic ritual as actual face-to-face contest, this conflict was also about government policies and local practices that favored the interests of the iron manufacturers and other aristocrats of the Ariège Pyrenees.[45]

While oak was the primary wood of ships made in Europe, pine and fir were also important components of oceangoing vessels, not because they were used for structural components in the body of the ship—they were thought too weak for that—but because the tall conifers that grew in abundance in Scandinavia, France, the southern Baltic region, and North America made excellent masts and spars—light, strong, and flexible enough to withstand all but the heaviest weather. The structure of the trees is a key to their function. Oaks and most other deciduous trees to which the Europeans had access did not produce extremely straight tall stems, but rather provided the branching needed for compass wood. Pines and firs, on the other hand, send out layers of horizontal branching from a stem that continues to grow straight toward the sun unless "topped" by weather, after which a dual stem will grow.

Before they gained access to tall Scandinavian trees, shipbuilders obtained tall masts by piecing together shorter lengths, joining the sections with complicated joints and pins, and binding them with iron bands. This was, however, labor-intensive and costly work, easily forgone when one could simply fell the tree, remove branches and bark, and trim the stem with a plane specially made to round large-diameter pieces of wood. [FIG. 53]

Political instabilities, outright warfare, and piracy impelled imperial powers to look to the south and east as well as to the north for the big trees they wanted. In the eighteenth century Britain obtained the majority of its masts from the Baltic Sea region or, after 1760, Norway.[46] In the early nineteenth century British timber traders sought wood for ship-

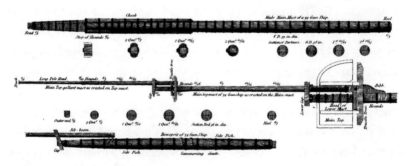

FIG. 53
"Ships Mast." Plate 8 from Abraham Rees, Cyclopedia, or, Universal Dictionary of the Arts Sciences and Literature *(London: Longman, Hurst, Rees, Orme and Brown, 1820). Pieced-together masts were functional but weaker and hence less desirable than the strong straight masts available in the woods of Russia, the Baltic region, and North America.*

building in the eastern Mediterranean, hoping to get their hands on the timber resources of southern Russia as well as those of the Ottoman and Austrian empires. They were thwarted in their larger goals by the power of the French military in the area, as they began to pay the price of the country's failure to enact a program to replant the woodlands swept away by the shipbuilding boom of the previous century.[47]

Promise and Peril in America

The uncertainties of the Baltic and Mediterranean timber and naval-stores trade made the allure of the Americas great, but it was not without complications. Almost from the beginning of European contact and exploration, voyagers to the northern parts of North America remarked on the abundance of enormous conifers growing within sight of the shore. It took no genius or vision to recognize that on this vast continent were the raw materials of empire that, if secured from the invasion of one's enemies, could be the source of riches for centuries to come. The Spanish, armed with a great armada of galleons and trading ships, sailed south, looking first for a path to the East Indies and second for the gold about which they had heard tales. The British, Dutch, French, Swedes, and Russians tracked northward, the first two finding not the gold of shining metal but the "green gold" of the forests.[48]

Unfortunately for English imperialists, the colonials who had made the voyage across the North Atlantic beginning in the seventeenth century seem not to have shared the Crown's hopeful vision that they would willingly provide all the masts and naval stores that the Royal Navy desired. Settlers in the southern "plantations," as they were called, preferred to make their fortunes growing the "sot-weed" (that is, tobacco) and later corn and other grains.[49] As yet untapped were the great forests of southern pine of the upland South, which were thought inferior to northern pines for shipbuilding, and that were in the hands of Native Americans who forcefully resisted white advancement into their lands. Similarly, the British did not favor American live oak (*Quercus virginiana*), which grew abundantly along the southeastern coast of North

America. In the northern colonies much of the settlers' activities were centered around agriculture and Indian relations, even after the Dutch and Swedes had exited as imperial powers. The English Civil War and the Puritan "Long Parliament" had something to do with this focus—or lack of it—as Oliver Cromwell turned his attention to the Baltic, determined to keep open the trade in mast woods from Scandinavia, and oak from western Russia.

By the end of the seventeenth century, after the restoration of the monarchy and the brief disruption of the "Glorious Revolution" of 1688–89, the English monarchy returned in greater earnest to realizing the possibilities of the North American woods. While never abandoning the Baltic trade, the British nonetheless began to try to control those woods considered the king's domain. As early as 1691, the Crown acted to take possession of all trees suitable for masts by including a clause in the Massachusetts Bay Charter that reserved them for its use. The persistence of the colonists in flouting the law is evident from the enactment of subsequent laws to the same end. Thus in 1729 Parliament passed a law that stated that "no white pine trees are to be cut without license." The act further reserved for the Royal Navy "All the Masts . . . exceeding Twenty four Inches Diameter, All Trees that exceed Fifty-four Feet in length in the Stem, All young thriving Pine Trees that seem promising to grow to Masts." The act also called for a survey of trees and stipulated that the king's trees be marked, and allowed that "All other Pine Trees or Trees now dead (altho' long) may be cut down."[50]

The main message is clear: The best trees were to go to the Crown for the use of the Royal Navy. Clearly the Crown was convinced that the colonists were looting the king's woods for their own purposes, and thereby not only acting as disobedient and disrespectful subjects but also behaving in ways that were undermining the Crown in imperial matters. Parliament continually tried to exert administrative control (complete with record keeping) on the timber trade in America. It was only the beginning of a larger struggle. In addition to traditional peasant claims on wood, general pilfering was a constant problem among shipwrights, who had been traditionally granted (or had simply presumed) the perquisite of keeping "shorts," pieces unfit for shipbuilding use by virtue of their size or otherwise flawed condition. Accusations that the workers simply stole wood or cut it to make it unusable abounded, and Crown officials simply assumed corruption in the shipbuilding trades.

In 1733 the Surveyor General published an announcement: "Whereas Paul Garrish, Esq., being imployed [*sic*] and licensed to cut white pine trees, for masts, yards and bowsprits, for the use of the Royal Navy, and having accordingly cut some, . . . some evil disposed person and disaffected to His Majesty, has about the 14th instant, cut and render'd unfit four of the said trees. . . . [I am] Offering a reward of forty pounds, New-England money, for identifying proof of the perpetrator."[51] Even in "New-England money," forty pounds was a hefty sum. This was both a serious offense and, by inference, one that may have not been rare. The act itself appears to be one of vandalism, though it is possible that the perpetrator was hoping to abscond with the timber at a later time. In any event, the official reaction, which was important enough to appear in print, indicates that the Crown's claim on the best timber was nettling the colonists considerably.

The English colonists in North America continued to liberate trees from the forests of the region throughout the eighteenth century. A Crown broadside dated January 1, 1770, announced that "all White Pine Logs cut and hauled out of the King's Woods in to the *Connecticut* River, or elsewhere, will be seized to his Majesty's Use and Trespassers dealt with according to Law." The argument for the seizure was to "prevent such fraudulent practices, and to preserve the Innocent from the Evil," but that noble gesture was an incidental consequence of the real goal of the decree, which was to stop "Persons [going] into the King's Woods and thence haul[ing] White Pine Logs into the *Connecticut* River without License, and against the Laws made and provided for the Preservation of the King's Woods."[52] Time and again the Crown and the colonists clashed over wood; edicts were issued and occasionally people were prosecuted for poaching. Historians' attention to the hostility between

Whitehall and the colonies has traditionally been directed to conflicts over stamp duties, tea taxes, and the like after 1763, often ignoring the abrasive relationship over the resources of the hinterland that had been festering for more than a century. Perhaps this is because the tea tax and stamp duty encounters were urban battles that involved crowds whose actions were recorded in colonial newspapers and broadsides, thus leaving a trail of description for succeeding generations of historians and mythmakers. Battles over wood in the forests were less likely to have involved large numbers of people or to have been recorded, save in the less spectacular form of the printed edict. A "tea party" makes for more sensational copy than hacking up or swiping some tall trees when no one is looking.

The navies of the major seagoing empires also faced constant competition, opposition, and harassment from their own merchants, despite the merchants' need for naval protection. In 1771 a bill was brought before Parliament to limit the tonnage of East India ships because of the competition for oak. That there was a shortage of the material was not in question, but whether the East India Company was responsible for that situation was debatable. The bill passed, both to control the use of the material and to wrest control of the East India Company from private hands and place it into those of the Crown.[53]

The wrangling over wood in North America should not blind us to the success the British achieved in extracting enormous quantities of it from their colonies, thence into world trade networks. Extant customs records are incomplete, but these shards can provide a window into the trade. An account book of the Commissioners of Customs in America for the period January 5, 1771–January 5, 1772, in the Massachusetts Historical Society, reveals a thriving commerce in wood in and out of North American ports. Over one million "feet"[54] of mahogany from "the British and Foreign West Indies" entered ten ports in North America in that year. Exports of oak from twelve ports in Canada and the southern colonies totaled more than three-quarters of a million feet, and exports of pine totaled nearly 2.5 million feet. Included in the customs records are figures for the number of feet of "oars" (about 350,000) and "staves" for making barrels (6.25 million). These figures were for boards, rather than "timber," or logs. In that form 261 tons of cedar left Philadelphia (a major port for that species), in addition to the approximately 2,700 tons of oak and 5,300 tons of pine that departed from this and other ports in North America. Most of the logs left the ports of Halifax, Falmouth, and Boston, while the variously sawn or otherwise processed wood left from ports scattered up and down the east coast of British North America.[55]

The overall volume of the wood trade between North America, the Caribbean, and England in the eighteenth century was enormous, especially when you consider that the preceding data represent only a part of the trade of *one* imperial power with its colonial possessions, and from North America to Great Britain in *one* year. Trade between Britain and the Caribbean increased these figures considerably, as did the commerce of French, Spanish, Swedish, Danish, and Dutch traders, all of whom sought timber and sawn boards for everything from shipbuilding to furniture making.[56]

Trade in African wood to Persia, Arabia, and India flourished as well, for example in red mangrove (*Rhizophora mangle*), which was used for buildings in the Middle East because of its durability, attractive coloring, and resistance to some insects. Those powers with ties to tropical areas—Spain and the Netherlands in particular—conducted a flourishing trade in woods such as teak from India, southeast Asia, and central Africa and goncalo alves (*Astronium fraxinifolium*) from Central America. More localized trade in woods such as chanfuta (*Afzelia quanzensis*) has gone on for centuries in the Indian Ocean basin, since it is a material favored by the makers of dhows. In southwestern India, merchants and traders in the hardwoods native to Malabar, Kanara, and Travancore provided raw materials for coastal shipbuilding, local heavy construction, and export throughout the Indian Ocean trade network.[57]

World trade in wood products included shingles, clapboards, blocks, hoops (probably for barrels), handspikes, oars, staves, trusses (for heavy construction), laths (to support plasterwork), spars, woodenware (plates, spoons, handles), and even house frames (an admittedly small number of those are recorded, but their appearance suggests that these houses were cut to order in one place for assembling somewhere else, even in the eighteenth century). Small amounts of lignum vitae and logwood (*Haematoxylum campechianum*) are noted, the former because of its unusual physical qualities of hardness and self-lubrication and the latter because it was a source of purple dye.

Trade between the United States and foreign ports resumed actively soon after the signing of the Treaty of Paris in 1783. Between November 1786 and November 1787, the port of Charleston, South Carolina, exported slightly over one million feet of planks and boards, listed as "Lumber" in the custom house records; it was likely to have been pine from the upland forests now open to exploitation. Flowing from the port outward as well were more than 3.5 million shingles, more than 1 million staves, and 2,700 cedar logs. More than 4,000 barrels of tar and pitch were exported and almost 4,000 barrels of turpentine. South Carolinians imported little in the way of woods, save the 220 tons of logwood, 3,000 unspecified logs, 18,600 feet of mahogany, and 50 tons of lignum vitae.[58]

American trade in wooden goods and products continued to expand in the early nineteenth century, in spite of the dislocations caused by the War of 1812 and the uncertainty of sea trade that resulted from European conflicts and skirmishes. Between October 1814 and September 1815, 16 million staves, 25 million shingles, 3 million barrel hoops, and 51 million feet of boards and planks were shipped out of the United States, as well as more than 7,000 tons of hewn timber. About 57,000 barrels of pitch, tar, and rosin were exported, as well as 76,000 barrels of turpentine.[59] By 1840 North Carolina dominated the country's production and trade in naval stores, turning out 593,000 of that year's 619,000 barrels. The change reveals both the depletion of South Carolina's relatively limited pine forests and the opening of the vast pine woods of the much larger North Carolina upland areas. It also indicates the South Carolina concentration on the slave-labor industries of rice, indigo, and cotton.[60]

Historians and their readers are justifiably impressed—even overwhelmed, if they think about it—by the magnitude of the imperial and industrial revolutions that altered the lives of almost all the world's people. But big ships and large cannon are, for all their bulk and appetite for wood, not the whole story of watercraft and wood. An armada of small boats, with a litany of names corresponding to specific shapes, lengths, and numbers of masts facilitated the marine life and in some cases the livelihood of more people than did the great warships and merchant vessels. The latter pair may well have had a larger, but indirect, influence on the daily lives of ordinary people as the political and economic changes wrought by war and commerce played out over the long haul than did the day-to-day decisions about whether to build or otherwise acquire a watercraft, or in which area to cast the nets. But for all but the last half-century, smaller boats made of wood or wood and bark transported goods and people on a local scale, with little or no reference to the larger issues of war and empire that concerned the political leaders and economic heavyweights.

Engineering technology for the nautical uses of wood may well have developed in a more complex manner than we might think. Rather than solving their particular engineering problems in isolation from others, shipwrights working on large men-of-war may have acquired technical knowledge that the architects and builders of great land buildings had developed. The engineering advances brought about by the special problems involved in building such huge craft for long voyages in dangerous waters may have in turn transformed some of the techniques of those who designed and built sloops, dinghies, and rowboats. The extent and nature of this technology transfer is difficult to ascertain. It is also possible that the scale and demands of the two types of craft were so different that they quickly separated from each other and have only occasionally crossed intellectual paths since the fifteenth or sixteenth century.

Overland

The importance of waterborne transportation might suggest that the most important things, ideas, and people traveled by water. If you weigh the amount of goods and numbers of people conveyed, river transport on rafts, barges, and other boats and sea transport on ships large and small were certainly the most efficient ways to move cargo. Sea vessels were obviously the *only* way to get from Eurasia to most of the rest of the world and certainly provided the easiest passage from Europe to Africa.[61] But for most of human history people and animals carried goods on their backs or pulled them on wheels or runners. These more mundane aspects of moving from one place to another are in the end more important for examining the places of wood in the history and culture of the world than are the engineering feats involved in moving the massive weights for the Great Pyramids of Egypt.

The conveyances that required the least amount of wood processing were those composed of small-diameter poles and branches that were lashed together. Resembling a ladder, these could be harnessed to a person or trained animal and dragged overland with goods tied to them. Nomadic indigenous peoples of the Americas commonly used what French traders and explorers termed the *travois* to move tepees and goods between winter and summer settlements and to new lands. This form was probably used all over the world in areas where water transportation was untenable or impossible. In parts of Europe and Africa a narrow tree fork, overlaid with planks pegged into the surface, functioned in much the same way as the travois, though generally it was a heavier structure.

In winter climates the basic mode of wooden transportation did not rely on dragging the ends of two poles on the ground but rather made use of the friction-reducing qualities of snow and ice. Regional variations of these vehicles abound among the polar peoples of North America and Eurasia, but most rely on a system of flat runners surmounted by narrow platforms onto which cargo is secured. Teams of dogs—most often samoyeds, malamutes, and huskies—pull the sleds, with a driver at the rear to control the sled and team. Ash is the favored wood for dog-sleds because of its lightness, resilience, and strength. The Sami of northern Scandinavia for centuries used reindeer to pull the *pulka,* a wedge-shaped boatlike sled made of sewn pine or fir planks that slides particularly well over softer snow. Similar tublike sleds made of joined bark pieces were used by Native Americans. [FIG. 54]

Wheels were not particularly useful in a climate in which snow covered the ground for much of the year and short-stem plants such as moss and lichens were common during the summer. In drier and hotter climates, however, the wheel was an innovation of revolutionary propor-

FIG. 54
Wooden sledge. Spruce or pine. Northern Finland, ca. 1900. A traditional winter vehicle of the Sami peoples, who used reindeer to pull them along the snow and ice.

tions. Archaeologists theorize that it has been in existence for about 7,000 years. The earliest wheels were simply round discs, probably crosscut from a tree stem, with an axle of some sort joining them. One can think with some amusement (hindsight being what it is) of the first bright idea that joined wheels on an axle, since it was doubtless compromised by the inevitably crooked or bent axle limb or the uneven or radically unequal diameter wheels. "Good for moving things in a circle," an early human cynic might have thought.[62] But the history of technology, as the engineer and historian Henry Petroski has argued, is the history of failure, improved upon.[63] Once people had figured out how to make a circle (which can be done with two sticks and piece of rope or rawhide), and that equal-sized circles were the only useful form for wheels on one axle, the rest was fairly simple. Wheels from the Near East from about 4,500 years ago were made of planks joined together, usually with dowels.

Whether one-piece or joined, these wheels worked, but they were heavy, thereby increasing the load to be moved. Wheelwrights figured out that they could remove most of the wood in the center of the wheel, leaving a rim and cross-members to support the load. At first simply carved away, the spoked wheel of constituent parts was developed and in use in the Middle East and China about 3,500 years ago. Making these lighter-weight wheels requires knowledge about the physical properties of woods and their grain structure as well as great skill in manipulating those materials. The hub of a wheel had to be of great strength, ideally of a wood with interlocking grain. Elm (the *Ulmaceae* family) has such characteristics and was the wood of choice in both Europe and North America. Spokes also had to be strong—to withstand the force of the vehicle load for that moment when they were in contact with the ground—and were therefore split from the straightest-grain wood of trees of great strength, such as oak, shagbark hickory, and ash. Trimmed with a spokeshave, they were mortised into the hub and then into the rim, which was made up of sections of ash or some other resilient wood that could take the shock of uneven roads and the force of braking. It was crucial that the arcs that made up the rim (called *felloes* in the English tradition of the craft) be composed as much as possible of complete annual rings so that the stress and strain of movement did not split or fracture the wood. Hence the ideal felloes were already curved in the tree, and failing that, cut and shaped to take no more than two spokes each, since longer arcs would mean exposing more end grain on the edges of the arc of the wheel, providing a potentially dangerous point of fracture. Ideal woods for the rim combined dense layers of late wood with porous spring wood in the tree's annual rings, as do beech and ash. Felloes were joined with dowels at the end grain end and the whole wheel secured with a heated iron band that shrunk as it cooled, drawing the wheel together tightly. [FIG. 55]

Wheels were the just the beginning of the wagon or cart builder's challenge. Even at slow speeds, most wheeled vehicles traveled over roads that were rough at best. In cities and towns stones may have been laid down to keep streets relatively mud free and from deteriorating in other ways, but these hardly provided a smooth surface. And in the

FIG. 55
"Charron" (Wheelwright). Plate 2, volume 3 of Denis Diderot, L'Encyclopédie *(1751).*

countryside—that is, most of the land—roads were barely maintained, if at all, and likely to be rutted and full of holes. (Occasionally they were paved with stones, such as some ancient Roman roads, but this was the exception rather than the rule.) Log and plank roads, for the most part a phenomenon of the nineteenth century in the United States, were bumpy as well. They were, moreover, a temporary solution, since even the most resistant woods rotted in constant contact with the soil, and they did so more quickly in the troublesome areas in which they were likely to be laid.[64] So important were wooden roads to the transportation of men and matériel in warfare that during the American Revolution the British general John Burgoyne had a wooden corduroy road laid for his march from Lake Champlain to Lake George. Whether it was a wise expenditure of time, effort, and raw materials is debatable, since his army was defeated.

The shipwright was challenged by water, wind, and heavy seas. The challenge for the cartwright, wheelwright, coach builder, and other tradesmen who built vehicles was that the glues commonly available and in use—usually made from animal hides—would not stand up to the jostling and the vibrations of standard wheeled transport. Hence highly skilled woodsmen and woodworkers made their products durable by using dry or cinched joinery—mortises, tenons, dowels, and wedges. For thousands of years four-wheeled wagons and two-wheeled traps and carts were the essential tools for transportation in rural areas in which rivers, lakes, and streams were scarce or easily avoided. Carts and wagons in particular had to be constructed both to carry heavy loads of manure, hay, stones, grain, and other goods and to be maneuverable, sometimes in very tight spaces. Like ships and wheels, these vehicles were built of strong hardwoods whenever possible, especially in the undercarriage, which carried the weight of the load to the axles and wheels. In Europe and the United States the same woods in demand for ships— oak, ash, and elm—were the most sought after. As in shipbuilding, the trees with the straightest grain and the fewest knots were ideal; wheelwrights often picked out trees themselves and certainly examined closely every piece of lumber that timber merchants tried to sell them.

The consequences of this confluence of demand become profound by the seventeenth century. Imperial wood requirements for ships and small craft were joined by a considerable demand for these same woods to build land conveyances. Long-standing preferences for specific wood species for building parts (oak for beams, for example) and heat also complicated the implementation of new policies. Finally, while there is some evidence of concern for the survival of the forests as early as the fifteenth century, little actual replanting occurred until the eighteenth century.[65] These efforts met with only half-hearted enthusiasm in England, since the return on the investment was likely to take at least one hundred years. This was an especially difficult idea to sell to landowners who could realize much more lucrative gains in a relatively short time by engaging in the trade and commerce that, ironically, the navies and the imperialists had brought about.[66] The pressure brought to bear on forests throughout Europe, the Indian Ocean, and China was immense. Although China and Japan ceased major seaborne imperial activity before most of Europe became engaged in colonial adventuring, their massive ships and large river craft had already drawn off considerable forest resources.

Other wooden conveyances drawn by animals—carriages and coaches—were to carts and wagons what parlors were to barns and other farm outbuildings. Refined in their finishes and trimmings, carriages and wagons were also distant from each other in numerical terms. The 1870 census of wealth and industry of the United States lists 44 "carriage trimming" establishments and 11,847 wagon making businesses.[67] While carriages often used some of the same woods as the wagon or cart for support and structure, they were pared down in the size and weight of the container for their cargo—people who for the most part weighed a good deal less than did a load of manure or grain, but who demanded considerably more cushioning and comfort.

Coaches and carriages were enclosed, the "cab" made of thin walls of mahogany and other fine-grained woods. Smoothness was an important characteristic of the exterior of these vehicles, and many were painted to add shine to their surfaces as well as to protect the wood. The interior finishing required the skills of coach trimmers, who outfitted the seats with cushioned upholstery and other accoutrements that clients had come to expect. These rolling parlors and their specially adapted furnishings screened out the rough and tumble character of the urban streets and the coarseness and dust of the country road, as well as inclement weather. Nearly all of these vehicles had a suspension system that eased the effects of the bumps in the road. Bound metal straps or leather slings shielded riders to some extent and turned a bump into a bounce. This model was almost directly adapted to the automobile, which around the turn of the twentieth century was in fact as well as in name simply a "horseless carriage."

At the outset automobile manufacturers just replaced the horse with an internal combustion engine, maintaining many of the wooden parts of the carriage. The terminology of some of the early automobile parts—dashboard, running board, floorboard—attests to this, as does the use of traditional wooden spoked wheels. Early trucks used wood for their cargo-carrying areas, simply adding a cab and engine to the front of the wagon form. In the twentieth century wood paneling continued to be used in some automobiles, and wood is still used in at least one car, the Morgan, which has for more than fifty years used English ash for its frame.

Railways

The railway locomotive may have been the "iron horse," but it ran on wood. Long before the mid-nineteenth-century advent of long trains of cars, short railroads powered by horses and humans were built for mining operations. They were built almost entirely of wood—cars and rails alike. Mines were often tunnels excavated into the sides of hills, from which material was removed in small cars or carts that ran on wooden wheels. The wheel-and-rail connection used in these aged mines is essentially like that used today—the flanged wheels fit on the inside edge of the tracks. (In central Europe a different sort of arrangement was common. There a small centrally placed wheel rode between the thick wooden rails while four additional wheels rolled atop the rails.) In any incarnation, these cars were essentially crude boxes on axles and wheels that were pushed or pulled, drawing their heavy loads out from ever deeper and steeper tunnels into the earth.

By the beginning of the nineteenth century, horse-drawn railcars were used to haul ores to waterways, where barges and boats carried the load downriver to smelters. In areas in which water transportation was impractical, early rail lines made use of stationary engines to haul cars up slopes too steep for early locomotives to handle. Eventually the steam-driven railroad commandeered the movement of coal and other materials over land, supplanting some of the water-borne transportation of the goods, just as trucks would supplant railcars for all but the heaviest and largest loads.

Coal mining was an especially important stimulus in the building of human and animal powered railways. As early as the Roman era in England, Britons burned coal, obtaining it by digging from outcroppings rather than by mining below the earth's surface.[68] By about 1200 coal was mined along the river Tyne. When timber shortages in the 1500s pushed the English to look for other sources of fuel, coal mining experienced a flowering, if you can call it that. When the steam engine, powered by coal, was introduced in the middle of the eighteenth century, the combination of coal, wood, and power brought forth the railway, the locomotive, and the Industrial Revolution.

The locomotive, developed in England at the end of the first quarter of the nineteenth century, made it possible to pull a relatively large amount of cargo on a bed of rails, and therefore it made economic sense to construct tracks over long expanses of territory. The first railway in the United States was the Baltimore and Ohio, which opened in 1833. The prospect of high returns made investment in the railroads an attractive venture for both individuals and governments. By the 1860s there were networks of railways throughout the northeastern United States, eastern Canada, and much of Europe. In 1882 Li Hongzhang, an important Chinese political official and entrepreneur, opened a steam-powered railway that served his Kaiping coal mines, replacing the pack animals that had been hauling coal and other heavy goods.[69] Rail lines were built in Central and South America and entrepreneurs, dreamers, and con artists made plans to develop continuous rail lines all over the world. Supplies for these projects ideally came from areas close to the route, but in some instances it was cheaper or only possible to import wood. Some late-nineteenth-century builders in East Africa, for example, imported Norwegian pine and spruce to construct rail lines from coastal cities to the interior.[70]

The hundreds of thousands of miles of track laid since the 1830s consumed billions of railroad ties, most without much thought to replenishing the world's supply of timber. About 20 million were used in 1840, 491 million in 1880, 798 million in 1920, and 333 million in 1960. These served the 66,000, 215,000, 669,000, and 758,000 miles of track respectively.[71] Finding an abundant wood species that would stand up both to weather conditions and to the pounding of heavy loads was a challenge, but railroad-tie suppliers had an advantage over other wood suppliers. Their product did not have to have a particularly finished appearance or have the potential to achieve it. Moreover, ties were relatively short, so tall trees were not necessarily prized. In addition, ties were rectangular in shape, and the wood only had to be milled to rough specifications, rather than the more demanding needs of spokes, undercarriage, or marine uses. Finally, ties could be treated to resist rot and insects with chemicals, at first with tar- or turpentine-based materials (as in the ship industry) and then, by the later nineteenth century, materials developed by the petroleum industry.

As railroad building became more diversified throughout the world, builders sought out native woods that filled their requirements. Ekki (*Lophira alata*) is a coarse-textured, high-density wood native to central Africa from the Ivory Coast to the Congo River basin. Also known as red ironwood, African oak, aba, bakundu, and esore, it is, as wood expert Jon Arno describes it, "an absolutely punishing wood to work with hand tools," but "can be dressed to a smooth enough surface."[72] The tree grows to great height (over 150 feet) and girth (five feet) and resists acids and to some extent termites, making it an ideal wood for heavy construction and railroad ties. Red mangrove (*Rhizoflora mangle*) is a similarly textured wood, found in brackish coastal swamp areas throughout the tropical world, and it too became a favorite wood for railroad ties in those places. Jarrah (*Eucalyptus marginata*), a mahogany species native to western Australia from Perth southward, was also used to great effect.

People living in industrial and postindustrial societies in the early twenty-first century do not think much about railroads, and if they do, they do not associate them with wood. In some parts of the world, concrete has replaced wood as the material for railroad ties, and metals, plastics, and synthetic fabrics constitute the trains themselves. Those who know something about industrial design think of the streamlined steel locomotives designed in the 1930s. It is easy to forget or fail to recognize that these materials only recently displaced wood as the major material in tracks and many of the cars carrying freight and people. For much of the railway's first century, passenger cars were built primarily of wood, with the undercarriage made of iron and steel. The Pullman Palace Car Company trimmed even middle-of-the-road smoking, sleeping, and dining cars with Andaman padauk (*Pterocarpus dalbergioides*), a crimson to dark reddish brown wood native to a string of tropical islands in the eastern Bay of Bengal. Flatbeds, boxcars, and other freight cars were also primarily of wooden construction until the mid-twentieth century.

Now, as tracks are pulled up and the steel discarded or recycled, the old wooden ties (also called *sleepers*) are piling up. Conscientious environmentally concerned gardeners and landscapers and a few furniture makers have for two or three decades been making use of these hardy pieces of wood. They brace and support hillsides that want to erode or otherwise move where they are not wanted, and some people use them to surround raised flower or vegetable beds, or to make walkways. This is not, however, a complete blessing. The ties survived their hard life under the rails and the trains because of those chemicals that preserved them from rot. In most cases the chemical is creosote, a derivative of coal tar that was used to treat wood for about 150 years. But creosote may be a carcinogen, so in many areas of the world creosote-treated sleepers are restricted to outdoor use in which there is minimal contact with the skin or foodstuffs. Those working with creosote-treated timber are required or encouraged to wear gloves and a dust mask. In the past two or three decades more treated-wood manufacturers have switched to the pressurized application of CCA (which we have seen has its own disadvantages), a process that produces the greenish lumber popular for decades. The small number of sleepers treated with ordinary salt to resist rot are safe for garden installation, though the salt may well leach into the soil and kill plants. A few woods used as sleepers are harmless to plants and people, since they were used in their natural state.

When wood began to disappear from the railroad car, so too did the domestic affect of the conveyance. Today we wistfully equate the era of train travel with some sort of gentler, slower, less frenetic pace of life, often creating a sort of reverie for a time in the past that was allegedly simpler and less complicated. But for decades after railways were introduced, *they* were the fastest means of transportation available, save the galloping horse, and no horse lasted very long running full tilt. Politicians and railroad men may have enthused about the "ribbons of steel" that would "conquer of time and space" (a phrase used by, among others, John C. Calhoun, the ardent American defender of slavery), but many of their contemporaries hated the railroad. The locomotive took people and things to places in a hurry and with so much noise that contemporary writers as early as 1840 complained of the "restlessness and din of the rail-road principle, which pervades its operations, and the spirit of accumulation which threatens to corrode every generous sensibility," a sentiment that hammered at both the speed of the trains and the restless and ruthless investment and speculation that accompanied the advent and spread of the railroad.[73] The age of the railroads—in fact any age—was not "simpler" or "less complicated." It was, if anything, more difficult, more complex, and at least as stressful as any succeeding age.

Bridging the Past

The expansion of rail lines throughout all sorts of terrain would not have been possible without wooden bridges. The idea of a bridge is obviously not a new one. The Romans built stone structures of immense height and length. In rural areas in which there was little in the way of lumber (the country areas of Scotland and Ireland, for example), small stone bridges still dot the landscape. The British built cast-iron bridges in the eighteenth century, but these and later bridges built of steel were a relative rarity until the twentieth century.

These structures survive out of all proportion to the total number of bridges built over time because they do not decay as do wooden bridges. The earliest type of wooden bridge was simply a log laid or felled over a stream. Deep gorges were traversed in some parts of the world on rope bridges, but these were restricted to people and small animals. For the most part wide and deep gorges were managed either by going around them or, failing that, climbing down the face of one side and up the other. Wide expanses of water accessible from a shore were "bridged," as it were, by boats, as they still are in some areas.

Conveying large amounts of goods across a gorge or ferrying them across wide waters is not practical or profitable for tasks such as supplying people with goods on a regular basis. In its heyday, the railroad was the most efficient way to move goods and people overland, and only profitable when the immense investment of building it could be recouped by constant use. For much of the first century of railroad and road building, the wooden bridge carried trains, traps, wagons, and carriages over obstacles, gobbling up hundreds of thousands of board feet of lumber. [FIG. 56]

Timber railway trestles built in mountainous areas of the world employed a system of joined vertical and horizontal beams and diagonal cross-braces, the entire system roughly trapezoidal in cross-section. They were, in engineering terms, simple post-and-beam structures in which weight was carried directly down and outward to the broader

FIG. 56
*"Pack Train Crossing Bridge,
Jasper Park, Alberta, Canada."
Keystone View Co., ca. 1920.*

base. They were also less of a challenge to actually construct than other possible bridge forms, since lower levels could be used as staging areas and a rough "ladder" to the next level.

Suspension bridges made of wood and bamboo were built in Asia for centuries before the railways came. In China twisted bamboo functioned the way steel cables do in modern suspension bridges, anchored on each end by capstans and supported amidstream by wooden towers. In Europe engineers for centuries used curved timbers to build truss-and-arch spans that in some cases extended more than one hundred feet, although these were uncommon. Often these bridges were covered with a wood-and-shingle roof to protect the roadway and the support structure from the weather. In the United States covered bridges were less common, and the few that still exist are classified as architecturally significant and endangered, and the subject of preservation efforts. An even smaller number still carry automobile traffic.[74] The materials used, as in nearly all forms of transportation-related wood construction, were the hardest, most resilient, and resistant to rot. In the United States and Europe, this was the familiar triumvirate of oak, ash, and elm, though softwoods and other hardwoods such as maple were used as well, especially if they were locally available. [FIG. 57]

Waterways and Water Ways

Railroads did not sweep away other forms of land transportation because of their greater speed and their superior carrying capacity compared to that of the cart and wagon. The persistence of other forms of transportation is not necessarily a function of stubbornness in the face of new technology or of the huge initial investment that railways required. Even as the railroad was being developed, workmen dug canals to float cargo and people from one place to another. Connecting two rivers or lakes or some combination of waterways was hard and expensive work, but it was not necessarily a great engineering triumph—as long as the

FIG. 57
Covered wooden bridge. Cornish, New Hampshire. First constructed in 1866, completely renovated in 1954, 1977, and 1989. The bridge is 449 feet 5 inches long and consists of two spans of 204 and 203 feet. It has an overall width of 24 feet, a roadway width of 19 feet 6 inches, and a maximum vertical clearance of 12 feet 9 inches. Of extra heavy framing, the advantage of the covering is that the roadway is less affected by weather.

bodies of water were on roughly the same elevation at each end and the intervening landscape was not full of hills and valleys.

But most of the earth's surface is not like that. Most canals, from the earliest narrow ditchlike affairs to the Panama Canal, either traverse uneven countryside or connect bodies of water of different elevations. The engineering feat in canal building is in solving the problems of elevation. This is done by means of locks, small areas of water controlled by gates that open and shut, isolating the watercraft briefly while the water level is adjusted; the craft then proceeds out the side of the lock opposite from where it entered, traveling at the new water level. Locks were and to some extent still are made of wood, either treated with some substance to resist rot and marine borers or of a species that can withstand submersion or constant wetness.

Although built as early as the fourteenth century in Britain, canals with locks were relatively rare for more than a century after that. The enormous weight of water that even a small lock must withstand mandated massive timbers and precise control mechanisms, and early locks often used timbers at least one foot on a side. Wooden levers attached to the lock to open it were also enormous (as much as twenty-five feet long) because of the force needed to move both the lock and water. Greenheart (*Chlorocardium rodiei*), a dense, resilient, water-resistant, and flexible wood native to Suriname, Guyana, and northern Brazil, had by the twentieth century replaced the woods first used in canal construction, oak and elm.

Greenheart and other water-resistant woods such as mangrove and some species of cypress are also utilized in docks, piers, and pilings. Steel piers are recent developments in the maritime industry and have by no means replaced or supplanted wood. While aluminum, plastics, and synthetic textiles have indeed surfaced as the materials of choice in many of the maritime-related industries, replacing natural materials such as wood, cork, rubber, bamboo, and cotton, there remains a remnant of the waterborne culture that is still based in wood and its products. Strip canoes are still made for an admittedly small market in the northern United States and Canada. Even the venerable birch bark canoe is still made by a few Native Americans and other craftsmen who trek through the woods in search of the perfect bark and make their crafts entirely by hand. Small boatyards specializing in wooden boats remain active throughout the world, serving those who opt for the traditional materials even though they usually come at greater expense than most plastic and metal craft, and require more maintenance. Others continue to use wooden boats because the new materials and technologies are unavailable or too costly.

Wired

On October 21, 1861, the first transcontinental telegraph connection was completed. Three days later, the Pony Express, that romantic experiment that got a letter across the Great Plains and western mountains to Sacramento, California, in about ten days, ceased operation. In business for about eighteen months, the Express was no match for wires and poles. The end of the run was probably a great relief to hundreds of tired horses and the end of a job for their riders, but the success of the telegraph meant that millions of trees would leave the forest to carry the lines of the telegraph, telephone, electricity, and fiber optics across the globe.

Hard, heavy woods such as oak, elm, ekki, and bulletwood (*Manilkara bidentata*) were in demand for electrical and telegraph poles. Remnants of poles installed in the American West during the 1860s indicate that old-growth redwood was also used.[75] California redwood (*Sequoia sempervirens*) resists insect infestation and water damage, and grows to great height and girth over a span of hundreds of years. For these reasons and because almost no one cared to think in terms longer than a few hours or days or beyond the next fat profit, it was almost exterminated from the band of land, ten to thirty-five miles wide and five hundred miles long, in which it grew in California. Cutting restrictions are now in place, but they can be undone as fast as they were put in place, and maybe faster, given American presidents' penchant for executive orders (some of them positive, some not) and the American Congress's perpetual dithering and partisan bickering, as if the fate of the Republic and its resources did not matter more than the next election or the next lobbyist's contribution to their political "war chest."

Imperial Ironies

We commonly think of empires as being great expanses of space and people controlled and exploited (and also benefited in some ways) by a foreign power that because of its technological and military superiority is able to set it and its supporters in positions of power even though their numbers are tiny in comparison to the colonized peoples'. But empires are not necessarily those that have spread across oceans. The United States, Turkey, China, Persia, and Russia all have had imperial designs that included ocean travel, but they were fundamentally land-expansive empires.[76] In all these cases, whether land-based or transoceanic, empire depended on wood more than any other substance for transportation. Metals were more important in the weaponry used to subdue indigenous peoples and other settlers who did not necessarily want or need either "care" or subjugation.

Wood underlay (often literally) the various modes of transportation that moved goods and people throughout the world and within their immediate surroundings, which is where most people throughout history spent their entire lives. Whether on land or sea, the builders of almost all of these forms of conveyance were after the same small constellation of hard, flexible woods—oak, elm, and ash in the Western temperate climates, and ekki, teak, mangrove, and other such woods in the tropical areas. In some cases these demands caused environmental degradation almost immediately after the European imperial expansion began in the fifteenth century. By the late sixteenth century Britons were fouling their air with coal smoke because their woodlands were being depleted with such rapidity. In other countries it took longer to reach a crisis stage, as the ravaging of forests took place on other peoples' lands. One conclusion seems inescapable. Empire was built on and carried by wood, but empire also destroyed that which made it.

5

Artifice: Furniture, Faith, and Music

T he same qualities that made wood so useful for all manner of practical goods also provided artists with a myriad of aesthetic possibilities. Furniture is a necessity first, but it quickly becomes something more—an expression of art, design, and craft that represents a level of refinement and social organization that goes beyond simple need. Owning furniture is often an indication of wealth, power, and status, save for the occasional fashion of minimalism among a small coterie of aesthetes. Wood is the preeminent material for furniture because it can be worked, even with the simplest tools, yet is hard enough to take on the shine and the smoothness associated with civilization. The variety of tree species and wood's infinite variation in color, grain, and figure make it an ideal field for artistic endeavor as well as practical application.

Seating Is Power

The New Shorter Oxford English Dictionary devotes nearly two-thirds of a page to the word *seat*. Rooted in Old Norse, Old High German, Middle Dutch, and Middle English, its meanings encompass everything from the part of the body that rests on a thing or place (also called a seat) to a legislative body, "a manner of sitting on a horse," or the location of disease.[1] The multiple and interlocking meanings tell us something of the importance of the word. While modern people simply take sitting for granted, it has long been a special sort of activity, associated with wealth and power. Think of thrones, the seats of the mighty. Technically it is not essential to have a back on a piece of furniture to make it a seat. Stools and benches, or chests and tables, can function as seats, but the lack of a back and armrests are signs of lower status, rather than, for example, a back stronger than the person's in the chair or throne. It is the opposite relationship in fact—greater vigor and power enables a person to be surrounded, as it were, in wood, textiles, and leather. Multipurpose furniture—chests and tables that can function as seats, for example—historically has been an indicator of lower status, not clever efficiency.

This cuts across cultures and time. Elders among the Dan people of the Ivory Coast are reserved low joined chairs; powerful members of the Gurage or Jimma of Ethiopia often rest in chairs carved from a single piece of wood; Angolan leaders sit on ornately carved thrones.[2] In some eras of Japanese history, chair use was discouraged, forbidden, or rejected, and sitting in them was reserved for the emperor, empress, and a few other dignitaries. Imperial chairs were made of precious woods, such as ebony and rosewood, and carved and decorated with elaborate motifs and materials.

The earliest surviving wooden furniture is from ancient Egypt. The best of the lot survived because they were entombed with elites, who for the most part were the only people who could afford them. Chairs from as early as 1600 BCE demonstrate well-developed carving and gilding skills, as well as a thorough knowledge of mortise-and-tenon and doweled joinery. Some of this furniture is veneered as well. Egyptian chairs suggest that they were part of the accoutrements of power, resembling thrones. For ordinary people it is likely that stools sufficed, although some, for use by the wealthy, were ornately decorated. Whatever their decoration, stools were still a form of submission, albeit less so than sitting or kneeling on the floor. Egyptian furniture from at least the eighteenth dynasty (1567–1320 BCE) has all the sophistication of furniture made since that time, though the tools to produce it were more cumbersome and in most cases less effective than those of the modern cabinetmaker. In addition to veneering (in practice since at least 3000 BCE) and joinery, Egyptians made use of marquetry and inlaying of metals, ivory, and ebony, which was highly prized for its hardness and rich black color.

The ancient Greeks and Romans did not normally bury furniture with their dead, but—usefully for us—pictured it in wall paintings and on ceramic wares. Like the Egyptians, the Greeks and Romans made a broad spectrum of furniture, from simple folding stools to great thrones, tables, and beds. Both Greece and Italy are rich in a variety of wood species, including regional variants of beech, maple, holly, yew, and lime. Moreover, since the Romans in particular spread their empire across a considerable part of the Mediterranean and southern Europe, their traders were able to secure special woods from remote parts of the world, such as citron from North Africa.

The most important innovation in classical furniture is probably the klismos chair. Developed as early as the fifth century BCE, this graceful chair form included a curved back and curved legs that flared outward from the seat. While some early Egyptian chairs included a sloped back, making some accommodation for the comfort of the body, the klismos chair probably represents one of the earliest attempts at ergonomic furniture design. Although it ceased to be made for centuries after the fall of Rome in the fifth century CE, it resurfaced in the nineteenth century, particularly in the United States, during a time of renewed interest in the classical world.

Joiner's Chairs and Other Forms

Furniture was relatively scarce in northern and eastern Europe during the medieval era (about 500–1400 CE), even in the homes of the great landowners. While their houses were often large, much of the interior was devoted to a great hall or some other ceremonial space. These expansive rooms often included a large open central fire and were the public gathering places for those who served the powerful.[3] Most of those sitting around the great banquet tables in castles and manors sat on benches. In churches most of the seating furniture consisted of benches and pews, and status was measured by proximity to the altar.

Most medieval European furniture was either frame-and-panel construction, the frames held together with pinned mortise-and-tenon joints, or of butted boards nailed together or joined with metal straps. Oak was the most commonly used wood, owing to its sturdy and stable qualities and also because it could be effectively carved into high relief, a favorite mode of decoration in northern Europe. Farther south, where softwoods were used in furniture, painting the surface in some cases replaced carving as decoration. Even in the best of circumstances this furniture was, in comparison to furniture made about the same time in China, crude. Thick planks and pieces were used throughout, and the range of furniture forms was narrow, including stools, tables, beds, cupboards, chairs, and chests.

The weight and construction methods may have had something to do with the instability of daily life in medieval Europe, which was circumscribed by constant threats from organized warfare and freelancing marauders. In an environment of limited communication and fragmented and diverse military and political power, maintaining personal and material security was more important than fine furniture. Perhaps the more lightweight furniture and other wooden arts of classical Greece and Rome were not only expressions of the people's expansive appreciation of and education in art and design but also emblems of confidence in their safety and their culture's permanence, however mistaken or deluded that turned out to be.

In contrast to medieval Europe, the manipulation of woods in architecture and furnishings in China of about the same era suggests that the Chinese had a more confident stance with regard to social organization and security. China presents a useful comparative case study in the evolution of furniture and the use of woods for it. Furniture excavated from tombs dating to the time of the Chu kingdom (475–221 BCE) shows a sophisticated grasp not only of complex joinery but also of finishing techniques using colored lacquers.[4] Particularly impressive are the folding beds, low tables, storage chests, and armrests that have survived from this era. For the most part these furnishings were only slightly elevated above floor level, consistent with the practices of kneeling and sitting on mats. Nearly one thousand years later, in about 500 CE, even the more well-off in Chinese society still made little use of seating furniture.

Social practices began to change during the Jin dynasty (265–420 CE), at about the same time the Roman Empire began to decline. By the Tang dynasty (618–907), more elevated seating had become common among the wealthy. Economic growth and relative prosperity during the late Ming period (1368–1644) expanded the demand for furniture, and the resumption of trade in the late sixteenth century—after a period of voluntary isolation from other parts of the world—brought more tropical and other foreign woods to Chinese craftsmen, who by that time had developed highly sophisticated joinery and designs.

Ming-period furniture makes European furniture look heavy, crude, and amateurish by comparison. Completed in 1607 and printed in 1609, the *Sancai Tuhui,* an illustrated encyclopedia of furniture, depicts several everyday and ornate chairs that appear to be made of bamboo, and others of indeterminate woods. What is striking to the Western eye is the economy of form and the sophistication of engineering and design.

Pierced chair backs are common, as are pared-down stretchers and light-weight seat frames. Graceful curved chair rails wrap around the few vertical members in the back and sides of armchairs. Curves are everywhere in these furnishings, in rails, stiles, and brackets and other supports, but they are almost always restrained. The thinness of the chair parts is in stark contrast to that of heavier European forms. [FIG. 58]

Huanghuali was one of the favorite woods of Chinese furniture makers. Akin to or in many cases actually a wood of the padauk group (genus *Pterocarpus*) or the *Dalbergia* rosewoods, huanghuali is a close-grained yellowish to purplish brown wood that can be smoothed to a high luster. It is native to Hainan Island and to Southeast Asia, from which it was imported. Some furniture was made of zitan, a dark brownish wood that was dense and close-grained and dark enough to look black as it aged and acquired the patina of use. While there were some native stands of zitan in China in Guangdong Province, it too was more plentiful to the south. Chang-mu, or camphor wood, was a favorite for chests, among both Chinese and traders and travelers who came along later. Less commonly used in fine furniture were jichimu and tieli, which

FIG. 58
Side table. Hua-mu (flower wood, possibly birch), huang-huali. China, Qing period (seventeenth-eighteenth century). Gift in memory of Arthur F. and Mary C. Wright. Courtesy Yale University Art Gallery.

show alternating bands of yellow and purple wood. Box, ebony, elm, cedar, and cypress were also used, but few furniture examples using these species survive. Kublai Khan liked tzu-t'an (a purplish form of sandalwood) well enough to order it used in his palace, built in the late twelfth century.[5]

Joinery in this furniture was precise and effective, employing slotted and pinned lap joints and complex mortises and tenons. It seems likely that Chinese furniture masters developed these joints to enable them to accommodate the demands of lightness in furniture constituent parts, relying not on the weight of frames and panels or iron straps and nails to provide support and stability but on the tightness and interlocking nature of the joints. They also seem to have taken pride in concealing this work, making the meeting of legs and seats, for example, magically seem like one piece. In other cases pins are visible, but they form part of the entire decorative scheme, nonetheless concealing the complexity of the joint within. Glue was almost never used.

A similar sophistication can be found in Japanese furniture. As in China, Japanese domestic life for the most part did not include sitting on elevated pieces of furniture. The Japanese cultural historian Kazuko Koizumi argues that chairs and the practice of sitting off the floor were brought to Japan from China sometime between the Kofun, or Tumulus, period (ca. 300–552 CE) and the end of the Nara period in the eighth century. The popularity of chairs among elites waxed and waned as foreigners (the Chinese again, Portuguese and Spanish traders, and finally Westerners in the Meiji period from 1868–1911) brought them to the islands.[6] There are scattered references and even occasional survivals of Japanese chairs, but they are not at the heart of Japanese furniture. Furniture in general was scarce in Japan even after the beginning of the Kofun period, when some elites began to build housing elevated off the ground. Between the sixth and eighth centuries Chinese influence in Japan was strong, and the wealthy and powerful made use of low benches and tables and employed coffers for storage. From the fourteenth to the sixteenth century the literary culture of the Muromachi period brought with it associated forms, such as the writing box and table and the reading table. The finish of these pieces was at least as important as their function. Lacquering wood was a highly skilled trade and a painstaking process, sometimes requiring over one hundred layers. Grain and figure were less important than the deep shine and glow of the finish. Woods such as ebony and rosewood escaped the finisher's hand because of their rarity or special appearance.

Neoclassicism, Chinese Elegance, Plain Style

There are two traditions of chair making in Europe—joiner's chairs, derived from frame-and-panel furniture, and turner's chairs, made primarily with the lathe. Combinations of the two traditions became the norm for the furniture of ordinary people. By the seventeenth century, ladder-back (then referred to as slat-back) chairs were in production in the Netherlands, from which they spread as a form of "plain style" chair consistent with the Protestant Reformation's injunctions against elaboration and "fancy." North to Germany and west to England, craftsmen produced millions of chairs with turned legs and pinned mortise-and-tenon joints. Seats were in the main made of pliable plant materials such as rushes, woven together and secured by threading the ends through the weave and pulling them tight. Indigenous hardwoods were almost always used for these chairs—oak in England, maple, oak, and other hardwoods in the United States. [FIG. 59]

These chairs were relatively easy to produce and could be found in various incarnations throughout Europe and the United States. In the latter they have been made as "settin'" chairs in the countryside and as everyday "plain chairs" by, among others, communitarian groups such as the Shakers. The woven tape seat identified with the Shakers married their skill and productivity in textiles with both an interest in comfort and the realization that a comfortable solid wood seat required extensive work with an inshave. The ladder-back was also a more efficient way to produce a chair back. Although it entailed chopping six to ten

FIG. 59
*"Mountaineer's Home near Graphiteville, N.C., U.S.A." Keystone View Co., ca.
1910. Ladder-back settin' chairs grace the porch.*

mortises (which could be roughed out with an auger), that process involved less careful and less demanding work than did the multitude of
round holes that had to be drilled at precise angles to hold the spindles
of the Windsor chair's back and sides.

The basic structure of formal wooden chairs has not changed significantly from ancient Egyptian times to the present, although variations
in formal style have certainly altered the immediate visual impact of the
form. Vernacular chairs made by people untrained in formal aesthetics
have similarly showed little variance beyond their immediate visual impact. An exception to this is the difference between the Windsor form—
with its gallery of spindles—and chairs with flat pieces of wood set
between vertical members of the back. Here was an area for display in

FIG. 60

Slat-back chair. Maple, bird's-eye maple, cane. Probably upstate New York, ca. 1840–60. Although this factory-made chair is somewhat rough—the stretcher between the front legs has irregular edges—the lyre back splat and the use of bird's-eye maple for the back's crest rail refer to the neoclassicism and the "fancy" elements fashionable in the early and mid-nineteenth century in the United States.

figured or carved wood. Below the seat, however, it has still been stretchers and legs. [FIG. 60]

High Style

Most historians now agree that a "consumer culture" began to form in Europe sometime between 1500 and 1700.[7] Some argue that this economic and cultural phenomenon began only after international trade became more common, while others are convinced that it could only have occurred when goods were available on a mass scale. Still others maintain that competition between elites and for royal favor led to fashionable material display long before the Industrial Revolution, even as far back as the reign of Elizabeth I (1558–1603). Each of these explanations has great merit and strong evidence to support it. Consumers' ideas and desires predate the appearance of goods to substantiate them. For our purposes, the origins are less important than the way the impetus to consume and display became tangible in the goods.

Architects and artists studying the classical ruins of Greece and Rome in the seventeenth and eighteenth centuries are credited with spreading the knowledge of classical furniture to the rest of Europe, and furniture makers and designers readily adapted building motifs to their designs for furniture and furnishings. Exact copies were not necessarily their aim; instead they sought a visual effect that would identify them and their patrons and clients as sympathetic to the glorious or noble attributes they found in the literature and material life of the ancients.[8]

In England the classical revival is usually associated with architects such as Robert Adam, who designed the long columned facade in Bath, and with the furniture designers Thomas Sheraton and George Hepplewhite. Adam interpreted classical forms in the interior decoration of his buildings, using pilasters and columns in fireplace surrounds and mantels, as well as in the decorative paneling of formal houses. Sheraton and Hepplewhite incorporated classical motifs such as lyres and columns in the structural elements of their furniture, as well as inlaid classical designs in exotic woods such as satinwood. In continental Europe the neoclassical style is evident in the Biedermeier furniture popular in Germany between 1815 and 1848, with its use of columns and applied brasses. In the United States, the ideas of democracy and republicanism bonded Americans to the ancients in the form of columns on country and city houses, symmetrical plans or facades in buildings, Sheraton- and Hepplewhite-style furniture (and their vernacular, or "country," variations), and a brief clothing fashion of diaphanous gowns cinched under the bust.

Even as architects and furniture makers were gaining knowledge about the classical world and its designs and ideas, their sources of inspiration were expanding beyond the confines of the Mediterranean. In Spain the material culture of everyday and ceremonial life had been greatly influenced by the art and design of Islam, since much of Iberia had been occupied by North African powers for centuries before the 1490s. In the rest of Europe, for the most part isolated from Spain and the parts of eastern Europe into which the Ottoman Turks had penetrated, the major non-Western influence came from China.

Imagine the surprise of an eighteenth-century European who for the first time saw a chair or table made in China. Not only was the design of the piece radically different; the wood looked and even smelled different. Thomas Chippendale (1718–79) is the most famous of the designers who adapted Chinese forms and decorative elements to his furniture, in part because he was adept at it and in part because he wrote a cabinet-maker's guide to the practice, complete with drawings. Chippendale and his cohorts did not mimic Chinese furniture; they added their interpretations of Chinese forms to the tops of chest-on-chests, even as they maintained an essentially neoclassical form for the piece. It was in little accents, such as the brass escutcheon plates of drawer pulls, that you can find Chinese designs, as well as in the carved basket-weave motifs employed on the top of chest-on-chests and the elaborate curvilinear carved motifs used in chairs. Chippendale's chairs share something close to the shape of grand carved Chinese chairs, but there is little reference to the smooth round members and graceful curves of the classic huanghuali chair. "Chinese Chippendale" became a shorthand, not so much for representing replicas of Chinese furniture but for the adaptation of the motifs that Chippendale and others grafted onto traditional forms of Anglo-European furniture and furnishings.

This furniture is a result of the economics and politics of empire, as the China trade opened up to Europeans. The latter exploited Chinese elites' desires to remain apart from international affairs while simultaneously trading with the West for their own immediate benefit. In cases in which the aid of the local powers ceased or proved ineffective, European imperial powers resorted to violence and terror, overcoming their disadvantages in manpower by means of superior technology.

The people living in these far-off places were of only passing interest to those venturing into their ports, whether the voyagers and adventurers were Chinese, Arabian, Portuguese, Danish, or British. But the exotic materials that the explorers and traders encountered were of surpassing interest. Mahogany, which grew in various forms throughout much of the tropical world, became a sought-after wood not only because it was easier to work and more dimensionally stable than were oak, beech, ash, elm, or maple but also because it was different, odd and strange. Some of the species are a deep reddish color virtually unknown in Europe and North America. Crotch mahogany provides a startlingly beautiful grain pattern when book-matched, and this too added to its desirability, although it is no more visually appealing than the similar formation in the crotch of a birch tree, nor is it more spectacular than the striping effect of curly or tiger maple, or any other wood that can develop those striations if the conditions are right. Mahogany trees were huge, both because the tropical forests in which they grew had been untouched by the European hunger for wood and because of the consistently wet and warm growing conditions of the tropics. Thus boards wider than Europeans were accustomed to seeing could be had, and these too became status symbols.

Face and Field

Most furniture forms—especially tables, chests, armoires, and doors—provided the artist and woodworker with areas, or fields, of opportunity for carving or cleverly displaying grain and figure. Here architectural elements—the tracery from Gothic church windows, for example—wrought in large scale could be reduced in size, modified, and interpreted in wood. The harmony of design between a building and its furnishings advanced the owner's cultural values, such as piety in the case of the Gothic.

In Europe and in the Islamic areas of Africa and the Middle East wood was a desirable option for sheathing and decorating interior space because it could show both wit and skill. It could be made to look like something else or to mimic other art forms in a material that was easier to work (marquetry, rather than mosaics made of stone or ceramics). In China and especially Japan, wood as the interior choice has more to do with Buddhist teachings about the relationship between humans and nature and theories about the proper environment for contemplation.

The habitations of poorer folk—the vast majority of the people—in medieval Europe had little in the way of interior furnishing or decoration of any sort. It seems safe to conclude that until the Industrial Revolution in the West, most people never sat in chairs. But in the larger halls and manor houses of the well-to-do, wood paneling and carving gradually began to be used as a major decorative component of interior spaces. Such woodwork was more expensive to use than stucco or plaster, but it was also more refined, more distant from nature. It also had the advantage of feeling warmer in winter than did stone, especially in those rooms heated with fires in the center of the room.

Wood, and particularly oak, was also an important symbolic choice because of its associations with strength. Plaster, by contrast, made its frailty obvious when it cracked or separated from its subsurface support. While wood also cracked, it generally did not fall apart, and the technique of floating panels in grooves in the surrounding frame helped alleviate the negative effects of annual swelling and shrinking, so wood gave the appearance—and the reality—of superior strength.

In Europe paneling offered a field for sculpture, grain, and for showing off the aesthetic qualities of figure. From the fifteenth century onward, one of the most common motifs for sculpted panels is the stylized, curvilinear folded cloth, called the *linenfold*. Popular particularly in the Netherlands and England, even the most carefully carved of these treatments seem unlikely to have fooled anyone, since the "linen" is unsupported. That was hardly the point. It was a statement, a tour de force, a pun in visual terms that speaks to the owner's taste, refinement, wealth, and sense of humor. For the carver it is a manifestation of skill and patience, though the latter was probably of no concern to the client. Plain flat panel wall treatments were also common and these offered both a less expensive option for wall covering and an opportunity to show off figure and grain.

Ceilings provided the decorator and architect with both the opportunity to show the comfort of a support structure and a decorative field. Carrying beams and cross-beams offer large expanses of flat surface between them, often deeply recessed from the plane of the beams. Panels inset in these spaces form *coffers,* literally places where valuables of some sort are stored. In this case the valuables are art and decoration, either carved in relief from the coffer surface, carved elsewhere and applied to it, or inlaid. In some instances painted or molded plaster was applied to ceilings, especially when the architect or client wanted to present the ceiling as the sky, the stars, or heaven itself. Even in these latter cases the beams were often decorated, even if only slightly. Painting enhanced the decoration of the coffers, and chamfering the corners of a beam (cutting or planing the corner off to reveal a smaller flat surface) not only decorated the beam's edge but also made it slightly more difficult for the beam to catch fire, since sharp corners more easily ignite. Careful chamfering often ends with a decorative flourish, a curving motif known as a *lamb's tongue.*

Floors were much less likely to be decorated in intricate patterns, in part because they would obviously be worn down in use, and in part perhaps because the floor symbolically and literally was of a lower order of place than the ceiling. Great spaces with a highly ornate floor inlay or design in wood are relatively rare until the nineteenth century; for millennia before that, stone instead had been used to great decorative effect. More common in bourgeois houses in the United States and Europe were decorative wooden floors of more restrained parquet designs, what people in the nineteenth century called "wood carpets." While narrow pieces of oak, maple, and other such hardwoods were the mainstay of these floors, contrasting colored woods (walnut, mahogany, cherry) were used for accents. In Rochester, New York, on a short street between two major avenues there are two houses that include a room with striped floors, alternating boards of black walnut and white ash. In one house, an Eastlake, or English reform, style building probably designed by one of the city's first woman architects, the room so decorated is the dining room. In the other, a French empire style building, the striped floor is in the parlor.[9] Parquet floors—even if the decoration was only around the perimeter of the room or hall—were a common feature of middle-class housing and institutions such as hotels and business offices until wall-to-wall carpet supplanted them. Once viewed as part of the crusade for sanitation, wood carpets and hardwood floors in general have again become status markers, although the international trade in familiar woods, bamboo and hitherto unknown species from the tropics (such as jatoba), has made them less expensive per square foot than some carpeting.

Intricate pictorial or decorative inlay and veneering in the form of flowers or other designs (marquetry) can be found on expensive furniture produced as early as the seventeenth century. International traders found exotic colored and figured woods from hitherto inaccessible parts of Africa, India, Southeast Asia, and the East Indies archipelago, and craftsmen skilled in both drawing and cutting the small pieces transformed both the palette and the expectations of consumers. Eventually the exotic became more accessible as trade expanded to Central and South America and the mechanization of veneering in turn made available to consumers of more modest means a way to make their humbler furniture resemble the solid-wood pieces of the wealthy, just as inexpensive plated silver, available by the middle of the nineteenth century, gave poorer people the appearance of solid silver on top of their veneered mahogany or rosewood table.

Intricate geometric patterning in wooden floors and ceilings may have roots in the Islamic world, where such fine woodworking has been ongoing for centuries. By the tenth century CE the tradition of making elaborate and decorative panels composed of very small pieces of wood was well established in Egypt. Though less favorably endowed with native timber, and perhaps because of this, Egyptian woodworkers excelled at making these panels, which resisted splitting caused by the daily temperature extremes of the Egyptian climate.

Early Egyptian woodworking demonstrates the thriving commerce in wood in the eastern Mediterranean. Much of this work was made of woods imported from Turkey, such as oak and pine, as well as more specialized woods from the south, such as ebony. Egyptians continued to import cedar from Syria and Lebanon, and their supplies of teak, used primarily for structural members in buildings, came from India. The spectacular intricacy of Egyptian architectural woodwork from this era is probably no better shown than in the Ibn Tulun mosque, built in Cairo in 876 and restored several times thereafter. The *minbar* (pulpit) of the mosque contains nearly all of the elements of North African and Middle Eastern craftsmanship in wood. The railings are composed of extraordinary latticework (*mushrabiyah*) made of thousands of tiny turned ovals linked by rods. Mushrabiyahs were also used in windows, their complex patterns diffusing the hot sunlight.

The sides and entry door panels of the minbar are composed of irregular polygonal panels that are combined in an asymmetrical pattern. There are repetitions within small areas of the panels, but there is no repeat of the pattern overall, which may suggest that different craftsmen worked on the panes of the panel, each leaving his own signature.[10] That they were held together with no nails or glue (which protected them from splitting) also suggests that they were either built in place or fitted together as the several workmen completed individual pieces all the way across the bottom, then moved up to the next level. This free hand with design left open the possibilities of decorative accident and opportunity, an affirmation of what the author David Pye termed "the workmanship of risk."[11]

Carving and Covering

Carvers used the woods native to their areas—pine in Scandinavia and northern Europe, oak in France and England, walnut in far southern France and Italy, teak and sandalwood in India, pine and nanmu, a form of sandalwood, in China. Virtually every wood surface was decorated in some way in these cultures, either as tracery in screens, or as painted sculpture. In Europe carvers and clerics often conspired to add humor to the arrangement, standing pulpits on columns carved to look like composite parishioners or characters from the Bible. Statues of the deity or deities, whether of Christ, Buddha, or deities who protected, inspired, and guided people were common in sacred structures (save

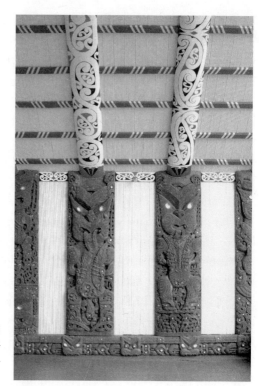

FIG. 61
Carved and painted walls, raf-
ters, and roof interior of a ma-
rai. Rotorua, New Zealand.

those faiths that consider such images to be unholy), as were decorations
on and around roof beams. Many of these were painted in a wide palette
of bright colors and fortunately escaped the depredations of various "reform"-minded sects who terrorized people and their houses of worship
over the past two thousand years.[12] [FIG. 61]

The great age of the woodcarver in the ecclesiastical architecture of
Europe began in the fifteenth century. Given the right wood, such as
oak or walnut, carvers and clients discovered that much more elaborate
and detailed work could be achieved in wood than in stone. A parallel
change was occurring as well in the nature of interior carving, moving
from the more formal linenfold and flat planar style to a more naturalistic one that sought to imitate nature and fool the eye. This impulse

toward realism also gradually supplanted the more abstract style of attenuated bodies and other-worldly faces that those of the faith had thought appropriate for representing the Divine.

The most impressive examples of intricate ecclesiastical carving are often in the choir stalls of great cathedrals, which frequently employ the forms of spires and steeples, bringing these inside the building as decorative elements. Churches were also places in which the painterly aspect of woodworking—usually associated with marquetry work on furniture—found expression. These realistic "paintings in wood" were actually the composition of minute pieces of different hues and tints of wood so carefully cut and inlaid that they appear to be paintings, complete with shading and perspective. This intarsia work—whether in wood or stone or any other substance—applies a new face to a surface, not of one form of wood as in veneering, but as a painter applies color to a surface, often in tiny amounts, in order to build an image. Intarsia does not use layers but tiny discrete points to create the whole. It is akin to the late-nineteenth-century movement in painting, termed *pointillism,* a technique most closely identified with the French painter Georges Seurat (1859–91).[13]

The decorative detail found in older European Catholic churches and the mosques of the Islamic areas of the world is for the most part absent from temples in Japan and China, though there is a considerable amount of carved wood in these structures. Aside from statues, many of which date from as early as the eleventh century, carved wood can be found in the ceilings and highly elaborated bracketing under the eaves of temple roofs. Unlike their Western counterparts, Buddhist temples are relatively free of furniture and built-in or recessed architectural detail, since such elaboration is inconsistent with a belief system that revolves around simplicity and contemplation. In some ways there would seem to be a sort of cross-cultural "convergence" between the systems of beliefs that characterize some Eastern religions and the simple life of prayer and daily labor of some of the monastic groups in the Christian

FIG. 62
Bell tower. Southeastern Finland, ca. 1850.

world, and even with the "simple" faith, architecture, and furnishings of the American Shakers or northern European Lutherans. [FIG. 62]

The tradition of realistic pictorial carving reached an apogee in seventeenth-century England. In about 1667, a young Dutch wood-carver arrived in England, settling near London. Here he got work carving decorative elements for the shipbuilding trade in the town of Deptford.[14] Within four years, while at work on a depiction of the crucifixion of Christ, John Evelyn, the distinguished author of *Sylva, or A Discourse on Forest-Trees* (1664), "discovered" Grinling Gibbons, and promptly introduced him and his work to the architect Christopher Wren and the writer Samuel Pepys. Through his skill and fortunate connections the carver soon found a patron in Charles II. This is as meteoric a rise as imaginable in the latter seventeenth century, perhaps all the more remarkable because England and the Netherlands were in an almost continual state of war between 1652 and 1674. Nonetheless, Gibbons continued on his upward trajectory, and was at work at Windsor Castle by 1677. Prior to that, in about 1675, he worked on the first documented limewood carving of naturalistic foliage, in Hertfordshire. From these beginnings Gibbons fashioned a career that, while not without the occasional bump or complaint from a difficult client, was a model for what all artists and artisans (and he was both) desired. He continued to receive royal commissions, eventually got an annual pension for his work restoring and cleaning the carved work in Windsor Castle, and lived to the age of seventy-three when most men barely made it into their forties.[15]

Gibbons succeeded not only because he was lucky but also because he was considered by contemporary critics, and by succeeding generations of connoisseurs, to be the best of all such artist-carvers. His work is intricate, complex, accurate in proportion and shape, and brilliantly executed. It is a constant marvel to look at, with a depth and attention to detail that remain mysterious. But Gibbons's work does not dazzle viewers merely because of its detail, accuracy, or breadth. It sets viewers back on their heels because of what he attempted to do, feats of the mind and the hand that are so beyond the boundary of normal humans that they can at first glance escape notice, as subtle aesthetic gambles will when from the hands of a genius. But soon after the first exposure the subtleties of his work emerge.

Obviously Gibbons was not the only accomplished artist in wood carving. Others, in other cultures with other tools, other training, and, importantly, other expectations from their communities have in different ways brought out of the grain shapes and forms that endure as art. Gibbons's specialty was high-relief carving, accomplished with a battery of sharp tools. His great talent was made manifest in part because the British steel and iron industries by the late seventeenth century had developed to the extent that he could obtain tools of great quality and sophistication. His work also continues to amaze Western observers because he worked in an aesthetic with which most of them are familiar—realism. His arresting overmantel decorations often consist of large "chains" of fruits, game, fish, and plants that were "draped" on the wall as swags over and on either side of a space intended for a painting.[16]

But carving in wood, whether for religious purposes or for secular decoration, is so diverse across so many cultures that it raises fundamental issues about the messages contained and conveyed by elaboration. Everywhere in the world—save perhaps Anglo-Europe and Anglo-America—those who look for it can find carved and sawn decoration so elaborate it is both a delight to the eye and perhaps a bit overwhelming. Westerners often have trouble negotiating these visual fields because they are covered with designs, a quite different aesthetic from the restraint their culture emphasizes. But "restraint" is a result of classical thinking (and, hence, neoclassical values) and a Protestant value system that condemned decoration and statuary as "idolatry" and decadence. However venal and otherwise corrupt the papacy might have been at times in its history, one can never accuse the Church of Rome of resisting decoration and finery. One can, however, certainly make that case for the followers of Oliver Cromwell, John Calvin, John Wesley, and others who sought what they called the "simple faith" of the original Christians by destroying or defacing decorations they could reach.

Most of the rest of the world does not think that "less is more," as the architect Mies van der Rohe put it about a century ago. They think "more" is evidence of greater devotion and of greater mastery of the material, as well as evidence of an eye that values intricacy and narrative in a visual field. *Narrative* implies stories or tales writ symbolically in wood. But narrative is not always what modern people think it is: history in a linear perspective or stories that are in the main the stuff of fantasy. It is easier for westerners in particular to think of human events in this way, in part because they and the idea of linear narrative are in many ways the heirs of the Reformation and the Enlightenment. Scientific understanding, as Isaac Newton pointed out, is based on the "shoulders of giants," a phrase that explicitly assumes a linear progression of knowledge. We don't build a house from the roof downward, to put the phrase in a more mundane setting. But it is not quite as simple as that. What some analysts might call the "uneven" and "internally contradictory" patterns of behavior among people of bygone or contemporary alien cultures may simply be a different way of knowing.[17]

Holy and Magical Wood

All religions and belief systems grapple with the nature and meaning of life and death: where life comes from, what the beginnings of the human race were, what happens when people die, what the nature of human consciousness is, whether humans are transformed into something else at death, whether there are other forms of life or power in the world or universe, and the nature of the relationships between humans and other beings.

Intricate patterning and stylized depictions of spirits and deities occupy much of the field of door surrounds and other panels in Norwegian stave churches, and scholars have rightly noted similarities to the carvings (most of which exist only in stone now) of the Celtic rim of Great Britain and of northwest coastal France. Entwined in the flourishes and interlocking tendrils of the designs are animal and spirit references, as are found, for example, in wood carving on Maori traditional weaponry and in their buildings [FIG. 63], in that of the Indian subcon-

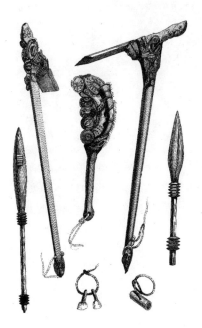

FIG. 63
"Nouvelle-Zeeland, Armes Pagayes."
Plate from G.L. Domeny de Rienzi,
Oceanie, ou Cinquième Partie du
Monde *(Paris: Didot Frères, 1836–37).*
Elaborate carvings on weapons pay
tribute to gods and spirits and were
thought to encourage bravery among
warriors.

tinent, in carved walking sticks made in the American South, and in the carved poles and decorated boxes of the Haida and other Native Americans of the northwestern coast of North America. These figures and faces, some part animal and part human, are visual shorthands, pictorial representations of a larger story that exists in oral form. [FIG. 64] It is a stylized and abstract rendition because the artist's intention was to act as a catalyst for memory, to repeat or restate a traditional narrative.

FIG. 64

"In the Land of the Totem Poles—The Indian Village of Kasaan, Alaska." Keystone View Co., ca. 1915. *Haida boatmen on Prince of Wales Island, Alaska, were recognized as some of the most skilled carvers and weavers (note the hat in the foreground boat) of their time and culture.*

FIG. 65
Burial markers. Northern Finland, twentieth century. Wood is obviously less durable than stone or metal, and these markers suggest that descendants may have regularly replaced them as they decomposed in the harsh climate of the Far North. Rather than impermanence, this suggests a continuing relationship with ancestors.

Common to many of these religious systems is the belief that ancestors must be respected and that they continue to have an existence in spiritual form that communicates with the living. [FIG. 65] Carved burial effigies of departed villagers throughout South America, Africa, the islands of the Pacific, and what is now Indonesia keep watch over the living residents; they help them connect with the dead or assist the recently departed on their journey to the unknowable beyond. The need to communicate with the dead for guidance and to honor them is also expressed by carved house posts, canes, staffs, and other spirit representations. [FIG. 66]

FIG. 66

Agbonbiofe (Yoruba/Nigerian, active by 1900, died 1945). House post, palace of Efon-Alaye. Wood and pigments. Ekiti region, Nigeria. Gift of William E. and Bertha L. Teel. © Museum of Fine Arts, Boston. Master of the Adeshina family of sculptors, Agbonbiofe carved this and two dozen other house posts to replace those destroyed by fire in 1912. The figure's headdress is symbolic of her status; the infants strapped to her back indicate royal continuity.

The largest carvings of this type are probably the Native American totem poles of the northwestern coast of North America. These poles, with their characteristic depictions of various figures stacked on top of each other, serve many purposes, such as honoring the dead, tracing a family's lineage, commemorating an important event, and showing respect to the power of the animal spirits that often define or describe a people's or a family's past. These carved and painted representations of the spirits help people exert some sort of influence over or gain some sort of information about the natural world and its forces. [FIG. 67]

FIG. 67
Edward Curtis, "Tlingit House with Painting and Totem Pole . . . ,"
Cape Fox, Alaska, 1899. Courtesy University of Washington Libraries.

Spirits are not humans, nor for that matter are they animals. Thus
their manifestations in wood or any other material are not intended to
appear natural or realistic. The traditional painted colors of Tlingit
masks and helmets are vermillion, green, and blue, and eyes are stylized
rounded almond shapes with prominent irises. Certain animals—the
grizzly bear, the sea lion and the land otter, for example—were central
repositories of spiritual power for the Tlingit, perhaps because of the
grizzly's and the sea lion's great size and potential for violence, as well
as their effectiveness in catching food that the Tlingit themselves cov-
eted. The land otter—wily and unpredictable—possessed enormous
supernatural power for the Tlingit and appeared in sculptured form in
many sacred objects, such as shaman's masks.[18]

Masks invoke the spirit world and are believed to transform the wearer when immersed in sacred rituals. The most spectacular ritual masks from the American Northwest Coast are the hinged raven masks, which are used in the Kwakwaka'wakw (Kwakiutl) Winter Ceremonial, the taming of the Hamatsa, which rids an initiate's body of dangerous spirits. In Japan, wooden masks of some of the players in traditional Nō drama define the character and its place in the dramatic narrative. Religious mask forms are likewise particularly important throughout sub-Saharan Africa. Among the Bembe and Kuba of Zaire, and the Yoruba of Nigeria, for example, masks are essential parts of rites of passage and invocations of spirits for purposes of aid and protection.

Carving on a smaller scale—amulets and charms—enables believers to cope with concerns about personal and portable protection from evil spirits, sickness, and demons. In some cases amulets allow people to communicate with their ancestors or to honor the spirits believed to be in all living things and even in some inanimate aspects of the human experience, such as the earth or the wind. Shamans used special rattles carved to suggest both their power and the nature of the effect they hoped to have by using the device. Oyster-catcher and raven rattles are probably the best-known of those created by Northwest Coast Native Americans, and they were frequently used to drive diseases out of the body. Many of the bird rattles were carved with a human figure lying on the bird's back, behind the bird's monstrous head. Other animals were also depicted on rattles, including otters and sea lions. In other parts of the world shamans used rattles in ways similar to those of the Native Americans, but with carved representations of animal spirits closer to their experience.

Certain woods and trees, such as ash, elder, rowan, and yew have been endowed with special properties, some of them curative, others magical. Among the Hopi, petrified wood pieces are sometimes considered important in bringing good luck to the bearer, while peach wood kept evil spirits at bay for some Chinese believers.[19] Ash was a desirable wood for Iroquois false face masks, and trees struck by lightning were often treasured, their pieces saved as amulets. Crosses made of the rowan tree were believed to protect people from the powers of witches, while ash and willow were once thought to cure a variety of afflictions.

The metaphorical connection between the tree as symbol of birth, death, and regeneration and the tree as symbolic of the heavens, the earth, and the netherworld can be found just about everywhere in the world. The foundation of Norse mythology is the ash tree Yggdrasil; Krishna, the Hindu god, is associated with the kadamba tree, which is considered sacred; the Acoma people of the American Southwest believe that two sisters, Iatiku and Nautsiti, planted four pine trees in the underworld under the direction of the spirit Tsichtinako, one of which penetrated the boundary of the earth above them, eventually leading to the presence of the Acoma on the earth.

The Tree of Knowledge occupies a central place in the Old Testament. Medieval legends connected that tree and the Christian faith through the belief that Eve planted a branch from the tree on Adam's grave, and the piece then grew into the tree from which the wood for Christ's cross was fashioned. The cross of Christ's crucifixion is Christianity's most powerful symbol and alleged pieces of the "true" cross are treasured as the holiest of Christian relics. That Christ was by trade a carpenter further links the faith to wood, both symbolically and as the material from which sculpture of holy subjects is made. Wood is also, perhaps ironically, the material of the stake, to which heretics and non-believers were bound before they were consumed by the fires set or ordered by the human "divines" who had assumed powers one might reasonably think were reserved for the deity.

Geometry and Pattern

The narrative and spiritual associations of decorated and sculpted wood are intellectual constructions that have embedded within them meanings and beliefs elevated above the mundane activities and values of everyday life. But the world of artifice also includes secular decoration, such as a chip-carved chest or chair or tool. The intricacy of these designs, in geometric low relief on all sorts of goods pedestrian and special, sets them apart from the more restrained work of the linenfold carvings of medieval Europe and the flat-panel and veneered surfaces of the neo-classical works of later eras.

Chip carving takes its name from the small bits of wood produced by the action of the small veining tools and knives that carvers usually use. Often chip-carved work is geometric, laid out with a straightedge and dividers. Dividers—an instrument consisting of two pointed legs joined at the top by a pivot—are useful for accurately transferring a circular design; navigators have long used them on charts to figure distance. For carvers they are useful not only for laying out circles but also for swinging arcs within a circle to mark out floral-like designs. It is a common form of decoration throughout the world. [FIG. 68]

The great arc of cultures from Islamic Spain in the West through India, Southeast Asia, and Polynesia to Japan can be thought of as an "arc of intricacy" that celebrates detailed pattern. Pictorial work finds no home in Islamic wooden arts, but marquetry and inlay have been highly developed in those areas, encasing boxes and occasionally larger forms in unending visual complexity. Patchwork patterning in Japanese marquetry allows for the presentation of larger surfaces in the overall

FIG. 68
Chip-carved plane. Beech. Eastern Quebec, ca. 1880.

pattern, thereby making it possible to include pieces of wood with un-usual grain patterning as well as color differentiation. This might be seen as a decorative bridge between Western enthusiasm for grain pat-terning and manipulation (as in book-matched veneered surfaces) and the intricate work of Islamic craftsmen.

While Islam's prohibition against figurative and representational de-pictions can explain why those elements were not part of the craft and art of Islamic woodworkers, it tells us nothing about why geometric pat-terning was popular. Did it resonate with the mathematical bent of the region's peoples? The ancient Egyptians were great astronomers and mathematicians; Islamic scholars contributed much to the learned world's understanding of mathematics in succeeding centuries. In addition, geometric patterning exhibited the makers' and clients' devotion, given the great planning and skill required to create the finished works. Fi-nally, as in other cultures, the high degree of artificiality that such pieces exhibit—straight lines, perfect circles, unusual combinations of wood and metals—are statements of the distance between humans and the rest of the living world.

In the part of the "arc of intricacy" inhabited by Buddhists, the exis-tence of visual complexity might seem surprising. There was little fur-niture in most of Japan until the twentieth century, to be sure, but there are a few surviving cabinet pieces with spectacularly patterned applied decoration of wood. In the most important form of Japanese furniture, the *tansu,* or chest for storage, the elaboration is in the multitude of iron fastenings common on these pieces. People first encountering *tansu* of-ten wonder about the amount of iron strapwork and hardware, as if it were there for functional purposes alone. One so struck might then think that Japanese cabinetmakers and clients were uncertain of the physics of storage pieces, or that they were overly concerned about security of the goods inside. The latter might bear some truth, but those conclusions are grounded in Westerners' assumptions about what makes for a pleasing and appropriate aesthetic. Wood's "natural" face was important in a culture in which nature was an ever-present part of the people's belief system. Iron bore no such resemblance or relationship, since it was the product of fire and human manipulation. It was emblematic of distance from the mine and from stone and earth.[20]

This is not to suggest hard and fast rules for cultural practices, but to identify tendencies and commonalities in the use of wood in the material cultures of peoples. There are examples of inlaid goods, marquetry, and parquetry (an inlay of geometric patterns) in European furnishings and buildings. There are examples of flat-panel interior treatments in the Islamic and Hindu regions of the world as well, and certainly in China and Japan, where that was a powerful aesthetic. But the larger picture of architecture and furnishings shows a substantial difference between northern and central Europe and the "arc of intricacy." The separation is linked to differing visions of the natural world and the duty or responsibility of humans to find a balance or harmony with it, or to know it and master it. None of the cultures considered above embraced entirely one vision or the other, but they valued different aspects of the human condition and the fruits of human art and craft.

Music in the Tree

Music is intangible; sound has no fabric or weight. But even if you cannot see it, much of what makes music still comes from the forests.[21] We do not know when people first made music by beating on something or blowing through a tube. Hollowing out logs and hitting them produces a sound that is the result of vibration in and throughout a void, the *sound box*. It is possible that early drummers discovered that they could make different sounds by varying the voids and sizes of logs. Both ordinary communication and music were the result.

It is even harder to pin down how humans figured out that they could make sounds by blowing through tubes, and that the tube's length and diameter affected the sound that emerged. What historians can follow, however, is both progression through time and diffusion in space. A small irony in the development of what were to be called the *woodwinds* is that they probably originated as instruments made of bamboo, which is technically a member of the grass family of plants. Bamboo has the great virtue of being hollow and of having species that grow to different diameters. We will never know who first pursed their lips and blew through a bamboo section to produce that slightly unearthly sound so different from the "whoosh" of those who had been simply blowing air through the tube, just as we will never discover who first found out that closing and opening a hole in the tube or that differing lengths and diameters of the tube changed the sound.

The bamboo *sheng,* which consists of different lengths of bamboo that are lashed together, has been played in Asia for millennia. Tone modulations are achieved by blowing through two or more tubes simultaneously, akin to the modern harmonica. Complicated *shengs* were equipped with several tubes into which holes had been bored, while simpler instruments relied on fewer tubes, sometimes with no holes in them. It seems likely that just about anywhere bamboo grew, this early wind instrument would have been in use.

It is not much of a leap from bamboo to bored-out wooden instruments. Advanced pipes and reeded musical instruments were in use in the Mediterranean region by the time of the pharaohs, and it seems likely that horns and other metal instruments have been around since the Bronze Age, since even coarse metallurgy dramatically expands the opportunities for making instruments. Drilling out short pieces of wood was possible as soon as metals were available and people figured out how to sharpen an edge. Moreover, people could also make a tube by sawing a piece of wood in half, hollowing it out and reattaching the halves by tightly winding them with damp hide or plant materials, as in early alpine horns. The outside of the instrument does not have to be round; it can be uneven or rough, though it is unlikely it remained so for long, given that the player would have worked the surfaces to suit his or her comfort.

The history of wooden instruments is once again an example of cultural convergence—similar changes taking place across cultures and across space with little or no evidence of contact. Wooden musical instruments were common artifacts in Africa, where kings and chiefs were escorted or serenaded by players of ornate wooden trumpets. Horns of bamboo or metal were played in imperial Rome, throughout North Africa, among the indigenous peoples of the Americas, throughout Oceania, and in much of Europe. Experimenting farmers and musicians from Scotland, Ireland, and eastern Europe seem to have discovered the bagpipes at about the same time, their pipes made by turners with pole or treadle lathes who bored instruments round on the inside and turned them on the outside.

Simple wooden pipes and whistles have been traced as far back as ancient Egypt. Players created sound by blowing air directly through a channel in the mouthpiece, the air rushing against the edge of an opening cut into the cylinder, just below the mouthpiece. The action of the air over the edge sets it into vibration, causing the characteristic sound of the pipe, which the player regulates by means of stopping or opening the holes cut into the barrel. Reed instruments generate sound by means of blowing air across a reed or reeds bound to the mouthpiece of a tube with holes drilled into it at precise positions to produce the sounds desired. More complex tones and harmonics required more holes and control of them, and enlarging the natural reach of the player's fingers was achieved by means of mechanical wooden, and later metal, keys.

All sorts of other less complicated wooden wind instruments were also available. Instrument makers discovered that certain woods produced different qualities of sound. Box, a dense and fine-grained wood, had a consistency to its tone quality, as did ebony, a species highly prized but devilish to work because of its density. It was also an imported wood for most of the world and that restricted its use in Europe until trade with Africa expanded during the Renaissance. As the machining of wood became more precise in the late medieval and early-modern period of European history, so too did the precision of instrument making evolve to the point where listeners and musicians could predict what they were supposed to hear from a particular instrument. Wood offered greater and more subtle modulation of tone than did metals, as well as the slight variation in tone from instrument to instrument that aficionados wanted in the tools of their trade.

Stringed instruments generate sound when a length of animal gut or, more recently, metal wire is plucked or stroked with a bow to set up vibrations. It seems reasonable to assume that this discovery evolved from bow-and-arrow hunting. There were stringed instruments in ancient Egypt, such as the *nanga,* on which strings are stretched perpendicularly to a sort of boat-shaped soundboard and attached to the end of a curving neck. Essentially a form of harp, the nanga's descendants and relatives are many, including the lute, sitar, koto, dulcimer, zither, and eventually the guitar and violin. Like the hunting bow and the harp, these instruments were initially plucked with the fingers; by about 900 CE musicians were producing sound by passing a bow across the strings. Making music by hitting the strings with hammers reaches back to at least ancient Assyrian civilization, in the form of the dulcimer.

Common to all these instruments, in addition to their strings, is the wooden "box" or "chest" in which sound reverberates. In the earliest cases the empty space over which the strings were stretched was the hollow of a naturally occurring plant form, such as a gourd, but in the main the story of wooden stringed instruments is that of worked wood formed in different shapes and pierced in different manners to produce distinctive sounds. By the tenth century stringed instruments played with a bow in Europe and Asia were essentially of two forms. One was the instrument made from a single piece of wood, carved to produce a short neck and gouged to form the sound box. The other form, which anticipates guitars and the stringed instruments of the modern orchestra, was the fiddle, which instrument makers fashioned by gluing and pegging together component pieces.

Guitars and other similarly made viols could be had in Spain by the fifteenth century. They were the work of highly skilled artisans trained in both the North African traditions of fine architectural and decorative woodworking and the ancient traditions of Assyrian and Egyptian plucked and hammered instruments. By the late sixteenth century an Italian innovation, the violin, became a popular instrument, at least in music played in the courts of the nobility. The great Italian makers of the "modern" stringed orchestral instruments—Amati and Stradivarius—are the most well-known to connoisseurs. The Amati family (active ca. 1500–1692) worked in Cremona and are considered the first great Italian violin makers; they often made violins in pairs, or consorts, and counted European nobility and royalty among their patrons and clients. Antonio Stradivarius (1644–1737), whose name has also long been synonymous with superb viols, also worked in Cremona.

Violins and the larger orchestral stringed instruments offer insights into the technical and artistic uses of wood in music. One of the first things you notice about fine violins—and even some ordinary instruments—is that makers often used figured woods, striped for the backs and tables (tops) and in some cases for the ribs (sides) and bridge. Certain species are favored for the back and table, such as sycamore and maple, while some makers swear by spruce for the table. European foresters to this day carefully tend plantings of sycamore trees grown especially for use in stringed instruments.

The process for making these instruments is complex, a combination of aesthetics and practicality. Since instrument parts are made of radially sawn pieces, or thin wedges sequentially cut from the pith to the bark, the trees must be more than twice the width of the back to be useful, since the pith and sapwood are normally discarded. Sequential pieces cut from these select trees are joined edge-to-edge, the flat surface brought to its final curved profile with gouges, as it is painstakingly tested for consistent vibration patterns. Artisans then cut the outline of the sound box with thin-bladed hand or powered saws. F-holes are cut in the table by drilling an entry hole and then cutting the curved openings with fret saws. Purfling, a thin strip of decorative hardwood, is then inlaid around the outside edge of the table and back, adding an edge resistant to the wear and stress to which the larger pieces will be subjected when played.

The ribs of the instrument are fashioned by bending thin strips—usually laminated sycamore and pine—which are glued to six blocks that are in turn glued to the back at the points where the outline of the instrument changes direction; that is, at the C-curves on each side, at the neck, and at the bottom of the sound box. Then the head, neck, fingerboard, tailpiece, end pin, and chin rest are affixed to the sound box, pegs for adjusting the tautness of the strings are inserted in the head, the bridge glued in place, and the sound post inserted into the box. Ebony and rosewood are among the favorite materials for pegs, the fingerboard, end pin, and chin rest; and sycamore for the head and neck, though in less expensive instruments hard plastic and other hard tropical woods might be used. The bridge has to withstand the enormous pressure of the taut strings, and maple is a favorite wood for this piece because of its fine grain and great hardness. [FIG. 69]

MUSICAL INSTRUMENTS.

FIG. 69
"Musical Instruments."
Plate 15 from Abraham
Rees, Cyclopedia, or
Universal Dictionary of
the Arts, Sciences and
Literature. *(London:*
Longman, Hurst, Rees,
Orme and Brown, 1820).

Makers of fine stringed instruments argue about the tone qualities of the woods they employ, and it is not clear that the "fiddleback" figure many makers and musicians prefer actually resonates better or more truly than other figuring or plain wood. But it does seem reasonable to argue that wave motions will perhaps achieve a color and depth when encountering wood with striations, though it may be the striations of annual rings and the alternate bands of new and late wood that do the trick.

It is also possible that the violin makers' gravitation toward fiddleback figure was purely for its decorative qualities. The instrument shape may well have evolved to its present form through experimentation to produce the richest and most distinctive sound, and there also may be something of the craftsman's tour de force about it. Certainly the scroll at the top of the head has its roots in decoration, especially since some instrument makers have taken the opportunity to elaborately carve that part to establish a signature in three dimensions.

Globalization came to the world of stringed musical instruments in the form of the pianoforte, or as it was known when introduced by Bartolomeo Cristofori in 1709, the *gravicembalo col pian e forte*. The "cembalo" part of the name refers to the instrument's roots in the dulcimer, a stringed instrument of long history that made its sound when its strings were struck. The "pian e forte" refers to the instrument's capability of producing soft or loud tones by modulating the force with which the player struck the keys and hence how hard the hammer hit the string. The inability to control volume was the drawback to the pian e forte's precursor, the harpsichord, which had preceded Cristofori's instrument by at least two centuries.

Producing this instrument, which we now call a piano, required an exacting knowledge both of music and of the technical characteristics of wood species from around the world, as well as exceptional skill in woodworking. There are several opportunities for potential disaster, or at least ineffectiveness, in building a piano. The keyboard has to be able to withstand the vagaries of humidity and dryness. Changes in shape distort the action of the keys, which also must be relatively free of material movement. Basswood is generally favored for the keyboard, although some makers favor sugar pine. Both are lightweight and relatively easy to work. The same cannot be said for the ebony traditionally used for the black keys, or the ivory-encased white keys. Ivory is no longer used because of the terrible toll ivory hunters and poachers have taken on the elephant populations of the world, and like ebony it is often replaced by plastics. In any case the goal was stability in dimensions.

The real complexity of the piano, however, is found in the mechanism that links the keys to the hammers that strike the strings. Enormously complicated, it is composed of very small parts that traditionally are made of wood. These parts not only have to be environmentally stable but also must be able to take multiple borings to seat the tiny screws that hold the mechanism together. The keys sit on a platform of a hardwood such as beech or maple, and the connection between key and mechanism is often of quarter-sawn rock maple or some similar hardwood. The small parts of the mechanism demand an exceptionally hard wood, such as European hornbeam, which does not swell or shrink much and can stand continuous use without distortion.[22]

The hammers, which do the work of striking the piano string, and the dampers, which release the string when it is struck, are generally made of hornbeam (*Ostrya virginiana*), west African mahogany of the *Khaya* genus (species *anthotheca, grandifoliola, senegalensis*), sapele (*Entandrophragma cylindricum*), which is native to central Africa from Ghana to Uganda, or some other such dense and dimensionally stable wood. The soundboard, which transmits the vibrations of the struck strings, is often made of slow-growing, vertical-grain, quarter-sawn pine, traditionally harvested from stands in eastern and northern Europe. The wood frames holding the strings of early pianos compromised the instrument to some extent. The most desirable tone requires that the strings be in high tension. The blows of the hammers exacerbated the strain on the frame caused by the tension of the strings. The iron frame, introduced in the nineteenth century, allowed the use of steel strings under greater tension, which in turn meant that the hammers could bang on them with little chance of fracturing the frame.

Cases for the piano's strings and striking mechanism offered practical and aesthetic opportunities and challenges for makers and players. Rosewood was used in the nineteenth century, perhaps because the forests of South America had become more accessible to traders, and because the great expanse of the piano's flat sides offered artisans an opportunity for decoration. Many wealthy clients demanded intricate and flashy treatments, since competence in music for amateurs (especially women) and evidence of wealth and taste were considered marks of refinement and an edge in the marriage market. Fancy veneers, especially burls of exotic species, were often applied to the exterior, as was black "japanning," a hard varnish surface often decorated with fanciful scenes reminiscent of those found on nineteenth-century Chinese tablewares. Clever decorators also painted scenes on the underside of the hinged top, which, when opened for playing, were then revealed to the audience. Modern grand pianos are usually encased in laminated materials, with mahogany or walnut on the outside over a core of woods such as African mahogany, obeche (*Triplochiton scleroxylon*), or agba (*Gossweilerodendron balsamiferum*), all of which are native to central Africa. Square and upright pianos, which became especially popular beginning in the later nineteenth century, were often encased in solid woods.

Wooden percussion instruments, such as the marimba and the xylophone, utilize the sounds made by striking different lengths of wood bars with a softened hammer. The marimba is probably of African origin, and reached Guatemala and Mexico as slaves were brought westward beginning in the late sixteenth century.[23] The vibrations of the strike resonate in chambers (first gourds, later pipes) below the wooden bars, the chambers being of varying lengths commensurate with the tone produced by the bar. According to a publication of the Latin American Folk Institute in Washington, DC, the instrument gets its name from the Bantu word *rimba,* which literally means "flattish object sticking out."[24] People have made marimbas out of whatever they could find, but the preferred wood for professional performance is a tropical hard-wood such as rosewood.

Alchemy and Artifice

Artifice is the alchemy of woodworking. Hustlers, hucksters, and true believers once thought they could find a "science" for transforming base materials into gold, but they were mistaken. But wood can be transformed, if not in composition then in appearance and shape. Smoothing and polishing is certainly the most obvious way to accomplish this, but the "alchemy" of the endeavor occurs when it creates, shapes, and transforms activities or forms of communication and it transforms spaces from voids or enclosures to areas of meaning greater than the elements of construction or manipulation alone. Obviously it is not the wood itself that does this, but what humans bring to the experience that endows or imbues the space, the object, and hence the material with meaning. But that is in some ways a reciprocal relationship: We carve or otherwise transform the wood into shapes and forms that in turn motivate further activity, whether it is contemplation, awe, or obeisance to the Divine. Sculpture makes belief systems three-dimensional, whether those of religion or of the more mundane, such as class distinction and power. The burl on the tree trunk—a lumpy roundish blob created by disease or some other deforming event—becomes the fancy and fanciful swirl of the book- or quarter-matched veneered surface. The tiny hornbeam components of a piano become the music of the strings; the gourd or the pipe of the marimba becomes the sound of the struck wooden slat.

Furniture is part of the accoutrements of living. It connotes both power and status and a drive for certain kinds of physical comfort, or at the very least rest from weariness. Because it is both a tool for the body and a presence among the constellation of other artifacts and belief systems of those who own or use it, furniture is both an extension of the body and its movements and a composition of ideas. Human comfort, as the historian Katherine C. Grier has ably demonstrated, has much to do with the use of textiles, which are softer and more pliable than wood, though comfort can be achieved by carefully shaping wood and linking constituent parts of furniture in such a way as to better accommodate the human body's irregular forms, as in the rocking chair made by Sam Maloof shown on this book's cover.[25] For the most part, however, when it comes to comfort, the artifice and the alchemy of wood in furniture are often not so much in the shape of the surface but in the visual affect of surfaces when we sit or recline and in our trust in the frame beneath the textiles. The "comfort" that wood provides is thus metaphorical. It is the comfort of the park as opposed to the uncertainty of the forest or wilderness.

Modern peoples tend to think of the wilderness in a positive light. Seen as the antithesis of life in an urban, industrialized society, wilderness is valued more in the mind than in reality. For most of human history, however, the wilderness was a place defined by the first four letters of the word, when *wild* was hardly a positive quality in either human beings or animals. Only a very few of its celebrants actually knew of it directly or lived in it, even for a short time. The wood around us—as opposed to the "woods" out there in increasingly smaller places on the planet—replaces the angst of the unknown and potentially dangerous with the comfort of artifice, the metaphor of both nature and the human ability to transform it without obliterating it, if we but know how to do that.[26]

6

Thinking Inside the Box

I n my office at home is a small, nineteenth-century traveling desk. I discovered it at a country auction in rural New Hampshire in 1999. It may have been a presentation piece; inlaid in the middle of the top is a small brass shield in which are engraved the letters "C. G." and "1876." [FIG. 70] It obviously did not come down in my family through generations; in fact none of my forebears were in the United States in 1876, though it is possible that the desk is not an American-made artifact. That, however, is irrelevant to me. What I like are its secrets.

Made of solid mahogany, its top and front are veneered in what appears to be Carpathian elm burl. Inlaid brass pieces protect and adorn the four corners, and L-shaped inlaid brass pieces strengthen

FIG. 70
Writing desk. Mahogany and leather, brass. England, ca. 1876. (Closed.)

the edges between the corners. The hinges and internal fastenings and hardware are also brass. Hinged on the long side, halfway down from the top, it opens to reveal a black leather-covered writing surface, the edges of which are trimmed in a narrow band of gold-leaf embossing. A small brass spring-loaded clip holds in place the half of the writing surface nearer the writer. This lifts to reveal a space suitable for storing paper. The far side of the box has at the top a long slender open compartment for storing pens and pencils, a small (one and a half inches) square open compartment next to the pen holder for erasers or other small accoutrements, and a brass-capped glass ink bottle set into a third compartment. The pen holder lifts out to reveal a small storage area. Even the invisible parts under the surface of the desk are smooth and varnished. The piece is impressively finished, even formal. [FIG. 71]

The top section of the writing surface, just below the pen and ink compartments, also opens to reveal a place to store things; it was probably intended for envelopes. The only other clues to the history of this piece come from a small violet printed paper label affixed to the underside of this writing surface. "Independent Order of Good Templars" surrounds an engraving of a lion lying down in front of a shield depicting

FIG. 71
Writing desk. (Open.)

a heart and an anchor, and behind that, two flags, one British, the other simply emblazoned with the words "Good Templars." Above the shield is the "all-seeing eye" associated with the Masonic orders. The reclining lion is a strong clue that it was made in England.

Upon a closer look, something does not look quite right. The pen and ink compartments are about one and a half inches deep, yet the vertical piece of mahogany in front of them when the top half of the writing surface is opened is about two and a half inches deep, nearly reaching the bottom of the interior of the desk. What is behind this, if anything, and how can I get at it? I lift out the pen carrier, but all I see is a small empty space, probably for storing pens. It's not deep enough. The joints are tight; pushing the walls or bottom does nothing. I remove the ink bottle; still no answer. Finally, after pulling up on the dividers between the compartments, I hear a loud snapping noise as the divider between the pen and the incidental compartment lifts straight up (with some difficulty), the front piece springs out at an angle, and two very small mahogany drawers with tiny ivory knobs appear. They are empty. No treasure, no secrets, no incriminating letters, but there they are. [FIG. 72]

FIG. 72

Writing desk. (Open, with secret compartment exposed.)

Why build them into the desk? What was there to hide and from whom? I shall never know the answers to these questions, but that does not trouble me, although it would have been wonderful to find a long-lost and juicy letter between lovers, gold coins, or some other treasure. More important is the clever way in which the apparatus works—a brass catch, released when the wood divider is raised, lets loose a narrow band of spring steel that pops out a panel, revealing the secret drawers, which are themselves very highly finished and decorated, a proper nest for things of value. The whole apparatus is theatrical every time I open it. I cannot help but think that it was meant to create a properly dramatic moment, appropriate for the magical revelation of secrets.

Boxes create security; they carry secrets. They have done so since humans first figured out that they needed or wanted to do that. The idea for a box probably first occurred to someone when they saw a naturally hollowed out tree, rock, or gourd. Stone and shell tools allowed people to hollow things out for themselves; the first boxes probably had a top of some other material. Only big, smart animals could get in—and humans clever enough to have figured it out as the maker had.

A box hewn from a single large log, its top hinged by a wooden rod, is an ancient form in many cultures. Using sawn boards to make boxes, however, opened up a new world of possibilities only dreamt about in the society of the axe and adze. The simple six-board chest is a form common throughout the world, but more likely to be found in places where people had excess things to put in them—textiles, for example—and where people tended to live in one place for a long time. The six-board chest was a common form in medieval Europe and in the United States from the seventeenth through the nineteenth centuries. In the United States in particular, wide pine boards—lightweight and easy to work—were readily available. These chests were commonly painted on the exterior, and their four "feet" were created by cutting out pieces from the bottoms of the end boards.

When carved and decorated, these boxes could convey status, wealth, refinement, and perhaps religious or other beliefs. They are often adorned with ebonized turned devices that have been split lengthwise and glued to the surface of frame members, the rounded devices on the flat surfaces of the frame providing machine-made testament to the safety of the house, as opposed to the rigors of the wilds. The panels between the framing members are often decorated with shallow relief carving. Complex designs of stylized flora—tulips, sunflowers, vines, and tendrils—appear with regularity on many of these chests. While not as complex and intricate as the interwoven designs of Celtic and Norse carving, or as geometrically inclined as the work of Islamic, Indian, Polynesian, or Asian carvers, they are nonetheless highly skilled works of traditional design. People make aesthetic choices for reasons, perhaps not always at the front of their conscious thought processes. Some of those reasons may indeed involve simply wanting goods that look like the possessions of other people with whom they want association. But such an explanation as the *exclusive* reason for consumer choice privileges the very few or even the single individual (it had to start somewhere with someone) and supports a theory of history that indicates a tiny minority makes everything happen, from chests to revolutions, while everyone else merely follows. There are plenty of failures in the marketplace to lay to rest the notion that consumers behave like sheep.[1]

We know these chests were containers of valuable goods because they generally have locks, usually set in the center of the front panel, just under the lid. Though sometimes highlighted by an escutcheon plate of different colored wood, ivory, or metal, locks were most often identified simply by the keyhole, an understated but yet noticeable indication that what was inside mattered. Although they were primitive by the standards of modern mechanical locks, these older devices nonetheless sent the message to the would-be thief or troublesome child that they could not gain access without either smashing the chest (difficult if it were of oak), picking the lock (possible with the right training), or carrying the whole thing off for later invasion (also difficult, since the wood and contents were heavy).[2] Smaller locked boxes, called coffers, were the places the well-off secured their jewels, coins, and other precious items. They could be hidden in odd places—barns, root cellars, or attic eaves—or taken along for safekeeping when traveling.[3] (See Figure 14.)

The Chest of Draw(er)s

Cased furniture pieces with drawers are at least as old as ancient Rome, but seem to have disappeared from Western life for centuries after the fall of the Roman Empire. They are harder to move than chests, they require more wood to construct, and provide less storage space than a chest of equal dimensions. But they do offer specialized spaces and the possibility of organizing possessions, an opportunity that did not seem to matter in much of the West, at least until about the sixteenth century.

The debate among scholars about when and where the "commercial revolution" occurred in post-Renaissance Europe, which was mentioned in chapter 5, does not reveal much about the advent of chests of drawers. Having access to more goods and more types of goods does not necessarily translate into the desire to have a greater number of discrete spaces in which to store them or to gain access to some things without disturbing or disrupting just about everything else. Modern people who have chests with a lid and one large space inside understand how irritating it is to search for something, knowing full well that it is usually on the bottom. Chests are thus reserved for long-term storage, either for seasonally used goods or for those things saved for reasons other than their day-to-day utility.

For some Europeans, the chest of drawers became a necessity as clothing became a consumable product of which diversity and decoration were demanded as the Renaissance began in Italy. Mediterranean traders brought finely textured long-staple Egyptian cotton to European tailors and the salons of the wealthy, and trade along the Silk Road had European royalty and courtiers angling for that fabric as well. These goods were refined in most of the senses of the word as understood then. Their physical characteristics of shine, lightweight body, and smooth texture made them the equivalent of polished metals and woods, akin to the manners and mores of their wearers, at least in theory. Clothing may not have "made" the man or woman, but it gave the impression that somewhere underneath it a virtuous or at least "polished" human being existed.

As the anthropologist and historian Grant McCracken has observed, Elizabeth I managed to convince the country elites that they had to be a presence in the court and that they should pay for the privilege.[4] In this world of fashion and finery a *lumpen* chest of undifferentiated space would have been not merely inconvenient; it would have been a social drawback at the least, too coarse in form and finish to be fashionable, and ultimately a catalyst for mockery among the courtiers and others currying favor. The clothing repertoire for both men and women grew increasingly complex, with layers of different fabrics of different weights. Separation of the goods was imperative and the chest of drawers— which the well-off could certainly afford—may well have met the need of the new elaborate fashioning of the body.

Drawers also suggest that the idea of specific and limited functions for clothing and its storage furniture reached court culture long before it permeated other forms of daily life. This idea, so closely linked to the mass manufacture of goods and to the scientific revolution, seems to have predated both in clothing forms and, in some of the trades, tools. The latter seems a more comfortable and logical place to locate this "modern" idea, since it is connected to technology and therefore science, but clothes may just as well have "made" the idea as well as the man—or woman.[5]

Portable and Precious

We can find a parallel evolution in the history of writing boxes and desks. The small traveling desk, like my nineteenth-century English piece, has been in use for centuries and across cultures. It was surely less common than large hinged-lid storage chests, because until the eighteenth or nineteenth century only a tiny minority of people in the world were literate. Even among those who could read and write, written communication was uncommon until the late eighteenth century. Paper was expensive because making it was labor-intensive; cotton rags had to be boiled so long that they became a thick soup, which was then pressed in wooden molds, each the size of a sheet, and dried. It was used sparingly, though more so than its more expensive alternative, vellum, which was made from thin, carefully treated animal skin. Textiles, a measure of wealth and refinement, moved from one protected storage place to another as they changed form and function.

Most writing boxes were small, and early forms usually enclosed mostly undifferentiated interior space. Usually made of the same materials as chests, they were often decorated in the same fashion, with low relief carving, paint, or applied ornament. By the seventeenth century the desk, like the storage chest, changed radically. The folding writing surface remained, but it was larger, and the storage portion expanded to include drawers below and, in some cases, enclosed shelves above. These latter pieces of furniture, variously called writing cabinets, bureaus, secretoires, and secretaries (note that the latter terms derive from *secret*), provide a glimpse into changes in cultural history that occurred during the centuries of imperial expansion.

Writing and record keeping became essential tools as economies expanded beyond local areas, requiring systems of finance and credit that were more complex and less determined by face-to-face communication. Personal exchange and barter require little in the way of records, but an imperial economy and expansive centralized governments are voracious carnivores of paper and demand storage for it. As Western Europe and China expanded their reach throughout the globe, record keepers and accountants followed them, at least metaphorically. More complicated government forms to ensure that the crown or the emperor got his or her cut of the profits meant greater specialization in government organization and a concomitant need to store records in a way that enabled retrieval. The small writing and supply-storage desk would not do for organizations of this scale, and a new form emerged as a practical solution to the new economy.

Hence the bureau, which *The New Shorter Oxford English Dictionary* defines first as "a writing desk," became the center of the run-of-the-mill activities of trade and government.[6] Eventually the word came to mean the equivalent of the chest or case of drawers for storing clothing, but its original meaning is probably derived from French, Greek, and Latin words associated with the color red, which may be connected to the choice of red cloth tape once used to tie together stacks of documents. The word *bureaucracy* literally means "rule by the desk," or more idiomatically, by those of the desk, the functionaries of the titular leader. Without desks and bureaus, capitalism and bureaucracy would have been hamstrung, as would have been other systems of organization and control of people, government, and commerce. Socialists and communists needed desks as well, but it is worth wondering, if only for amusement, whether libertarians and anarchists could live without them or their modern counterpart, computers.[7]

In Japan, trunks (*nagamochi*) and chests with drawers (*tansu*) have been in use since at least the late seventeenth century, and the latter in particular became one of the characteristic furniture forms in Japan in the eighteenth century. Fashioned from native varieties of woods such as paulownia (*Paulownia tomentosa*), zelkova (*Zelkova serrata*), cedar, box, ironwood, white mulberry (*Morus alba*), and plum, much of the woodworking and cabinetmaking of this era is of an intricacy and competence unknown before.[8] As we saw in chapter 5, *tansu* achieved their apogee of popularity in the Meiji period (1868–1912), when many furniture forms became economically accessible to a wider spectrum of Japanese society than in previous eras. Boxes large and small and shelves of varying shapes and dimensions had been a part of the storage furniture of middle-class and wealthy Japanese households for centuries, as had the coffer (*hitsu*), but the drawered chest ultimately came to represent Japanese furniture almost as much as the lacquered box.

Tansu often were a combination of drawers and cabinets combined in a single large piece of furniture, the whole arrangement sometimes mounted on wheels. While mortise-and-tenon, frame-and-panel construction often forms much of the carcass of these pieces, they were also routinely decorated and bound together with elaborate iron fittings. More mundane functional pieces, such as apothecary chests and commercial storage fixtures, were usually undecorated. In form (but certainly not decoration) they are reminiscent of the combined cabinet and drawer storage units of the Anglo-American secretary-bookcase or even the storage units of the American Shaker societies, although many of the latter were built into buildings, rather than as free-standing pieces. It may be that such an arrangement befits a culture entering or amid an economy that privileges or values organization and specificity of function. That was surely the case in eighteenth-century Anglo-Europe and America. It may have also been the case in Meiji and early twentieth-century Japan, when the Japanese empire began its great push for expansion of territory and economic influence.[9]

As important as the *tansu* and the *nagamochi* were in Japanese daily life, small writing and book boxes constitute probably the most common form of Japanese storage box known outside the country. They were small enough for foreign collectors and traders to bring back from their eastern travels, and the highly decorated lacquer surfaces and sophisticated joinery probably appealled to those smitten with Japanese arts and culture, as were many Westerners after the country was "opened" by Commodore Matthew Perry in 1854.[10]

Document boxes (*bunkō*) and letter boxes (*fubako*) were intended for storing loose papers and for presenting letters, respectively, and were of modest size (roughly two by two by eight or ten inches, with considerable variation). Writing boxes (*suzuribako*) carried the tools of Japanese written communication—the brush, ink stone, ink sticks, and water dropper. Some traveling *suzuribako* included storage areas for paper, but most were intended for use on a writing desk or table (*tsukue*). One form, originally intended for commerce, included not only writing gear, paper, and document storage but also a small safe and an abacus. Popular from around the middle of the Edo period (1600–1868), eventually *suzuribako* became common in the households of less wealthy families as well.[11] [FIG. 73]

FIG. 73
Attributed to Igarashi Doho I, Japanese, 1643-78. Writing box with decoration of autumn flowers over a fence. Lacquer on wood with gold and mother-of-pearl, maki-e. Japan. Edo period (eighteenth century). William Sturgis Bigelow Collection. © Museum of Fine Arts, Boston.

As in Japan, the small box afforded craftsmen in India, the Middle East, North Africa, and Europe a field for creativity and the exhibition of skill with wood on a minute scale. Whereas much of the exterior decoration of Japanese small boxes was done in lacquer, the artists in these other areas often worked with tiny pieces of wood arranged either in patterns or as representational art. Marquetry, intaglio, and complex figured veneers applied to the surfaces (sometimes all four sides and the top) transformed the box into the *casket,* a container for precious things, and a precious thing itself.

Why have people wanted to own these boxes, and why in particular those made of wood, when metal was available? The very wealthy certainly owned boxes made of precious metals, often encrusted with jewels, even as they kept more of the same out of view on the inside. The decorated wooden jewelry box, music box, writing box, and casket kept in the private and personal rooms of living quarters of the less well-to-do were an expression of restraint as well as narrowed economic station, assuredly *not* the showy and even gauche statements that the gold or silver jeweled box was. Though not plain, as were "plain style" chairs and other goods, they were nonetheless made of what people still thought of as a natural material, cleverly manipulated. Their designs were as well often tricks of the eye, witty puns on perspective and pictorialism, the painter's and the architect's craft and art slyly slipped onto the "humble" box's surface.

Native Americans of the northwestern North American coast have been making incised and colored kerf-bent boxes for centuries. A series of saw kerfs cut on what would become the interior side of the wood allowed the box maker to bend the material to form corners. The characteristic stylized decorations on the exterior were of great religious significance for the peoples who made and used the boxes, linking each one and its contents to the cosmology of their religious beliefs. Other peoples did much the same with special boxes, carving them to alert the onlooker that the contents—relics, totems, or other special artifacts—were connected to far greater significance than were other things. [FIG. 74]

FIG. 74

Viola Garfield, "Tsimshian carver Bryan Paul uses [a] hand drill on carved rattle, Port Simpson, British Columbia, 1934." Courtesy University of Washington Libraries. Paul uses a hand drill, rotating it between his hands. Another rattle lies on the porch near an adze. Paul's toolbox, decorated with traditional carved and painted designs, is on the right.

Cabinets, Curious and Otherwise

The decorated box and the secret, often valuable nature of its contents were the common person's miniaturized version of a longer tradition of encasing rare goods and special prizes. Most people have at some time in their lives visited or stumbled across a genuine rarity. People search for four-leafed clovers or gawk at the freakish or rare in a museum. Herein are the two-headed calf, the large diamond, a piece of petrified wood. In some shrines there are relics, sometimes even "a piece of the True Cross."

Those unfamiliar things are objects of wonder, desire, and sometimes fear and loathing. Whatever their effects, accumulating the rare, strange, and different has been the object of collectors for centuries.[12] With a collection, however, inevitably comes the question of display and protection. Some collectors merely want to amass, and the display of their goods is to them of little concern—better it be hidden away and safe from thieves and the envy of the neighbors, especially if what really matters is not the ownership and display but the "chase."[13] For others, however, the things gotten, the rarity and the completeness of the collection are of little interest and reward unless they are protected and artfully displayed for the amazement of acquaintances and for the private pleasure of simply looking at the collection, recalling the details of acquisition, and experiencing the satisfaction of owning what others want but do not have.

For hundreds of years serious collectors of small things have favored a highly ornate framed box—a "cabinet of curiosities" full of small drawers and compartments—to store their treasures. The cabinet itself was part of the drama of revelation; opening it revealed only a small fraction of the contents. The initial bank of drawers could be slid out and opened outward to reveal a second, third, or even further hidden layers of goods.

These evolved, in function at least, into the china cabinet and other display pieces of more modestly endowed people's homes, wherein were housed goods meant for dining or mere show. By the seventeenth century, these heavy "cubberds" (cupboards) commonly stood on a two-door locked cabinet base. Above that, many had only a small closed cabinet in the center, with open space on either side. The open areas show that people were willing to display some valued pieces all the time (Why bother showing inferior goods?) while simultaneously giving the impression that what is behind the closed and locked doors is even more valuable. The promise could be a bluff, but the outsider had no way of knowing that until the cupboard was opened. These pieces teased with display and connoted a sense of security—that the owner could show off some valued things without fretting about theft, or that those things, however valuable, might not mean that much to the owner. The box itself might have been imbued with status, but what was inside of it got people to thinking.[14]

Cabinets of curiosity, court cupboards, and their cousins, treasure chests or trunks, also formed the intellectual basis for the natural history museums and similar institutions founded all over the world during the past three or four centuries. Charles Willson Peale's 1822 self-portrtait shows him pulling back a curtain of his Philadelphia museum to reveal row upon row of little boxes full of things presumably rare and odd, or at least taxonomically arranged. Taxonomy itself, while seemingly mundane to the modern eye, was hardly that when the first naturalists began to collect and exhibit their specimens. The diversity of the species was a wonder to people who traveled and read little, and proof to some of the immense achievement of Creation.[15]

The cabinet as a container for special things has evolved into the box that contains anything. In contemporary English usage the cabinet-maker still retains his or her status as a furniture maker, designer, joiner, and artisan in wood. Those who produce plywood containers to store household gear seldom take on the title of cabinetmaker, but they are not strictly carpenters either. Books on this form of cabinetry normally call them "boxes" and discuss the ways in which the front can be covered with finer wood and accommodated to doors and drawers. This terminology is probably the result of the virtual elimination of joinery as a component in these pieces. Even dovetailed drawers can be made by a router and a jig that produce the serviceable half-blind dovetail. (There are machines that allow for making "through" dovetails, but they are in general more expensive, and most consumers cannot or do not see the difference.) The distinction may also be a function of the extensive use of plywood in this work as well as the places in which it is used, usually the kitchen and bathroom.[16]

Caskets and Coffins

For most people whose faith allows it, burial of a corpse tends to be in a wooden box. Ancient Egyptians began to bury their dead about 2500

BCE. Peoples whose lives and livelihood were closely associated with the sea often chose a boat or a boatlike container for burial, both because of their connection to the water and because they viewed the transit of the soul to the afterlife as a voyage on a watercraft. In regions as far apart as the South Pacific islands and Viking Norway, some of the dead are sent off in boat-shaped coffins, or in the case of the ancient Norwegians, buried in actual boats.

Most people were too poor or not important enough to end their days in a fancy wooden container. Their fate was a plain six-board box, nailed shut. It was a coffin, not for precious goods, as the word *casket* would imply, because the flesh of the body was not considered precious, though the bones and the spirit or soul were. Flesh was transitory; it rotted and smelled foul when it did so. Scavengers ate it, leaving the bones behind. In Judeo-Christian theology the flesh was corrupt, the seat of vile or at least degraded behavior. The coffin was the appropriate place for it—a box closed tightly that would isolate the foul from the pure, that would itself dissolve once safely in the ground, leaving only the skeleton as the traces of the body that once was. In central Africa the Fang and in northern New Zealand the Maori store and protect the skull or all the bones of the dead in boxes, often honoring or protecting them with carved figures. The difference between the Anglo-European coffin and other peoples' box for the dead was emblematic of the cultural chasm between Europeans and many of the other cultures of the world. It was particularly apparent during the age of imperial expansion, and perhaps no better illustrated than in Herman Melville's *Moby-Dick* (1851). Ishmael, the protagonist, stays his last night ashore in Peter Coffin's tavern, where he meets Queequeg, the South Pacific harpooner. Aboard the *Pequod* Queequeg begins to make and decorate with carving his burial box, a far cry from the pine box that the word *coffin* connotes, and in the end it is the means by which Ishmael survives the great white whale's sinking of the ship.

Ultimately, however, Queequeg's vision of the appropriate container for the dead triumphed over Europe's and America's plain box. By the mid-nineteenth century in Anglo-America, the coffin was becoming the casket, the enclosure of precious contents. Historians of death and mourning practices have linked this change to the extended period of public grief that Queen Victoria observed for her beloved Albert, who died at about the same moment Melville's epic was published. There is certainly some reason to believe that the black-and-lilac-clothed monarch had some influence, but the change is more complicated than that, especially since there is ample evidence that it was occurring while Albert was alive. In the United States, at least, transformations in the Protestant faith, from a stern and forbidding interpretation of the New Testament to one emphasizing gentleness and love, had been ongoing since at least 1830.

Rejecting or at least modifying the notions of Original Sin and damnation for all of the unbaptized and unregenerate, and accepting the perfectionist idea of the human potential (and responsibility) to create a "heaven on earth," Protestants of this theological stripe turned toward a conception of mourning and the afterlife that was more fully grounded in the emotional concerns of the Romantic movement that had taken hold of European and American literature and art. By the later decades of the nineteenth century, undertakers had been superseded by "funeral directors," who operated "funeral parlors." Rather than being prepared with shroud and coffin in the residence of the deceased, then to be buried in "graveyards," "bone yards," or "churchyards," the dead were prepared out of the house and "laid to rest" in an elaborate box of fine wood, lined with the richest textiles a family could afford; the dead were also made to appear "asleep," or "at rest," as many of the gravestones of the era intoned. In this area embalming became a commonplace treatment for the corpse, and caskets were often lined with metal plate to prevent the decomposition and disappearance of the remains. Modern psychologists may be correct in thinking that this phenomenon was a denial of death. It was certainly a transformation of thinking, one that linked white Euro-Americans with other peoples of the world in a way that they could have hardly imagined, given the imperial adventuring that they most often justified as the inevitable triumph of superior civilizations over peoples considered inferior by virtue of race and culture.[17]

Queequeg knew what the wood of his coffin meant when Ishmael only barely understood it. It was not merely a material to be worked but a field of endeavor for the carver that became both the repository and the substance of belief. When Westerners examine carvings of spirits, whether those of the Tlingit, Haida, or Kwakwaka'wakw of North America, the Maori of New Zealand, the Bembe of western Africa, or the peoples of the South Pacific islands, they usually see only representations of nonhuman, often monstrous-appearing forms. But others see wood as an appropriate substance for such representations because the material comes from one of nature's most awe-inspiring manifestations. Trees are larger; they are older; they are in an interlocked and symbiotic relationship with other creatures that they conceal, shelter, and nourish. And if you are careless or unlucky, they can kill you.

Barreling Through History

In about 1800, one of the many pottery manufacturers in Liverpool, England, produced a jug with what seems to modern eyes to be a curious decoration. On one side a young man in an apron stands pounding the hoop of a barrel that is only partly drawn together and that is clearly on fire, or at least there is a fire burning inside it. On the other is a coat of arms. Two animals—both with bits in their mouths and covered in round spots—have reared up and put their front legs on the central area of the arms. This device is full of formal rococo curves. In it are depicted three of the tools of the cooper. A ribbon below the arms proclaims "Prosperity Attend the Integrity of Our Cause." It is the coat of arms of the coopers' guild, and a testament to the importance of coopering as well as an indicator of the trade and class consciousness of its craftsmen. [FIG. 75] [FIG. 76]

Customshouse records in colonial Anglo-America reveal much about the nature of trade and commerce of the era. The list of commodities and the volume of trade demonstrate that oceanic commerce was immense in quantity and diversity, from entire tree stems to fine textiles. Wood played a major part in this trade, as both raw material and finished

FIG. 75
Earthenware jug. Probably Liverpool, England, ca. 1800. Pounding the hoop to draw together the staves during the first firing, done to add flexibility.

FIG. 76
Earthenware jug. Probably Liverpool, England, ca. 1800. The coopers' coat of arms.

goods—for example, ship- and house-building materials such as spars and shingles. At least as important were the quantities of "staves" and "shooks" recorded; these were parts of what for centuries was the most common container for goods to be shipped and stored—the barrel. If ships and, later, railroads were the engines of conveyance of the international economy, and factories and machines were the engines of production, then barrels (and their relatives, hogsheads and pails) were the containers without which trade would have been hamstrung, and in some cases impossible. Without coopers there would have been far less need for shipwrights, wheelwrights, and machinists.

Let's examine a few trade and manufacturing statistics from the United States, a century apart. For the year beginning January 5, 1771, export commissioners in North America reported that among the various goods exported to Ireland were nearly three million "staves," plus a further two million exported to Europe, and over ten million to the "West Indies."[18] A century later, well along in the Industrial Revolution, *The Statistics of the Wealth and Industry of the United States* of the Ninth Census (1870) reported 4,976 coopering establishments employing 23,314 workers, as opposed to 815 establishments employing 6,232 hands that manufactured wooden packing boxes.[19] Only eleven other industries listed showed more businesses engaged in trades.[20] Wet and dry goods may have moved *on* rails, roads, or water, but they moved *in* barrels, casks, and hogsheads until metal and plastic drums and heavy-duty lifting apparatus made the wooden barrel all but obsolete in the twentieth century.

Why were barrels so important and so widely used? They are much more easily moved when full than any other form; they can be rolled, whereas boxes have to be lifted and hauled on some other conveyance on wheels. In an age before devices for lifting large crates and weights were common, rolling barrels was the best possible way to transfer commodities. They can be stacked on their flat ends or on top of one another on their sides, each succeeding layer resting its wide dimension in the crook created between two adjacent barrels below. They are extremely sturdy containers, made (generally) of oak, with strong hazel (or similar flexible wood) or iron bands that when shrunk by drying or cooling pulled the staves together. Unlike other materials that could hold liquids, barrels were sturdy yet cheaper and lighter than the alternatives. These working goods were not the stuff of the table or the parlor; for those purposes shining metals and glazed ceramics were a more refined option.

Coopered containers date at least as far back as ancient Egypt. The step from the hollowed single-piece container to the staved bucket or barrel is probably linked to the discovery that you can bind together a cracked wooden vessel with a pliable sapling that shrank as it dried, drawing together splits in the wood. With the advent of edge tools in the Iron Age, and the development of axes and adzes, it became possible to shape pieces of wood to make barrels and buckets. Whatever their initial use, barrels and other coopered vessels were in widespread use throughout the Roman Empire, and probably in any other civilization with sufficient metallurgical technology to make edged tools.

Coopering and shipbuilding are related skills, although shipbuilders had the advantage of being able to use tars and other substances to keep water from entering a vessel through joints, while coopers could hardly do that for goods that people were meant to eat or drink.[21] As master shipwrights began to excel at producing watercraft with minimal stuffing between the pieces of the hull, so too were coopers more effectively achieving their opposite goal—to keep liquids in, rather than out. It seems likely that water-faring peoples developed shipbuilding and coopering at about the same time. Sound watercraft carried people and goods over the water without losing them to the deep; barrels preserved the food and drink that were essential to long-haul seaborne travel on the open seas. Together they were the engine of exploration and empire.

In *The Cooper and His Trade,* the English cooper Kenneth Kilby lists eighty-three different tools used in making coopered vessels.[22] Coopering tools are large, heavy, and often rounded, as opposed to, for example, violin makers' tools, which are small and delicate, but also curvilinear. In general the size and number of craft-specific tools indicate both the complexity of the job and the difficulty in working the woods used. What distinguishes the cooper's job from most others in woodworking is the necessity to make containers that are rounded, joined, watertight (or at least tight enough to hold fine dry materials such as flour), curved in at least two directions, and strong enough to resist the shocks and stresses of carrying heavy goods in rough forms of transportation. Thick oak—the favored wood of coopered vessels—is difficult to work; hence coopering tools must be large enough to take considerable pounding and to exercise significant leverage or mechanical advantage. But parts of casks and barrels must be cut with enough exactitude to hold water when assembled, so that many of the tools must also be capable of precise work. [FIG. 77]

Casks and barrels most often had a larger diameter in the middle than at their ends, while buckets generally had a larger open end. Thus each stave (the vertical piece of which the sides were formed) had to be cut at an angle along its long edges to accommodate the staves on both

FIG. 77
Cooper's tools. From left: adze, croze, topping plane, jigger, driver.

sides of it, as well as shaped to accommodate the changing diameter of the finished vessel ("dressing" and "listing," as coopers term it). Barrels also had to be bent to come together at their ends, trimmed on the outside, and hollowed out on the inside. Finished barrels have a considerable amount of tension and strain on the staves, so coopers have to be expert in selecting wood for clarity and straightness in the grain, and thus they prefer quarter-sawn oak. As in shipbuilding, the demand for oak made it a commodity of great value and demand, and countries that had begun to exhaust their supplies by the seventeenth century cast an adventuring and avaricious eye on the stands of the trees in North America, Russia, and Germany. Memel oak, so named for the port from which it originated, was favored because it was a bit easier to work than American oak, had a less resinous quality (and therefore imparted less alien odor to the contents), and was just as hardy.

As if cutting, shaping, and jointing were not enough of a challenge, "raising up" the barrel required considerable physical strength and manual dexterity. Ideally, the process went something like this. After aligning the staves within an iron or sapling hoop, a large hoop is set at an angle at the opposite end to hold the staves together. Then the barrel, cinched on one end and splayed on the other, is placed over a small wood fire to increase the wood's flexibility. A couple of larger hoops are then driven down the widening staves to begin to draw them together. The cooper then turns over the barrel and drives down the hoops to begin the stave-bending process, gradually drawing the vessel's splayed end together. Then the barrel is fired again, or "toasted," over a small wood fire on the shop floor, which provides "set" for the vessel and helps bring out the characteristic flavors to be imparted to the wine or other liquids to be stored or aged. After the bands are hammered in place and the vessel has cooled, the next job is to shape the ends of the staves— beveling and routing out a groove for the barrel's round top and bottom "heads." After shaving the inside smooth, the tricky task of inserting the heads remains. This requires the cooper to slacken the hoops, slip in the

FIG. 78

"Tonnelier" (Cooper). Plate 1 of volume 10 of Denis Diderot L'Encyclopédie *(1751). Jointing and shaping are in the right foreground, and the barrels in this case are held together with wooden hoops of hazel or another flexible wood.*

heads, and rehammer the hoops in place.[23] Completed barrels were often brined to reduce the strength of tannins in the wood, thereby "sweetening" the oak for its intended contents. [FIG. 78]

Standard sizing for barrels, casks, tubs, pails, and other vessels evolved over time, and a table of sizes and nomenclature for these containers provides some indication of how broadly the cooper's products permeated everyday life. White coopers made vessels without swollen sides such as pails and tubs. [FIG. 79] Kilby lists four different pails and buckets as well as wash, cook's yeast, grog, scalding, oyster, and other tubs, and splayed storage vats, some large enough to hold 5,500 gallons. The latter were generally used in the brewing industry.[24] Dry casks were made in specified sizes for packing products such as soap, seeds, white lead, butter, pepper, putty, gunpowder, preserved meats, ink, mustard, apples, grapes, and paraffin. [FIG. 80]

FIG. 79

"Boisselier" (White cooper). Plate 1 of volume 2 of Denis Diderot, L'Encyclopédie *(1751). White coopers made all sorts of smaller and sometimes cruder goods. Note the shoes, or* sabots, *in the foreground.*

FIG. 80

"Busy Scene in the Ozark Apple Region of Missouri: Picking, Sorting and Packing in Barrels." Keystone View Co., ca. 1920.

Dry coopers began to lose their share of the packaging market in the nineteenth century as wooden boxes, together with the winches and cranes to lift them, and railroad boxcars made rectangular containers a practical alternative for the efficient shipping of goods. Their counterparts in the wet coopering trade, however, were able to maintain their position as the primary producers of containers for beer, wine, spirits, vinegar, tar, turpentine, and just about any other liquid for industry or human consumption that had to be hauled or stored in bulk. A whole nomenclature evolved for the different sizes of barrels. The following are English measurements:

Pipe	108–116 gallons
Butt	108
Puncheon	72
Hogshead	54
Barrel	36
Kilderkin	18
Firkin	9
Blood Tub	7.5
Pin	4.5

There were also large vessels identified simply by their size, such as 250-gallon ovals and the brewer's 500- and 1,000-gallon bulge vats. The above capacities were in widespread use, but for some goods, and especially wines and spirits, measurements sometimes varied from these semiofficial standards. Wine casks were produced in a great number of sizes, ranging from tiny one-gallon casks to huge sherry butts of several hundred gallons.

The spectrum of sizes for coopered containers offers a clue to the special nature of the relationship of the cooper to the beer, wine, and spirits trades, even to the present day. In one respect this reveals the acute sensibility and concern humans have had for the quality and quantity of their favorite tipple. Armies, navies, and entire civilizations are said to have "moved" on their stomachs, but what appears to have moved their hearts and minds were fermented and distilled liquids. Coopers certainly took great care when producing containers for goods such as herring or flour or apples; the barrel had to stay in one piece, keep the goods inside, and not impart foul flavors. Herring coopers favored cold-climate spruce, while coopers for dry goods used the cheapest woods that could survive the rigors of shipping. But those requirements were merely the beginning of the job for coopers who made containers meant for the alcohol trade.

Part of the intoxication of woodworking lies in the smells and the tactile qualities of the material, especially after it has been worked. All woods have a characteristic aroma, especially when they are newly split from the log. But some woods impart to goods an odor, and hence a "taste," that people would rather avoid. Some foods smell so strongly that the wood chosen for storage barrels makes less difference than it does when more sensitive materials, such as wine, whiskey, beer, and ale, are packaged. In addition, some woods are more receptive to subtle intermingling of their aromatic qualities than are others, especially if left in the rough or if heavily fired when the container is made. Finally, some woods are so structured internally that they allow for a certain amount of breathing by the goods inside, a process that can alleviate pressure buildup in the case of fermented liquids and help along the development and maturation of the drink, a trait that is especially important in the alcohol trade.

Thus while the growth of the box and crate industries in the nineteenth and twentieth centuries and the supplanting of wood as a barrel material by plastic, steel, and aluminum have put an end to the majority of cooperages in the world, the wooden cask and barrel still reign supreme for wine and spirits. The beer and ale businesses have abandoned the wooden coopered container, largely for economic reasons, but also because the brewers have managed to convince consumers that pale ales and lagers with little of the taste, texture, and complexity of wooden casked beers and ales actually taste better. Even traditionalists who favor "real ale" still get the stuff from aluminum kegs and barrels. Changes in alcoholic content and standards have also allowed brewers to forgo the wooden barrel, as the general weakening of commercial beer and ale that has transpired since World War II ended has meant that brewers no longer have to worry much about the increase of pressure from fermentation in the keg; thus was breathability rendered unimportant.

Oak's cellular structure provides the breathability vintners need and, equally important, American white oak and European oak (in particular oak from the Alsatian Vosges region or that of the Allier and Nièvre regions of central France) absorb just enough of the flavor "notes" of its contents and dispense just enough of their florid aroma to produce the complex flavors and overtones that wine connoisseurs and spirits enthusiasts treasure. [FIG. 81] New oak wine barrels introduce tannins and vanillin to wine, and the porosity of the wood allows for the mild oxida-

FIG. 81

Oak wine barrels in storage. Moana Park Winery, Hawkes Bay region, New Zealand, 2005. Variously "toasted" by the final firing of the barrel, these storage barrels impart varying overtones to their contents in the aging process.

tion that many think imparts depth to the drink. French oak forests have been carefully tended for over three hundred years, with attention paid to the density of the forests (denser stands tend to yield trees with tighter grain and less porous wood).

Whatever the flacks for the aluminum wine-barrel trade would have consumers believe, the inert nature of the material adds nothing to the goods. It may not subtract anything either, but that is not the point. While chemically nonreactive and nonporous materials are desirable for cooking and food storage, thousands of years of wine-making suggest that this is not the best idea for producing and storing it, though the costs are surely less. In 2005 a new wine barrel from a reputable cooperage cost about $600, which put it out of reach for many winemakers. In France, there have been laws against infusing wines with oak flavors by using chips of the wood, but these restrictions are being lifted, thus imperiling an already dying craft.[25]

Whiskey makers still use oak barrels for aging their goods, in part because the aging process alters the distilled product over time, providing manufacturers with the opportunity to distinguish their liquor from that of others. The whiskey trade also provides for the recycling of wine, sherry, and other casks. Sherry casks add color to malt whiskeys, as well as hints of the taste of the sweet wine, while American bourbon casks are also utilized for some of the whiskeys, especially blends. Modern methods of cutting, planing, beveling, and hammering have mechanized much of the coopering process, so that a major factory such as François Frères of Saint-Romain, Burgundy, employs only forty coopers. Some elements of the process—selecting the wood for staves, and assembling and firing the barrels, for example—are still done by hand and eye.

It is difficult for contemporary people to conceive of how important and how numerous coopers were before the twentieth century. In every port of the world hundreds or thousands of barrels of different sizes were everywhere, full of trade goods from ceramic wares to salted fish to whiskey. Cooperages were common in all these ports and in those places where trade and production in substantial amounts took place. General stores were full of barrels, casks, kegs, and other coopered products such as buckets and tubs. Even on a small scale coopered goods were ubiquitous and important. A simple butter mold, for example, about six inches tall and stamped with the number 7 (indicating it was a "measure," or standard size) is a coopered hexagon. It has a removable bottom with a cow incised in it to produce a decorative effect when the butter is turned out of the mold for sale. [FIG. 82] Banded with lead and made (probably) of black walnut, it is merely one of a group of large and small coopered pieces commonly found either in retail operations (as in this case) or in the back of the household. So important were these storage vessels that these areas of seventeenth- and eighteenth-century Anglo-American houses were called butteries, not because butter was stored there, but because such containers, called *butts,* were. If we take the long view of woodworking in preindustrial and early industrial societies, it seems clear that there was a great triumvirate of woodworkers—the shipwright, the carpenter, and the cooper. No other trades seem to come close in importance.

FIG. 82
Coopered butter mold. Walnut and steel. United States, ca. 1885. Probably a measure of volume as well as a decorative device.

Boxed In

In 2002 the United States Census Bureau listed sixty-one categories under its heading of "wood container and pallet manufacturing." There are a few entries for barrels and other coopered vessels, but most of the entries refer to boxes and crates.[26] Much easier and cheaper to manufacture, crates and boxes use straight pieces of lumber, nailed or stapled together. They stack and store more efficiently than do barrels, but you cannot roll them around, an advantage of barrels that the forklift has rendered moot. On a small scale they began to replace barrels for dry goods by the mid-nineteenth century. By the latter half of the twentieth century huge crates were regularly moving across the oceans in gigantic container ships, hoisted in and out of holds by enormous machines. The barrel's round shape had become a hazard in large-scale transportation, since its potential for movement and momentum posed a danger on the docks and in the ships.

The fresh fruit and vegetable industry relies on special lightweight containers made of thin sheets of wood stapled to a frame and bound with wire. These boxes, which replaced older and more solidly built nailed crates and tops, in fact provided a quality useful to the growers and shippers of fruits and vegetables. Handlers risked damaging the contents in the lightweight crate if they tossed them around. A more solid crate allowed for rougher handling without visible damage to the fragile fruits and vegetables inside, much to the chagrin of the merchant who opened the heftier crate only to find bruised and smashed contents. Stronger crates thus mandated more packing material if the goods were to arrive intact. Herein is a small irony in the history of transportation and of the importance of specialized wooden containers. By choosing a more fragile container, shippers protected the goods more effectively by rendering it more difficult to hide damage while saving money on packing material. (The packing material in heavier crates was, until the twentieth century, either straw or fine shreds of wood called excelsior, which today is usually made from eastern cottonwood [*Populus deltoides*].)

The produce-related cousin of the wire-bound crate, the round bushel basket, is a standard measure for volume of many goods, and continues to be so, especially in the United States and Canada. Bushels are made of slats of veneer-thickness wood, fanned out in a circle and stapled at the center of the round bottom, that are then cracked and bent vertically to provide sides that are held together with bands of thin wood stapled to the slats. Two wire handles are attached to the top rim. This basket was probably inspired by coopered forms, but it is much less expensive to make, since it employs inexpensive machine-sawn wood of minimum thickness and machine-driven staples to hold it together. There is almost no handwork in the making of the modern bushel, and there was very little when the industry began. Bushels probably did not replace the barrel or the cask, since they were impractical shipping vessels, save for the small-market gardener's trip to market or a retailer. The container they replaced was the splint, cane, or grass basket, both in its role of carrier from the field or orchard and in its function as a container in retail display.

Baskets

Native Americans have been making and using baskets for centuries. Nipmuc, Abenaki, Iroquois, Salish, Mohegan, Mi'kmaq, and other native peoples made splint baskets centuries before Europeans made their way to North America.[27] They had discovered that when they debarked and then pounded freshly cut straight-grain oak and ash logs along their length and circumference, they could separate thin layers of wood, one annual ring thick, from the trunk. They split these sheets into thin strips, or *splints,* cut them to length and wove the bottom of the basket. They then bent the splints upward and wove additional splints horizontally, locking in the handles (traditionally of steam-bent ash) as they wove the last courses. Finally they bound the top course with either narrow splint or some other flexible plant fiber. [FIG. 83]

In Europe the splint basket tradition in Finland—using native pine—reaches back to at least the sixteenth century. In the Dalarna region of Sweden pine splint baskets were made at least as early as 1830, and by about 1850 splint basket making in Sweden became a large-scale, industrialized production in the Skåne region, near Stockholm. By the 1880s Swedish and Danish splint basket sellers had reached into markets in England and northern Europe, and by the middle of the twentieth century they were selling approximately one million splint baskets annually in Denmark, which then had a population of but four million.[28]

Some baskets were specially woven with large open spaces between the splints so that the vessel could serve as a sieve. Baskets used in cheese-making, for example, are generally of low height (six to ten inches) and

FIG. 83
"Indian Basket Weaving, Prince Edward Island." Keystone View Co., ca. 1915.

substantial width (sixteen to twenty-four inches), and are woven in a pattern of widely spaced hexagons. When lined with cheesecloth, they catch cheese curds while allowing liquid remains of the curdling process to pass through. Other sifting baskets, such as winnowers, were closely woven to keep the grain berries from escaping as the chaff blew away in the threshing process. Cylindrical splint traps are used to catch fish and eels.[29]

The Cherokee, as the historian Sarah Hill observes in *Weaving New Worlds: Southern Cherokee Women and their Basketry,* did not begin to use white oak for baskets until after they were removed from their homelands, preferring to use river cane even though they knew about white-oak basketry from their exposure to European settlers.[30] Cherokee basket weavers, using the simple plait weave, eventually developed patterns and designs using color (red from bloodroot, brown from walnut, for example) and different splint widths that we now identify with their work; they added the wooden handles that Europeans had long used and lids that could serve double duty as a tray when removed. For the Cherokee, white-oak splint baskets represent contact and interaction with white settlers, and more subtly, the defeat and ethnic cleansing of peoples native to the American Southeast.[31]

Native Americans began to make splint and other baskets for the tourist trade by the end of the nineteenth century, as the Indian wars ended in their defeat and sequestering on reservations and, eventually, on small plots of land inappropriate for agricultural activities or the people's culture.[32] The Cherokee exchanged "coarse baskets" for foodstuffs and other goods.[33] By the 1940s, when white oak had become scarce in Cherokee lands, basket makers discovered that they could make showy and shiny baskets from red maple. They began to enlarge their palette of basket forms and designs, primarily for the tourist market, since red maple is not as hardy as oak splint, but the baskets are smoother and hence more "refined."[34]

Their chief competition in the tourist handmade basket trade came from two sources. The Shakers developed a system of mass production of finely cut splint baskets, enlarging upon an activity originally developed to serve their own agricultural needs. Shaker goods—seeds, chairs, cloaks, patent medicines—gained a devoted following in the United States in the latter half of the nineteenth century, a result of the Shakers' reputation as makers of consistently high-quality goods and as people who were, to outsiders to the faith, quaint, romantic, and slightly odd.[35]

The other source of competition for Native American splint basketry came from the Appalachian "folk" whom folklorists, sociologists, and reformers "discovered" near the end of the nineteenth century. As part of the Appalachian folk revival that picked up steam in the 1920s and 1930s, splint baskets made by the mountain people formed part of the "craft renaissance" championed by such entrepreneurs as furniture manufacturer Wallace Nutting, educator Gertrude Flanagan, ballad collector Olive Dame Campbell, and social worker John Collier.[36] As early as the 1920s the splint basket had become an icon, a quasi relic, and an affective presence of vanished, superseded, and safely venerated cultures and peoples, passed by in the Industrial Revolution, seemingly simpler and more "genuine" in their poverty or isolation and, to some extent, more "American" than the urban and suburban dwellers who most fancied their goods, if not their lifestyle. By the later twentieth century basket making became an industry apart from the "folk," a manufacturing process like nearly all others.

Most of the splint baskets that have survived seem to have been used in the harvesting and temporary storage of hardy fruits such as apples. Large baskets were impractical for sensitive-skinned fruits such as peaches and apricots, since the weight of the fruit would bruise and otherwise damage the fruit on the bottom. But as apple producers began to breed for resistance to bruising and ease of harvesting in the later nineteenth century, larger and larger baskets could be used to pick the fruit. In some ways the big apple basket that is commonly found in antique

FIG. 84

*Splint and bark baskets. Ash, oak, and birch. United States, ca. 1920–
95. Native American and factory-made, these baskets range from the
rough apple baskets in the center to the more finely crafted Cree basket
at the top. Birch bark baskets were less flexible, but could be made
waterproof.*

shops and flea markets is an emblem of the homogenization of nature,
at least as far as fruit varieties are concerned.[37] [FIG. 84]

Handled wooden splint baskets represent a temporary, and usually
open, form of storage, one that does not attempt to conceal the contents.
They reference the world of the handmade, even if much of the process
of separating, splitting, cutting, and weaving the splints is done by machines. Although nearly all splint baskets on the market today are the
product of mechanized mass production, they look to be the work of
the hand, and are loaded (some might say freighted) with the values of

the preindustrial world, of small, family-farm agriculture, and of the refulgence and the reverie of the autumn harvest. Splint baskets, whether newly made or patinated with the dirt and mystery of age, are in some ways unlike boxes; they are an open and unabashedly nostalgic form, ruing the end of summer as they resonate with the rhythms of the romance of the orchard and farm or of the native peoples whose cultures struggle to survive.

Fence, Palisade, and Picket

The folklorist Henry Glassie long ago argued that certain regions of the United States can often be distinguished from one another by the shape, form, and construction of the fences people make. Fences are not strictly speaking "boxes" in the sense of having six sides (or five if there is no top) that enclose space. But they are enclosures, and for most of human history they have been made of wood.[38]

The oldest fences were meant to keep people and marauding animals out of gardens, domesticated animal areas, houses, castles, small villages, and fortified strongholds. Walls and moats were the most formidable lines of defense in warfare, but obviously more expensive and time-consuming to construct. Early fortified villagers used earthworks and palisades of pointed or carved logs (some carved with fierce faces to frighten attackers). Conjectural diagrams of seventeenth-century North American settlers' dwellings drawn from archaeological research suggest that these first houses and farmyards were fenced in, and most of the livestock and crop fields left open. Unlike their contemporaries and antecedents in northern Europe, these farms did not enclose a yard with timber or stone buildings but used fences to set off their living spaces.[39]

The wealth of trees in North America may have provided the raw materials to expand fencing to enclose grazing animals. Clearing trees for planting or pasture meant not only hewing and hauling the trees from the land but also ridding the area of stumps. (Not everyone did that, however, and foreign travelers and domestic critics lambasted American farmers for their sloppy, stump-ridden fields, the antithesis of the neat, geometrically aligned areas delineated in such guide books as John Worlidge's *Systema Agriculturae*.)[40] Those who removed the stumps discovered that they had the makings of an effective fence, although surely not one with the neoclassical regularity of posts and rails.

Eventually, however, either the stumps rotted away or the supply was insufficient to enclose the land. Europeans had for centuries marked off plots of land with stone walls, hedgerows, and woven fencing, often made from coppiced trees. In North America, however, fencing took on different forms, using heavier pieces of wood. Apart from the New England stone wall, the three major forms of fencing were the post-and-rail, the board, and the "snake" fence. The abundance of trees with long straight stems made it feasible to split them along their length into substantial sections. These rails were then let into the posts, which were set into the ground. The ends of rails were often pared down so that two could fit into mortises in the posts. It was a labor-intensive operation to make post-and-rail fences, and the intelligent fencer made certain that the wood for the posts was a species resistant to water and insects, such as black locust (*Robinia pseudoacacia*), Osage orange (*Maclura pomifera*), or members of the cedar and cypress families. Totara (*Podocarpus totara*), a rot-resistant wood native to New Zealand, works well in this capacity, as does giant sequoia (*Sequoiadendron giganteum*), which is no longer available commercially due to its scarcity. Board fences were less labor-intensive to erect, since sawn boards were nailed to the posts.

Snake fences avoid some of the problems of rot because the posts are not sunk into the earth. While they are essentially post-and-rail in form, the fence remains upright by means of interlocked rails and angled braces that create a saw-tooth pattern as the fence snakes along the land. Snake fences take more timber to make, but their limited ground contact makes them hardier, and the labor of digging postholes is for the most part eliminated. They were more popular in the American middle and Deep South, although the form can be found occasionally in other regions of the United States, probably the importation of settlers emigrating to the north and west.

In the nineteenth century, wire fencing became an alternative to the wood- and work-intensive post-and-rail, board, and snake fences. An article in the Albany, New York, monthly magazine *The Cultivator* made the case for "good, strong, neat and straight fence" and criticized the "plenty of awkwardly built board fence all through the country." The process they advocated went like this: Postholes "in no case [should] be less than two feet and an half deep." To "set the posts perfectly straight," the article recommended driving other posts into the ground and stringing lines between them to keep the fence path straight and each post perpendicular to the ground. "The *pounder* [italics original], for beating the earth firmly about them, should be shod in the lower end with a cast iron head." At least four boards are then to be nailed to the posts and the whole fence painted or whitewashed.[41]

But in that same issue was an article entitled "Cultivation of Forest Trees." It drew attention to the problem of introducing trees to the vast grassland interior of the United States. "The propagation of forest trees is becoming every year more important," the editors wrote. "There is a large portion of the western country, in which the natural destitution of timber constitutes a great obstacle to its cultivation. . . .There is no reasonable doubt that those bleak prairies might in the course of a few years . . . be covered with various kinds of trees." The article makes reference to what was then generally accepted knowledge about the prairies and especially the Great Plains—that the Plains were a "great American desert," covered with sod so densely packed with roots that it was untillable. The steel plow eventually changed that, but the problem of treeless areas seemed as vexing as the grasslands had been. The article discussed the species of trees most useful and most likely to be success-fully propagated, urging sylvaculturists to grow trees from seed rather than transplanting saplings. Other articles in this issue make the case for importing and propagating Scotch larch and for using Osage orange for hedgerows.[42]

These articles note the growing problem of deforestation in the United States, but they do not identify how this problem emerged, although the stress on planting new woodlands suggests the solution. But an 1849 article by J. S. Skinner in a competing agricultural magazine, *The Genesee Farmer,* took a more aggressive stance. "The cost of building and repairing the Fences in the United States is enormous, almost beyond the power of calculation, and forces the inquiry whether Legislatures ought not to be called upon to compel every man to keep his stock to himself. . . . It is this enormous burden which keeps down the agricultural interest of this country, causing an untold expenditure. . . . The system of compelling every landholder to enclose his property, is peculiar to the United States, with only the exception of England, where the fence nuisance appears again under the form of the hedge."[43] Skinner was arguing for the open field system of grazing and the enclosure of crops that would be harmed by hungry livestock, rather than the Anglo-American system of enclosing pasture lands.

The emerging iron and steel industry of the mid-nineteenth century offered a potential amelioration of the fencing and forest dilemma. *The Pictorial Cultivator and Almanac for the United States, for the Year 1851* reported that "Wire fences are attracting much attention in all parts of the country. Where they have failed, it has usually been owing to poor iron or too small wires, in endeavoring to make them cheap. Where stone walls or timber is scarce, they may prove valuable."[44] The article then provided explicit instructions for building sturdy wire fences, and it seems impossible for readers to have missed noticing that the process was far less labor-intensive than erecting wooden fences. The problem of consuming forests and wood at an alarming rate did not disappear, however. In spite of occasionally alarming forecasts of dearth and devastation, public policy in the United States did not begin to address the problems until the late nineteenth century, and even then in sporadic and piecemeal fashion.

In the latter-nineteenth-century American and Canadian West, and to some extent on the grazing lands of Australia and Argentina, barbed wire changed the calculus of enclosure for ranchers and farmers. Wire replaced wooden rails and kept cattle and horses inside, on pasturage that lucky or wealthy farmers wanted to keep for their herds. It began the process of closing the open range that many ranchers had been using for grazing. Other types of wire fencing were developed for controlling the movements of sheep and other livestock, and the ease of installation and the low price of wire fence allowed ranchers and farmers in the tree-poor plains to put fences where the expanse of space and the cost of wood in the past had made it a fool's errand. Today, board and rail fences still surround some pastures and fields, but it is more likely that they are around the horse farms of the well-to-do. In some places, even these are being replaced by fences of strong white plastic, which neither rots nor needs to be repainted.

The Virtues of Enclosure

Coopered, nailed, wired, and stapled wooden boxes of various sorts were as much engines of economic development as were watercraft, the railroad, the steel and iron industries, water power, the steam engine, and electricity. Absent barrels and their kin, it is likely that no one would have sailed anywhere far from shore, a development that many indigenous non-Europeans might have wished for, had they known what was to come. Without packaging that protected and preserved goods, no economy would have developed much beyond the local level. Without caskets and baskets, the affective element of human culture would have been poorer perhaps, and certainly different. Critics, consultants, advisers, corporate cheerleaders, and all the rest who neither make anything nor stick around after they have dispensed their advice have in recent years promoted "thinking outside the box," as if the box were some sort of prison. But the box and what is inside of it are often what matters most to people, whether literally or figuratively. Neither the box nor what or who is inside of it should be dismissed so cavalierly.

What do boxes tell us about human culture? As caskets or coffins they enclose precious materials or the remains of those no longer living. As houses and greater structures they provide shelter, inspiration, and comfort. They are complex entities whatever their size because they are artifacts that show both an exterior and an interior (whether humans are actually inside them or merely peering into them). They conceal secrets and, with the proper mechanism, can keep them, unless violence is brought to bear on them or they have aged and deteriorated until they fall apart. Boxes let people order their existence, whether it is putting books on a shelf or bookcase or clothing in a chest of drawers. Boxes let people take things out of public view—embarrassing things, precious things, things that reveal too much about them.

The box also presents an opportunity to embellish, both its exterior and, less commonly, its interior. Advertisers jumped at the chance to paste paper labels on the outside of boxes and crates, especially after chromolithography and other cheap color reproduction methods were introduced in the latter nineteenth century. Cigar manufacturers plastered brightly colored labels on the inside of the lid of their boxes, nearly all of which were once packed in Spanish cedar (*Cedrela odorata*), full of the perfume of the wood and the tobacco, an aroma that, whatever one's convictions about the "sot-weed," is unmistakably elegant and sweet.

In the hands of artists, boxes offer opportunities to experiment with and exhibit mastery of the technical details of drawers, compartments, and the visual interaction of figure and form on a small, even minute scale. Whether veneered or of solid wood, the sides and top of a box can be the setting for the concentrated visual magic of pattern in woods and wood figure that are too rare or dear to find in larger case pieces. While

FIG. 85
Mark Sfirri, Mini Truffle Chest.
Mahogany and gold leaf, 2002.

some artists stay close to the rectangular or round form, others have moved to more sculptural treatments in which the containment function, while present, is manifested in surprisingly shaped and placed compartments, and is secondary to the overall effect of the wooden box as art.[45] Whimsical forms and shapes have entered the domain of the artist box and will likely remain, as well as the practice of using rare woods and complex shapes to call attention to the box rather than to what is or what may be inside. [FIG. 85]

Are these new aesthetic creations the equivalent of the jeweled box of precious metals that graced the chambers of the wealthy in the past? Meant to be displayed, as the jeweled box was, these too seem to be a new order of interpreting wood, a signal not so much of the goods stored away from sight but of the savvy aesthetic sense of the owner, and perhaps even a comment on the environmental wisdom of celebrating small pieces of a renewable resource rather than the shiny exuberance of the extracted mineral that grows again only over millions of years, if at all.

7

Little Things with a Point

W hen I was coming of age in a small town south of Buffalo, New York, I was, like so many others, enchanted with baseball. Minor league baseball thrives in this area, and children and adults follow their team with dedication and the limited expectations and humility that minor league enthusiasts learn through experience and lore. Younger players on the way up to the major leagues caught our attention, but not so much as the former big-league players on the way down. For some reason I took an unusual interest in one pitcher—a big, quiet man, Sam Jones. He had been a successful player for a time, throwing the first no-hitter by an African American on May 12, 1955, for the Chicago Cubs. He went by the nickname

"Toothpick" because he was almost never seen without one in the corner of his mouth when he pitched.

Sam "Toothpick" Jones was the embodiment of someone who flouted one of the most common and dire warnings of our parents—never do anything active with a toothpick in your mouth, and never be seen in public so equipped. They invoked a vision of a toothpick lodged in the throat, stuck in the stomach—or worse. Jones would have none of that, carelessly or rebelliously planting his toothpick in his mouth for whatever oral or other gratification it provided. Not for him the public embarrassment associated with long-held traditions of proper behavior, or "manners."[1] He retired intact.

Hygiene and Manners

Before parents had begun trying to scare their children with these menacing stories, humans had long been wrestling with one of life's minor annoyances, food that stubbornly lodged between their teeth. Even people privileged enough to easily obtain only the tenderest cuts of meat cannot escape the annoyance that toothpicks ameliorate, and no one who eats delectable foods such as corn on the cob or spare ribs escapes. In spite of technological advances in dental hygiene wrought by manual and electric toothbrushes, the tiny smooth sliver of wood is still one of the most efficient ways to remove interdental detritus. But in many societies this raises the cultural problem of where and when to conduct the maintenance. Almost without exception in much of the world, using a toothpick in public is considered unrefined, yet there is little in the way of an alternative solution, at least in the immediate surroundings of a meal.

Centuries ago people deduced that failure to clean the teeth caused painful decay and foul breath. The anthropologists Christy Turner and Erin Cacciatore have found evidence of habitual rubbing between teeth in ancient Pacific Basin peoples through their study of dental remains, which they argue indicates the oldest intact evidence of the use of tools by humans or humanoids.[2] Firm-stalked grasses and twigs were probably the first toothpicks, though porcupine quills and the stem end of small or pared-down feathers probably did the job for those who had access to them.

Each of these picks has a disadvantage related to their form or substance. Twigs small enough to work between teeth are weak and splinter. Grasses break too easily, and quills and feathers can be too large, too abrasive, or too fragile. Splintered wood has rough edges, but people already familiar with smoothing tools could rectify that. Wooden toothpicks have been used by people above the level of subsistence for centuries, although pickers were urged to use them discreetly. The fifteenth-century *Boke of Curtasye* (ca. 1430) urged people to "At meat cleanse not thy teeth nor pick / With knife of straw or wand or stick." Hugh Rhodes's *Boke of Nurture* (1577) similarly advised, "Pick not they teeth with they knife nor finger-end, / But with a stick or some clean thing, then do ye not offend."[3] Silver, gold, and ivory picks were a stylish wealthy person's choice for centuries, until the toothbrush became a popular tool for teeth cleaning in eighteenth-century Europe and North America.

There are a wealth of interpretations, stories, and outright hogwash about the origin of the modern, mass-produced wooden toothpick. In the United States the tale usually begins with Charles Forster, who is alleged to have run across a supply of hand-whittled Portuguese toothpicks while at work in the Brazilian offices of an English shipping company.[4] Forster moved to Boston, Massachusetts, in 1865 and began working on a machine to mass-produce toothpicks, having decided that white birch was the optimum wood for his project. Birch had the right combination of flexibility and resistance to splintering, was relatively common, and nearly white in its raw state. The latter quality became even more important by the end of the nineteenth century, when the "sanitary crusade" was in full force, and whiteness was associated with cleanliness.

The New England toothpick tale asserts that Forster began his trek to success by convincing (probably hiring) several Harvard undergraduates to demand them after a meal at a Boston restaurant, the Union Oyster House. Eventually the establishment's owners decided it was worthwhile to provide them and Forster's fortunes and the toothpick industry took off. Whether or not that is the way it happened is not important. What is certain is that in 1872 Silas Noble and James P. Cooley of Granville, Massachusetts, received a patent for a machine that made toothpicks out of a block of wood. But Forster was not left out of the picture. He established a toothpick factory in Strong, Maine (near to both birch trees and cheap labor), in the 1870s. The firm was evidently a success, staying in business until bought by Diamond Match Company in 1947. The latter produced 90 percent of all toothpicks in the United States in 2000.[5]

The toothpick, important though it may be in oral hygiene, has other functions and meanings, at least in Western culture. The modern genius who thought of dipping them in aromatic oils, flavoring the little sticks with peppermint or some other mask for halitosis (or the fear of it) probably had no idea that the ancients had already figured that out.[6] But in the twentieth century clever marketers of insecurity among the American well-to-do and middle class managed to convince people that "their best friends won't tell" them if they had the dreaded condition. If your "best friends" won't tell you, then who would? All who suffered life's inevitable setbacks were thus left to wonder if they didn't get the girl or boy or job or promotion because their breath betrayed them.[7] With dentists and oral hygienists telling people that they risked all sorts of diseases should they neglect to brush, pick, and eventually floss their teeth, those little white paper packages of flavored toothpicks at the cash registers in restaurants and diners (not purveyors of haute cuisine, surely) were a small-scale gold mine.[8]

One way to gauge the use of toothpicks is through their accoutrements. Toothpick holders were common gear in the Victorian household. They were, in their simplest form, little cups made of glass, ceramic, or wood. The simplest forms are often mistaken for toys or egg cups or some other household artifact since most twenty-first-century people have ceased to display toothpicks anywhere near the dining table. But a century ago, toothpick holders were found in nearly every household.[9]

Children and some adults have made use of toothpicks in ways probably not imagined by Forster or the American Dental Association. With a little paste or glue, craft classes in schools and at camps taught the little nippers how to build themselves miniature cabins and other toys, and maybe helped develop the tots' manual dexterity. Some enthusiasts go much further than miniature houses and fences. Wayne Kusy of Evanston, Illinois, and guitarist of the rock group Flannel Tubs, has made a toothpick replica of the English passenger liner *Lusitania,* a sixteen-foot long construction of 194,000 toothpicks.[10]

Toothpick art and crafts aside, the noble splinter has also played a major role in the foodways and drinking behavior of a small but growing band of people in industrialized nations. At dinner parties and receptions in embassies, art galleries, and other public events, toothpicks bind together small portions of food variously termed canapés, hors d'oeuvres, or snacks.[11] The toothpick, which in its wooden form could be heated with the food, allowed cooks to wrap bacon's slithery substance around water chestnuts or the even more slippery chicken livers before roasting or baking. Toothpicks have become the flatware for stabbing and seizing small bits of fruit, vegetables, and other little victuals on trays while standing around making polite conversation and imbibing during predinner cocktail hours.

In the world of drink alone the toothpick has played a small and sometimes essential part. The martini, originally a combination of gin and vermouth, was often presented with a small pickled onion or stuffed green olive that had been skewered with a toothpick. The pick enabled the drinker to retrieve the vegetable part of the meal without distress or having to wait until the end of the drink. But the serious drinker's form of jolt to the brain cells was not the exclusive preserve of the toothpick.

Sometime about the middle of the twentieth century, Asian restaurants (at first Cantonese, then "Polynesian," then the rest of the region) began to proliferate in the United States and parts of Europe. In the United States these were probably the outgrowth of enterprising restauranteurs branching out from urban Chinatowns, while the Polynesian fad may have been a result of Americans' increased consciousness of Hawai'i after World War II, the popular musical *South Pacific,* and the Islands' quest for statehood. Australians were of course closer geographically to these areas than were Americans or Europeans, and their Asian and Pacific immigrant populations were doubtless the wellspring of their culinary diversity as well. But whatever the source of the Asian cuisine vogue that is still very much alive, many of these restaurants introduced still another drinking use for the toothpick—albeit a bit "tarted up"— the toothpick-stemmed parasol stuck in the top of a fruity, sweet, and potentially very useful class of alcoholic drinks. Plastic has replaced the

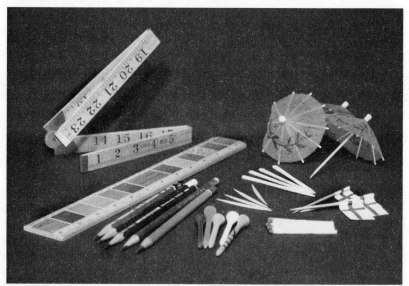

FIG. 86
Toothpicks, drink decorations, sandwich picks, matches, pencils, rulers, golf tees.

toothpick in some establishments, sometimes embellishing a drink with a little sword stabbed in the heart of the onion or olive, or as the stem of the parasol in those (usually) reddish drinks. Something adult seems lost in this transition. [FIG. 86]

"Americans," Theodore Roosevelt is alleged to have said, "like big things." This is doubtless true, as any observer familiar with the country can attest.[12] Big toothpicks in the foodways of Americans are as sure a thing as big food. The essential sandwich, allegedly invented by the English earl of the same name, required no complex joinery or attachments. But in the hands of the purveyors of big food, the sandwich, or more specifically the club sandwich, with its three slices of bread (not the measly two) and various vegetables and meats, required something to hold it together while people tried to eat it in something resembling a refined or at least nonswinish manner. Hence the long toothpick, now often with tiny plastic streamers attached to the top, became a necessity. The plastic part is doubtless considered an adornment to the food. There are even specially shaped toothpicks for the big sandwich. Club toothpicks are flat, one millimeter thick and five millimeters wide on the rounded large end that tapers to a pointed one-millimeter end, and (in the case of the box I own) made in China for a Canadian company.[13]

Manufacturing toothpicks is a sophisticated process in which (usually) birch trees are transformed into tens of thousands of polished, uniform, and tiny pieces of wood. Once a tree has been selected, hewn, and brought to the factory, it is cut into short lengths (about twelve inches), debarked, and conveyed to a steam room where the wood absorbs enough moisture and heat to become flexible. Then a veneering blade slices the cylinders into sheets the thickness of one toothpick, after which the sheets are cut into the proper width and length. Thereafter they are dried in a heated chamber, tumbled against one another to achieve a modicum of smoothness, then pointed and packed. A tree can yield thousands of them.

"Treenware" and Foodways

Woodenwares are probably the oldest of human eating utensils. Sharp sticks were effective tools for spearing bits of food when eating by hand was uncomfortable or difficult. As metallurgy and more complex social organization began to characterize human society, eating utensils became common among the well-to-do while the poor continued to eat with their hands. How one ate came to be one of the ways people distinguished between classes, as the sociologist Norbert Elias and the anthropologist Claude Lévi-Strauss have demonstrated. Initially people ate with knives and then combinations of knives and spoons. Forks came into widespread use somewhat later for most people, since they added relatively little additional convenience to the task of conveying to the mouth the stews and pottages typical of most diets.[14]

Wooden spoons served as both cooking and eating utensils. While they could be made from small gourds simply by splitting the fruit and removing the seeds, spoons could also be made of wood with an edge tool such as a plane or spokeshave, and the bowl hollowed out with a gouge or scorp. Spoons and bowls also can be turned on a lathe. Hardwoods such as maple and beech worked best since they withstood liquid immersion well and did not impart flavors to the foods.

Spoons for eating were a small part of the generic category of goods referred to as *treenware*. The *n* at the end of the "tree" portion of the compound word is a reference to an older form of case determination in English and other northern European languages. In this instance it is a possessive form, indicating that "ware" is "of the tree," or "from the tree." Treenwares associated with food included trenchers, plates, and low bowls. Trenchers were large bowls turned out on a lathe or scooped out of a single piece of wood like a miniature dugout boat. Everyone at the table ate from them directly, and they were the mark of people of low social standing. Someone described as a "good trencherman" packed away great quantities of food at a sitting, behavior unbecoming the refined, who presumably did not have to worry from where or when the next meal was coming.

In many Asian countries, the "flatware" of choice is a pair of chop-sticks, cylindrical or square in cross-section, used between the thumb and first and second fingers. Individual pieces of food can be grabbed by pinching the food between the sticks or they can be used together to pull rice or other small pieces from a bowl to the mouth. In many areas these sticks are made of bamboo, the long stringy fibers of the plant providing strength to the long and thin utensil. Other hardwoods are used in some regions, depending upon their local availability, while in urban areas plastic is replacing the traditional raw wood or lacquered stick. For those who are deft with the sticks, the fork seems a primitive and unre-fined instrument.

Woodenwares were also an integral part of the sales and measure-ment of foods. "Measures" were sometimes made of pewter or other metals, but often of wood, especially for dry materials such as vegetables. In many parts of the world today vendors at markets use boxes stamped with the insignia of the state or local officials in charge of standardizing weights and measures to ensure an honest volume. While unscrupulous vendors might well be able to doctor the volume if they were clever woodworkers, the penalties for such activities are stiff. Measure mills were built throughout many countries to provide the enormous volume of measures that retailers needed, and their reputations rested on the accuracy of the containers, as did that of coopers.

There was a special class of wooden gear that accompanied the dairy trade, especially the retail trade in butter. In a world that was for the most part devoid of vegetable oils, butter and rendered animal fats such as lard, beef tallow, and chicken, goose, and duck fat were the common shortenings for pastries (the dough rolled out with a wooden rolling pin) and other baked goods, and the main lubrication for and flavoring of fried and other cooked foods.[15] Americans had established a reputation for themselves as great lovers of fried and greasy foods by the nineteenth century, much to the chagrin of social and health-minded critics.[16] Until the late twentieth century, when medical research demonstrated links between saturated fats and heart disease, animal fats were treasured commodities, considered the font of flavor and an essential substance for building strong bodies.[17]

Most prized of these was butter. It took considerable investment in time, labor, and resources to produce it, and the flavor was prized by many as the most delicate of all the fats. Unless you were well off, it required raising cattle, milking them, separating the cream, and then churning it, arduous tasks that necessitated much hand labor until mechanical separators and churns were developed in the latter half of the nineteenth century. Once churned, butter had to be kneaded with wooden paddles to free it of water and air to achieve a smooth consistency.

Butter had a status that lard or other animal fats never gained. It appeared on the dining table, while lard and the rendered fats of fowl did not, except in the houses of the very poor. So special was it that a whole subgenre of wooden implements was developed to embellish it for sale or for the table. Called butter *molds* or *prints,* these tools were forms into which butter was pressed, with one facet of the form carved in some decorative manner so that when the form was removed a decoration in butter was left on the top. One of the most common images was, not surprisingly, a cow, though floral designs and the occasional coat of arms were also employed. Butter molds came in many sizes and in two basic shapes. The rectangular mold, often an official measure, was a box that was filled with butter, the lid pressed down, and the box then opened by undoing a hook and screw clasp on one of the corners, which allowed for the sides of the form to be taken apart, freeing the slab of butter. Round molds are cylinders into which the butter is pressed, the mold then turned over and the handle that protrudes from the top then pressed down in turn. Since this handle is attached to a carved disc inside the larger cylinder, when pressed down it produces both a uniform

FIG. 87
*Butter molds. Walnut. United
States, ca. 1885.*

cylinder of butter and a design on the top. Larger molds, or presses, of this type (three to six inches in diameter) were useful for retailers for measuring or presenting their goods, and small molds (one to two inches in diameter) were used for decorating the bulk butter for the table. Special ceramic and glass butter "keepers" attested to the food's special place in cuisine and culture. [FIG. 87]

Small bits of wood also became part of the ice cream industry, chiefly in two forms: the stick onto which an ice cream "novelty" was attached and the flat "spoon" that came with small paper cups of the product. (These were once known in some parts of the United States as the "Dixie cup," associating the brand with the generic form of the item.) "Popsicle sticks" (another example of brand-name application to the generic) were even more popular as craft raw materials for children than were toothpicks, in large part because they were potentially less harmful to the tykes since they were larger and rounded at both ends. Physicians use a longer and wider version of these sticks to depress a patient's tongue for a throat examination. White birch was usually the wood of choice for these goods, its relative tastelessness and whiteness as important for the ice cream industry and medicine as they were for the toothpick makers. While physicians still use the wooden tongue depressor, the ice cream industry has opted for plastic spoons in order to save money (since plastics are hardly an environmentally sensitive choice).

But while the plastic flatware offers the convenience of a tiny bowl for liquids or near liquids, they are less effective for hard ice cream because they tend to break easily.

Wood was everywhere and in nearly every way linked to food processing and consumption. It fueled the cooking fire (more about that in chapter 9); it was the major component of tables, chairs, benches, serving pieces, and utensils. Wooden tools worked the soil, cultivated and harvested the crops, stored the goods, sheltered livestock, penned them in, and carried the produce to market.

"Congreves" and "Lucifers"

Before the twentieth century one of the most important responsibilities in houses involved maintaining the fire. One compelling reason for keeping it going was that once extinguished, starting another fire was difficult, unless live coals could be obtained, either from another fire in the house or from a neighbor. Failing that, the job entailed striking a flint to produce a spark that, if it were a lucky strike, hit tinder that with a properly directed breath of air burst into flame.

Producing safe and inexpensive matches was one of those problems that energized inventors and dreamers for decades. The problem was not how to make flame erupt, but how to control it so that it wouldn't burn down the house or explode in the face of the everyday Prometheus. Legend had it that the ancient hero of pyromaniacs suffered mightily for his gift of fire to humans, the gods sentencing him to the agony of having his entrails eternally eaten by scavengers. No such fate awaited the match bringers, however. Explosives had been around since the Chinese developed them in about the tenth century. But an explosive is exactly what was *not* needed in the household; rather, the desired result was a slow burn after ignition. Once lit, the match had to have a cheap fuel to keep it going, and that fuel was wood.

In 1680 the English chemist Robert Boyle discovered that rubbing phosphorus and sulfur together resulted in a burst of flame. Boyle's discovery, while too dangerous for practical public use, was nonetheless a breakthrough in the attempt to chemically and mechanically produce fire. All sorts of frightening attempts were made to make a safe fire starter, including "fire bottles," which sought to contain the explosive potential of the phosphorus-sulfur combination within glass. The French chemist Jean Chancel in 1805 tried a combination of asbestos and sulfuric acid in a bottle, producing flames by inserting and quickly withdrawing a piece of cedar dipped in a combination of sugar and potassium chlorate. But sulfuric acid is about as dangerous a compound as one can find, and carrying off fire creation in a glass bottle is hardly a comfortable process.

Success in working out the science of the reaction and in making a safe and convenient match fell to John Walker, who is credited with developing the first friction match in 1826 or 1827.[18] Combining potassium chlorate, antimony sulfide, starch, and a gum binder, Walker's "Congreves" (after Congreve's rocket, which was invented in the early nineteenth century) were popular, but Walker did not patent his innovation and so did not profit greatly from its widespread use. Samuel Jones, a London chemist, reproduced Walker's matches, calling them "Lucifers." He marketed them widely throughout England, and in 1832 Richard Bell opened the first English match factory in London. These matches worked well enough, but produced a strong unpleasant odor. By adding phosphorus to the chemical mixture, Charles Sauria developed the friction match that would strike on any rough surface, though the resulting odorless fumes were toxic.

In 1855, the Swedish inventor Johan Lundstrom substituted safer red phosphorus for the toxic white form that Sauria had used, and further refined matches by infusing the phosphorus in the striking substance on the outside of the package so that they could successfully be struck only there. These were the first "safety matches" and they became a standard form soon after they were introduced. These matches, like their predecessors, were poisonous. The first nonpoisonous matches were patented in the United States by the Diamond Match Company in 1910.[19]

White pine was at first the preferred wood for matchsticks in the

United States. Aspen is now the wood of choice, perhaps because white pine prices have increased as the demands of the building and other trades have outpaced supply. Matches are made in much the same way as are toothpicks. Logs are cut into twenty-five-inch blocks, sliced on a rotary veneering machine into sheets, then cut into sticks, or "splints." After soaking the splints in monosodium phosphate (which prevents the splint from becoming a glowing ember when extinguished), they are dried, polished, dipped in hot wax, and the tips coated with a paste of chemicals that when dry will ignite when struck on the side of the box. People consume matches by the billions. Americans light about 500 million annually, the British more than 100 million. The Diamond Match Company manufactured about twelve billion matches in 2000.

Matches are manufactured throughout the world,[20] and constitute one of the most important labor saving devices developed in the past two hundred years. They made fire portable and accessible to nearly every person. They relieved people of the burden of having a fire burning at all times and saved countless trees from being consumed for that purpose. In hot climates and in the summer the end of the continual fire made life more bearable. The easy accessibility of fire probably made people careless with their use of matches, to the extent that it became something of an achievement to start a fire or light several candles with one match. In the end matches probably encouraged less care in organizing fuel to get a fire started, perhaps demonstrating how technological innovation can lead to the disappearance of skills, as the electronic cash register that shows the correct change has made obsolete the ability to subtract.

The portability of matches also had unintended consequences. Arson, a favorite strategy of the pyromaniac, murderer, avenger, and insurance fraud, was now far easier than it had been before the match came along. Matches also made it less difficult to smoke tobacco in places once inaccessible (there being no flame) and thus allowed people to partake who hadn't been able to do so in public before. Matches also helped make the cigarette a wildly successful product, especially among women and children. The latter found them convenient for sneaking a smoke behind some outbuilding or in school lavatories, much to the consternation of their critics.

Pipe and Stick

The burning match and the wooden smoking pipe provide striking examples of the importance of variety in wood's physical characteristics. While the match of pine, aspen, or a similar wood burns away in flame, pipe tobacco merely smolders, and does not ignite the bowl. Density and compositional differences enable a properly "carboned" pipe bowl to resist burning; briar is extremely hard and its grain and figure complex, while at the same time porous. These characteristics, and tobacco's smoldering tendency, have made wood the favored choice of smokers throughout the world, whether they smoke for ceremonial reasons, to ingest drugs, or for other reasons combining oral gratification and the pleasing nature of the experience, despite the habit's well-established dangers.[21]

Most briar pipes are made from a shrub native to the Mediterranean region, the white tree heath (*Erica arborea*). The word *briar* is descended from the Old French word for heath root, *bruyere*. In its raw form the root is bulbous and gnarled, and the grain and figure reflect that exterior. Making a pipe bowl involves sawing a blank (a rectilinear rough shape) and then turning it in two separate processes to produce the bowl and stem section. The area of the pipe base where the bowl gradually becomes the stem is generally finished by hand and the piece then polished with sanders and cloth wheels. Smooth-bowl pipes are then stained and finished, while those destined for decorative carving first go to the carver's bench.

In the United States, mountain laurel (*Kalmia latifoli*) root burls are used in the modern pipe-making industry; its close- and straight-grained characteristics make it an especially good choice for small turned items.[22] Wooden Native American ceremonial pipes (calumets) like those encountered by Lewis and Clark in their exploring expedition of 1804–6 were made of semi-diffuse-porous and semi-ring-porous woods and decorated with bald and golden eagle feathers, porcupine quills, sinews, and other references to animals the native peoples considered sacred or powerful.[23] These pipes served important ceremonial functions among Native Americans, for whom the acts of smoking and passing a pipe were instrumental in religious practice and interpersonal relations.[24]

Throughout much of Asia, opium and tobacco pipes were made of the roots of native trees similar to the white heath. Elaborately carved and decorated pipes demonstrate the importance of smoking tobacco and other intoxicants in cultures throughout the world. Functional pipes need only have a bowl and stem that resist ignition. The American corncob pipe is surely an example of this minimalism. But the field of artistic opportunity (in this case a small one) was seized upon by carvers because of the social and cultural importance attached to the act of smoking.

Carved decoration similarly reveals the importance of another class of everyday and ceremonial wooden artifacts—canes, staffs, and sticks. In the industrialized world canes are today associated with the infirm, but before the twentieth century they were considered an essential part of the gentleman's accoutrements and the rural carver's testament of skill. In other cultures they continue to be part of the vital gear of the clan or tribal leader.

Coppiced hazel is a favorite wood for cane and walking-stick makers in England. Its strong straight grain makes it effective for walking and for the carver, while its tendency to fork allows for a naturally config-ured handle at the top. Sticks of elite and precious woods, such as ma-hogany or ebony, were often further embellished with silver or gold tips and handles, and connoted both social status and power; they may have been the urban elite's way of fending off potential adversaries. In the hands of healthy men, the formal walking stick, while refined in ap-pearance, embodied the potential violence of the cudgel in the shape and finish of the refined. It was an announcement of power and the willing-ness to use it.

Historical references to caning reveal what all who carried the stick knew. In 1856, Massachusetts Senator Charles Sumner, a vehement opponent of slavery, was beaten so severely with a cane by Congressman Preston Brooks of South Carolina that he could not attend the Senate for three years. Brooks's actions outraged people (mostly in the northern and midwestern states), but his actions were extreme only in the place they occurred, in the class of his victim, and in their severity. Poor men carried no sticks unless infirm, their physical condition marking them as no threat.

These sticks had other powerful meanings. Folk cane carvers often used the form as a message carrier in ways similar to the much larger totem and spirit poles of Native Americans. In the nineteenth and twentieth centuries, the United States was replete with cane carvers. Folk carvers usually adorned their work with densely packed and complex motifs, some playful and others that referenced mythical and folk themes. One of the most commonly seen renditions on American folk art canes is the snake coiled around the length of the cane. Whether this is a witty reference to one of the favorite woods of the cane carver in America, snakewood (*Brosimum guianense*), or to the villain in the Garden of Eden is unclear, and it may be both.

Snakewood is a relatively difficult wood to find and harvest in its habitat, Panama, Suriname, Guyana, and the Amazon area of Brazil; it grows very slowly and insects readily attack the sapwood. It has been exported in only modest amounts and has found no mass commercial use. Cane carvers favor it because the fine-grained dark brown heartwood is exceptionally hard (it rivals ebony for hardness and durability) and polishes to a high degree of smoothness.[25]

Large decorated walking sticks, which might more accurately be called staffs, were once a near-global symbol of power and are still so in many cultures, though many are no longer made of wood. [FIG. 88] These may be in some manner descendants of weaponry—pikes and spears—but their ceremonial use has long since drifted away from the military. The Greek god Hermes carried a snake-entwined caduceus

FIG. 88
Unidentified artist, Shango staff. Wood. Yor-
uba/Nigerian. Early to mid-twentieth century.
Gift of William E. and Bertha L. Teel. © Mu-
seum of Fine Arts, Boston. This ceremonial
dance staff shows a woman in a pose that is
meant to indicate subservience when in the
presence of a king, or fon.

(which also may be the forebear of the folk carver's snake stick) and religious figures such as the Chinese god of longevity, Shou-hsing, are often depicted with long carved ceremonial walking sticks.

The staff is almost never associated with the infirm, but with the
strong and the hearty. Hikers and trampers have for centuries (at least)
made use of long sticks to ease their way through underbrush and thickets, or to steady themselves on treacherous ground. While most of these
were cut from limbs in the woods, the mass production of hiking staffs
began in the nineteenth century, among European Romantics seeking
woodland spiritual renewal and in the United States during the "rush"
to the Adirondack and other East Coast mountain areas.[26] These were
usually made of polished hardwood, turned on a lathe and fashioned
with a bulbous top. At around five to six feet in length, they were meant
to be held about halfway up the staff length and probably marketed to
urbanites on a quest for health in the mountains—whose hands were
unaccustomed to the rough surface of a staff cut while in the woods,
covered with rough bark and the small knobs left after the twigs were
trimmed off.

The staff's more pastoral form was the crook, a long pole with a large hook on one end. The function of the hook was primarily to snag the legs of sheep, which are notoriously skittish and quick at avoiding pursuers. Symbolically the crook came to be associated with Christianity, both in the figure of the shepherd and to some extent as a metaphoric extension of the equation of Christ with the lamb. In more mundane terms the crook is also considered by some to be the original golf club, first used by bored or mischievous shepherds who whiled away the time smacking stones, until one disappeared down an animal hole, thereby beginning the problem of workers sneaking off to play the game when they were supposed to be at the job or minding the flock.

Launching the Stone

In the centuries succeeding that first hole in one, those who found their way to the golf course had a difficult time getting the ball in a position from which they might strike the first blow on each hole. As the stone evolved into the "feathery" (a leather ball stuffed with feathers), the "gutty" (a real near-sphere made of gutta-percha), and the modern wound or solid ball, the problem of elevating it for the drive drew the attention of player-inventors. Most people simply raised a bit of sod with the heel of their shoe and placed the ball on it; a few whacked the turf with their club to do so, as the English professional Laura Davies does today. Still others used a little cone-shaped mold into which they packed sand or sod that they then unmolded on the ground, pointed side up, onto which they then embedded the ball.

In 1896 George Grant figured out a new way to set the ball above the turf. The Boston dentist made a wooden peg with his dental tools (retired ones, we hope), onto which he attached a rubber pad to set the ball. He obtained a patent for the device in 1899. Grant did not leave his patients for the lure of great riches wrought by his primitive tee.[27] A fellow dentist, the New Jerseyan William Lowell, may well have done so, however. Lowell figured out that simply producing the tee got him nowhere. He realized that to make a success of the wooden tee he needed name-recognition endorsements. He succeeded in getting the flashy and successful Walter Hagen to use it while barnstorming the country on an exhibition tour. By 1921 wooden tees were readily available, and the form has remained virtually unchanged since, surviving challenges from other materials (plastic, rubber, paperboard) and other designs that promised more distance and less resistance to the "launch" of the ball. Advertisers have discovered the tee as a source for hawking their goods, though it seems unlikely that many golfers pay attention to words on tees, even if they get them for nothing.

Golf tees are made in several countries, including the United States, China, and Taiwan. Birch is the most commonly used wood in North American tee factories, which are essentially turning mills. Given the international growth of the game and its unmooring from its associations with rich, whites-only country clubs, and loud, bizarre, and tasteless clothing, making tees is a small but potentially growing business. That is not a curious development, but what seems odd is the persistence of the wooden tee, a little thing that often breaks or is sent off to parts unknown by even a well-struck shot. Plastic tees seldom break, but are thought the gear of slashers and sod wreckers, akin to the embarrassing relative who shows up at the formal function wearing a lime-green polyester suit with white plastic shoes and matching belt. In an age when even young people use powered carts (a bane that is infecting even parts of the British Isles) and titanium technology has launched golf balls distances once only dreamed about, the wooden tee seems one of the last bastions of the game's conservative past. It endures more or less unchanged because it is a *safe* tradition; most players believe that it has no impact on the distance they can drive the ball, and thus the little wooden peg can serve as a harmless and perhaps quirky link to the past in a game that takes some measure of pride in its ancient lineage.[28]

Toys

Well-to-do Sumerian and Egyptian children living three thousand to five thousand years ago had toys not all that different from modern toys, at least in form. Their wooden dolls and moving toys were probably the one-off product of a gifted carver. Roman dolls of oak and other hard-woods dating from the first century CE had movable joints, achieved by using slot-and-tenon joinery with a dowel inserted through both parts of the joint to act as a pivot. Wheeled animal toys were simple affairs, the profile of the animal cut out of a plank, with an axle and wooden wheels attached to the bottom of the "body."

Toy making as a large-scale industry probably originated in Europe, in areas with abundant waterpower and supplies of cheap, easily work-able and neutral-tasting woods such as larch and spruce, and for more expensive toys, beech and ash. Central and eastern Europe, and particu-larly Germany and Bohemia, were centers of toy making by the four-teenth century. All sorts of woodworkers used slow work times—the winter months especially—to make toys. The southern German city of Nuremburg became one of the largest centers of the early toy trade, the finished goods distributed by itinerant peddlers. Oberammergau (a cen-ter famous for religious carving and Passion plays) and other small towns became the seats of the high-fashion doll industry, their carvers produc-ing technically sophisticated wooden dolls with exquisitely carved heads. British doll bodies were cruder than their counterparts on the Conti-nent, often made of a cylinder of wood and simply carved limbs. So su-perior were the Continental products that in the eighteenth century the English enacted trade laws to protect their native toy and doll industry. They succeeded in keeping the German products out of the British Isles for decades.

In the eighteenth century a special variety of toy was introduced, though in actual use the so-called doll's house may have been as much an adult plaything as it was a children's toy. Enthusiasm for the doll's house was likely an outgrowth of the interest in cabinets of curiosities. But instead of precious bits of stone, art, and floral and faunal materials, the wooden miniature house was a collection in tiny scale of things nor-mally found in European households. The game was to make the little house as accurate as possible, down to the smallest detail. When taken to the extremes it often went, the doll's house passion conflicted with children's desire to play with a world where they were giants, an all-powerful presence in total control of those who "lived" in the house. The very act of getting inside the house—by opening hinged doors that exposed the entire front façade—was unlike any house adults or children knew; it was the action of some fairy-tale behemoth. Most children in normal bourgeois and wealthy families actually got to play with their little houses, which were often too small for their "normal" sized dolls. Doll's house residents were tiny, no more than a few inches tall, living in rooms that were similarly minute. The scale, however, was not the point. It was the experience of power for those over whom adults maintained control by virtue of their size, physical strength, and command of the necessities of life, a situation buttressed by the child's need for love and protection.

Most early doll's houses were one-of-a-kind productions of carvers and woodworkers with time on their hands or clients with money. Eventually, however, they were mass-produced for a hungry market. R. Bliss Manufacturing Company, a late-nineteenth-century American firm, produced inexpensive wooden doll's houses with decal-printed façades that looked like brick or clapboard residences. The doll's house, like the twenty-first-century doll "system," was the opening salvo in what was surely a continuing drama of desire, pleading, and ultimately acquisition of an ever-increasing number of accessory goods—furniture, draperies, silverware, ceramic wares, more dolls, and textiles for the bedroom, parlor, kitchen, and dining room.

The boys' versions of the miniature house were the balsa-wood model and the toy log cabin. Balsa's light weight and light complexion, in addition to its even grain and flexibility, made it a sensible and profitable choice for the model trade, as well as the full-sized airplane industry, which used balsa in many parts of the fuselage.[29] Models provided parents with an opportunity to justify these and other toys as educational in the same way that doll's houses and dolls gave them license to indulge their daughters. They could be seen as things and pastimes that better prepared boys for adulthood, or at least provided them with the opportunity to develop intellectual and motor skills that all children would need later. Models and, in the United States, "Lincoln Logs" usually came packaged as component parts; the trick was to follow the assembly instructions included with the kit without breaking anything. Model airplanes in particular were a popular genre in the twentieth century, since balsa's lightness enabled children to make gliders that could actually soar, if only for a short distance. That a few children found a more dangerous pleasure in the glue they used to join some of the pieces was an unintended consequence.

In turning mills and on home lathes, woodworkers have over centuries poured out millions of toys, many of which were based upon the principle of the cut or carved plank attached to wheels and dowels. Tops were turned; puzzles cut from a board; more complicated toys, such as the once popular Noah's Ark assemblage of beasts and articulated figures suspended and swung between two uprights, were simply clever combinations of sawn and round wood. Sawn figures could be produced en masse by cutting the profile of a figure on a large piece of wood with a bandsaw and then cutting slices off the block. German toy makers pioneered a method of mass production using a lathe. They turned the figure's profile on a large cylinder, drilled out the center, and sawed small wedges from the doughnut shape. A quick application of paint and the toy was finished. Wooden blocks, which were a popular educational device in the late nineteenth century, were even simpler to produce, using powered carving machines developed in the furniture industry in the middle of the century. Pick-up sticks, short sticks that are thrown in a pile with the object being to remove them one at a time without moving another stick in the process, are simply large painted toothpicks. The key element in the success of nearly all of these toys was their exterior surface treatment. Paint and printed decals made the ordinary extraordinary, the bland colorful, the wood into something else— a bear, a wagon, a house, a rabbit.

Complicated wooden toys were still in production in the late nineteenth century, but were the minority in the world of the fantastic and the miniature. Albert Schoenhut, a German immigrant to the United States, was a master manufacturer of articulated wooden doll bodies and steam-pressed doll heads. His company also produced elaborate circuses, some with dozens of animals and other features. While dolls were aimed at girls, the most sought-after toy for boys was the rocking horse (although girls certainly made use of the creature as well). These were complex creations, many artfully carved and painted, held together with sophisticated joints, equipped with glass eyes, manes and tails of real hair, and decked out with a leather saddle and reins. Children certainly were capable of pretending that they had a steed with a stick topped with an abstraction of a horse's head, but once the would-be cowpokes and their parents caught sight of a rocking horse in all its glory, it was difficult for either the parents or the child to be satisfied with the pale imitation.

The rocking horse never actually went anywhere, whereas the stick did. Outside toys such as kites, hoops (rolled with a stick), and games such as mumblety-peg (in which a ball and cup are linked by a string and the goal is to toss the ball and catch it in the cup) were almost always made of wood until the twentieth century. Manufactured fishing poles for children were usually made of ash with a lancewood tip, though many poorer anglers simply cut a switch from a tree and tied a line to it. Toy guns and rifles made from wood were painted to look like metal, at least until inexpensive tin barrels and bodies became the standard in the late nineteenth century.

Wood remained the standard material for toys long after iron and steel had supplanted it in the adult commercial and industrial world, although iron and tin did make earlier inroads in coin banks and in moving toys, such as trains, wagons, and automobiles, the metals giving them a chance to survive the depredations of speed-crazy boys who reveled in crashes. Fine fashion dolls by the late nineteenth century were no longer made with wooden bodies but with stuffed kid leather bodies and bisque heads. The beautiful creations of the French companies Bru and Jumeau were for careful play, however. They were certainly well-made, even hardy, but their obvious excellence was also in some ways their undoing as playthings. They were surely treasured possessions, loved and respected, but also cosseted and protected, as parents hoped their children would be.

Wood endured as the material of playthings because it was safe (the occasional splinter notwithstanding) and durable in ways kid and bisque were not. It was, moreover, inexpensive, more easily worked, and took finishes readily, whereas bisque required painstaking hand painting to justify its expense and kid had to be tanned and smoothed. Steel and tin, while long-lasting, were hard, cold, and potentially sharp edged. Getting bopped with a wooden toy by one's sibling or playmate hurt and could draw blood; getting walloped with a cast-iron bank or a tin train could do a great deal more damage. Wood was friendlier and less industrial, at least in the minds of those who bought the toys for children. In the twentieth century brightly colored plastic has largely replaced wood and tin as the material of choice in toys. It is lighter, cheaper, and less likely to inflict wounds if a toy becomes an inadvertent or deliberate weapon. Is anything lost in the transition? Does it matter that children grow up without much physical contact with wood in goods that are theirs to keep, treasure, and destroy?

For some parents, it apparently does matter. In recent years wooden toys have undergone a renaissance of sorts in the "premium" toy industry. Companies such as TAG, Habermaas, Amenco, Automoblox, Beleduc, Brio, Tulip, Turner, Batavia, and Front Porch specialize in finely crafted, often educational toys, some designed for children with learning disabilities. The irony in this is that wood was the material for the toy of the ordinary child thirty to forty years ago; now it is the premium material.[30] The Web site www.kinderstart.com has links to the work of the psychologist Jean Piaget as well as to sites with information on breast-feeding, all of which indicates an equation or at least a connection between educational toys, a more healthy, even "organic" environment for children, and wood. Many of the wooden toy manufacturers most active in the United States market are based in Europe, where there has been considerable discussion about and commitment to the idea that wooden toys are more conducive to a healthier emotional start than are toys made of other substances.

It is probably impossible to know how different materials affect the growing child. The adult concern about the use of plastics rather than wood may in fact be simple romanticism or nostalgia for a past that may not have existed. It may be more a reflection of the parents' insecurities or fears about the present and the future of their offspring than a matter of the child's educational development. Wood may simply be a comfort for the adults in a plastic and metal world.

Pencil and Rule(r)

Skilled artisans have been drawing plans for millennia. Ancient Romans used the *pencillum,* a hollow reed into which thin animal hairs were inserted, to make fine ink lines.[31] This device was actually a brush, and the association of the word *pencil* with brush lingered long after the Roman Empire was a dim memory and a subject for history books and architects. When the photographic pioneer Henry Fox Talbot entitled his book of images *The Pencil of Nature* in 1844, he was referring to the artist's brush and not the wood and graphite artifact most people now know as the pencil.

The marking possibilities of graphite were discovered in Europe in the fifteenth century. Before that when people wrote or drew plans they used styli to scratch impressions in soft substances (such as wax) or they used brushes and points to convey inks onto paper and other surfaces. Henry Petroski attributes the first known and documented "lead" pencil to the Swiss naturalist and physician Konrad Gesner in 1565.[32] The discovery of large deposits of graphite, then called "wadd" as well as "black lead," helped stimulate the development of the cased pencil. At first enthusiasts for the new material simply put pieces in specially designed metal holders, reminiscent of the ancient stylus, or wrapped the graphite with a flexible material such as string.

The first documented maker of the wood-encased graphite pencil is

Friedrich Staedtler, who made them in the German woodworking center of Nuremberg in 1662.[33] A joiner by training, Staedtler made cased pencils by sawing out a groove in a small rectangular length of wood. Into the groove he inserted a thin piece of graphite, sawed it flush with the top of the wood and glued another piece of wood on top of it. He then planed or shaved it to produce a round or faceted pencil.

The first mother lode of graphite mines was found in Borrowdale, England, in 1565.[34] This deposit was thought better than graphite found elsewhere in Europe, and the British protected their supplies of the precious commodity. Pencil makers outside of Britain were forced to tinker with the composition of their inferior stuff and hence developed methods for making a practical substitute for pure graphite. In the 1790s Nicolas-Jacques Conté discovered that by pulverizing the graphite available to him, combining it with ceramic clay and water, and then firing it, he could produce a serviceable substitute for English graphite.[35]

Pencil making in the United States apparently began in Massachusetts in the first decade of the nineteenth century with the work of William Munroe. Within twenty years the pencil business linked a name now familiar to those who still use wooden pencils—Joseph Dixon— with a surprising name: Henry David Thoreau. The latter's father, John, established a pencil-making business in Concord, having learned the trade from Dixon. While the younger Thoreau is justifiably more famous for writing, his other careers, particularly in civil engineering and pencil making, offer clues to the connections among planning, drawing, clear thinking, and writing, practices that expand our appreciation of what appears to be an ordinary artifact.

The marriage of black graphite with wood poses several problems not immediately apparent. Graphite is brittle, as anyone who has tried to use it to mark stone or concrete block rapidly realizes, as point after point is either rubbed smooth or broken. Encasing it therefore means finding a substance that is practical for the job (workable without too much effort), yet rigid enough to prevent bending or torquing that would break the graphite, rendering the pencil useless because points would simply fall out of the wooden sleeve. Hardwoods such as maple and oak would prevent this, but they are impractical choices for the wood casing since they are in demand for other purposes, thus more expensive, and more difficult to pare away as the graphite wore down. By the late eighteenth century the ideal wood had been found—cedar.[36]

The wood pencil today is not much different from that of the nineteenth century. It was certainly popular in the United States—an estimated twenty million were consumed in the 1870s—and within forty years the supplies of American red cedar had diminished to such an extent that the United States Forest Service investigated possible substitute woods. Cedar varieties other than *Juniperus viginiana* and its relative *Juniperus barbadensis* (red juniper), such as western Port Orford cedar (*Chamaecyparis lawsoniana*) and incense cedar (*Calocedrus decurrens*), eventually took the place of the original southern cedar used in pencil making. In parts of the world where American cedar was unavailable or too expensive, alternative woods such as basswood, alder, Siberian redwood, Russian alder, and English lime were tried and found wanting. Until reforestation activities began in earnest in the middle decades of the twentieth century, the diminishing supplies of cedar helped push manufacturers to think of new wrappings (plastic, wound paper) or to strive to make superior mechanical pencils that used no wood but required a greater initial investment on the part of both the manufacturer and the consumer.[37]

Why does the wooden pencil persist as a consumer's choice? Mechanical pencils are inexpensive and no longer associated with fancy matching pens carried by rich people in suits or with "practical" metal instruments carried in the plastic-protected pockets of engineers and draftsmen. But most mechanical pencils cannot be sharpened, and the quality of the line is inferior to that of the freshly sharpened wooden pencil. (To get an acceptable fine line, drafting professionals use broad-diameter graphite in their metal and plastic holders and then sharpen the lead in specially designed sharpeners.) One seeming shortcut to a consistently fine line—the ultrathin 0.3 millimeter lead—snaps too easily for most hands. In the less rational realm of human activity and preferences, wooden pencils excel beyond their mechanical counterparts because wood gives way when chewed or carried in the mouth, habits (especially the first) in which people still engage, despite the disapproving looks of others, the embarrassing evidence of teeth marks in the wood, and sanitation concerns. Wooden pencils persist as well because they are viewed as inconsequential, inexpensive, and easily attainable, even though in reality they are not much less expensive than cheap mechanical pencils. As with wood in general, the wooden pencil may well be viewed by some as more "organic," in opposition to the industrial and technological world of metal and plastic. No less important is the psychological release the wooden pencil provides: There is little that feels better when angry than the guilty pleasure of snapping a pencil into pieces.

The point (if you will) of being able to sharpen a pencil may be aesthetic for some. For others, the aim is to be able to make erasable drawings that must be accurate. Builders and clients increasingly demanded detailed scale drawings as the Industrial Revolution proceeded in the nineteenth century. Older ways of doing things—trial and error, allowing fabricators and builders to make structural decisions on site and in the midst of the process—were less likely to produce the machines or buildings upon which greater demands were continually being made. The cost of failure increased exponentially as the complexity of machines and structures multiplied. Throwing out an entire machine because of faulty planning or execution became a financial disaster as steel replaced wood and casting replaced planing and shaping.

Marking and measuring are essential elements of accuracy and craftsmanship. One of the simplest ways to mark wood is to push a pointed instrument into the surface. Lines can similarly be made with a straightedge as a guide, although it is easier to use a knife edge when marking across the grain. Marking gauges—wooden instruments with points or blades—eliminate the need for the straightedge, relying on a movable

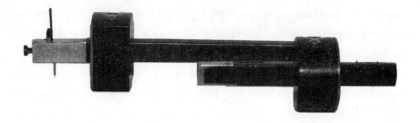

FIG. 89

Marking gauges. Ebony and brass, steel points. England or Scotland, ca. 1875–1900.
Ebony and brass were high-end materials for a tool, but a worthwhile investment for
the master artisan. Ebony's hardness made it possible to machine it to high tolerances,
and made more precise measuring and marking possible.

block, or fence, that can be pressed against the edge of the material to be
marked. [FIG. 89] The incised line, however, is permanent, and in some
cases difficult to see, even for people with perfect vision. The pointed
pencil, on the other hand, leaves a mark that can be removed and that is
easily seen on most light- to medium-colored woods.

We take rulers with accurate gradations for granted, but people who
had to measure things by transferring a dimension from one object to
another by first marking the distance on a stick and then marking that
dimension on the other piece thought rulers were a godsend. They might
have used dividers for measurements, rotating the tool from point to
point, but without a straightedge, accuracy was difficult. Standard measures, graduations in inches or meters, and the ability to produce very
accurate measuring tools were the industrial breakthroughs that underlay not only the Industrial Revolution and mass production but also the
ability of individual artisans to make more and better pieces. Metals
might well have been desirable for many of these tools had they been
less expensive to produce and mill with accuracy. Until the eighteenth
century, this was hardly the case, but there were species of wood that
could be machined to close tolerances and that resisted thermal and

aqueous expansion and contraction. The hardness and uniform light color and grain of boxwood were ideal for incising the fine lines of precise rulers.

Archaeological and pictorial records from Egypt dated to 2500 BCE demonstrate that artisans there were using stone and wood rules calibrated to standardized lengths related to the human foot and outstretched arm. Ancient Greeks and Romans used graduated rods of wood and brass, the latter more common in Rome. These measuring rods or sticks continued to be the most commonly used form of measurement through the eighteenth century, when rule makers began dividing feet into inches, quarter inches, and eighths.[38]

By the nineteenth century the carpenter's or joiner's workbench often had several wooden rules. In the beginning of the century the folding, two-foot, brass-bound boxwood rule was introduced, and it became the standard working rule. Many also incorporated a small graduated sliding brass bar let into one end to enable woodworkers to accurately measure interior dimensions. The variety of rules in use at any time is considerable, corresponding to the specific needs or common practices of tradesmen, but the distinctions are at best arcane to all but artisans and devoted collectors.[39]

Off the workbench and away from the workshop, most people are familiar with wooden rules in two basic forms: the one-foot rule, sometimes with a thin metal strip inserted along the measuring edge, and the one-yard or one-meter stick. The former is the straightedge of the school room and the home desk, the latter of the garage or other areas where such items of occasional amateur work are stored. Like many other implements of wood such as golf tees and pencils, the flat surface of the rule has been used as a field for advertising.

For much of the twentieth century the advertising function was enough of a boon that businesses (often in the building trades) gave away yardsticks and metersticks. Consumers gladly accepted the largesse, snagging a roughly accurate measure usually made of a light-toned softwood (the easier to stamp) in exchange for the vendor's advertising. But more recently the free rule has become scarce, if it even exists at all, and customers are actually asked to pay to advertise the business. This phenomenon is not limited to building-supplies merchants; it is even more extensive—and bizarre—in the clothing trade, where consumers actually pay a premium to hawk a brand and, in their minds at least, their savvy as connoisseurs of the momentarily stylish, if not necessarily superior-quality goods. Like money, brand names are best if discreet; flashing either indicates insecurity and perhaps the inability to recognize the qualities that constitute the well designed and the well executed.

This nervous bravado and social insecurity manifests itself as well in commonly held ideas and beliefs about craftsmanship. Some artisans have powers of intellect and mastery over materials that, when compared to the rest of us, are near to the magical and miraculous. They seem to instinctively know what they are doing and how to do it. The "instinct," however, is not some superhuman power; it is brilliance that the rest of us attempt to understand by attributing it to some power out of our control, the "luck" of the genetic draw. This attitude denigrates genius, but enables those with lesser skills to feel better about themselves. This interpretation of what seems "native" or "instinctive" skill is not merely negative. In the attempt to understand the work of the skilled who seem to have no tangible plan, people also forget or never consider that these skills may be the result of years of error and failure, lessons that have been learned and incorporated into the work of the mind and hand. Everyone plans; some draw on paper, some draw on memory and experience.

Hanging Things

Savvy entrepreneurs and inventors in the nineteenth century knew that one way to riches was through the workrooms of the household, in particular the laundry area. Backbreaking, repetitive, nasty work that scalded the skin, irritated it with harsh soaps, and was generally done weekly, laundry was the task women hired out first. No one liked it; commentators called it "the domestic dread"; the pleasure in getting clothes clean or ironing them smooth disappeared quickly, in the de-moralizing and certain knowledge that all those qualities would inevitably disappear as soon as people got dressed or slept on the sheets and pillowcases. Until the nineteenth century most people were largely un-washed, in spite of reformers' efforts to undercut the fear of bathing that had reigned in Western societies since the fall of Rome. (Such was not the case in much of Asia, where communal and ceremonial bathing had been linked to health and spiritual renewal, rather than to disease and licentiousness.)[40] Easing the burden of laundry was thus a way to success, even glory, at least in the minds of the women who did it at home and the professionals who did it for the more well-to-do. Patents for washing machines, irons, stoves to heat them, drying racks, and all sorts of other laundry gear are abundant in the United States and other countries with patent laws in place.

Drying clothes was not the labor-intensive travail that washing them was, but it was still a nerve-racking chore. Most people dried clothes by laying them over bushes, fences, or on the grass, either because they did not have sufficient lengths of rope or because they did not trust the line. Draping wet clothes over a rope strung between two trees was an effective way to dry them, but such plans could come to a bad end in a wind gust. Clothes got dirty again or simply blew away. Enter the clothespin. In the United States, credit for inventing clothespins (also called clothes pegs) as we know them has been assigned to the Shakers. But it seems likely that the country woodsmen and artisans who figured out how to make pitchforks, hurdles, gates, and other necessities of country living by partly splitting branches and twigs of flexible trees also ascertained that a small branch partially split lengthwise would clamp the garb to the line, utilizing both the suppleness of the wood and its spring to hold it fast. (It is also possible to make a clothespin by splitting the branch completely and binding one end with willow or hazel or some such wood or plant fiber.) It seems unlikely that the Shakers invented the concept, but it does appear that they developed the turned and kerfed clothespin.

At first the Shakers often did not try to patent their inventions (they altered that policy sometime around 1828), and they had lots of competitors in the race to develop the perfect clothespin. Between 1852 and 1887 the United States Patent Office granted 146 patents for "improvements in clothespins."[41] None of these revolutionized the clothespin industry, save one—the wooden pin of two sections bound together with a spring. This form has for the most part completely superseded the one-piece clothespin. Like the makers of toothpicks, matches, and a variety of other small goods made of wood, many clothespin manufacturers favored white birch, owing to the wood's light color (cleanliness, again), sturdy character when cut into small pieces by powerful machines, and relative cheapness, since it was not as much in demand as were pine or maple.

The passion for convenience and ease in the laundry process ultimately led to the diminution of the clothespin. With the advent of electric dryers and relatively cheap electricity in the twentieth century, hanging clothes in the open air declined precipitously, at least in the United States. Whereas people in city and country once strung lines between apartment buildings, trees, houses, barns, and posts, they now frequent the laundromat or use the powered dryer in their home. This has obviously made life easier and made drying clothes less of a problem in cold and wet weather. But the consignment of the clothespin to the back of the hardware store has broader implications. Using electricity, much of which is generated by fossil-fuel-burning power plants, not only drains consumers' finances (while wind and sun are free); it pollutes the very air that could be drying the clothes. The environmental irresponsibility of this tactic is only heightened by the astonishing bans on outdoor clothes drying that some municipalities have enacted, as if clean clothes on the line were some sort of affront to social mores or community aesthetics. Finally, the wholesale abandonment of outdoor clothes drying reveals the triumph of the advertising-induced value of "soft" textiles over the stiffer and, in the minds of some, more invigorating textures of air-dried clothing.[42] The clothespin is not merely a quaint tool, but a small part of a more environmentally responsible ethos and a stiff resistance to the ploys of chemically laden fabric softeners.

Does softness in fabric "feel" equal refinement and rank, as it once did? Stiff and coarse clothing and other textiles once connoted working-class status, while finely woven goods carried the message of wealth and power. But fabric softeners seem less a democratizing agent than another effort to carry people further from nature, even in some cases closer to a more polluted or at least altered natural world. The Industrial Revolution has brought people machines to refine natural fibers in ways only dreamt about before and it has also brought chemical fibers that have properties that natural materials do not possess. What is the price paid for these materials? Do they ever degrade when no longer used? Does hanging the microfiber fleece vest on the clothesline achieve an ecological balance?

In some ways the clothespin speaks to an earlier, less technologically advanced or dependent era. It is certainly a simpler solution to the problem of drying clothes, albeit one that is compromised by weather conditions. Its replacement by the powered drying machine is not an example of technology initially resisted by masses of workers. Initially only a small number of mill workers in the clothespin industry were downsized as the electric dryer took over. Part of the reason for the slow decline of the industry lay with those who bought dryers for bad weather but nonetheless used the clothesline when they could. But eventually cheaper foreign-made clothespins, declining use, and competition from plastic clothespins have sent the industry in the United States to near death. The West Paris, Maine, Penley Corporation, which had made clothespins from Maine birch since 1923, finally closed in December 2002, idling thirty-nine of the fifty-four people employed by the firm. The remaining workers now are employed by an importing and distribution company that deals in clothespins and matches, as well as plastic flatware and straws.[43]

Shoes, Clogs, and Sabotage

The Penley workers sent off from their jobs responded as do most people who are put out of work through no fault of their own—they carried on, finding other jobs or retiring if old or well-situated enough. Occasionally in the history of technology, differences between labor and capital have had a different outcome, however. Most people even slightly familiar with labor history know of the Luddites, those early-nineteenth-century English resisters to mechanization who smashed the new machines that replaced them. What they did in their blatant manner was not new; workers had in various covert ways been avenging perceived and actual wrongs of owners and bosses for ages, probably since the first time one person was able to command others to work, whether for wages or merely to avoid physical punishment or death.

The act of hindering or preventing work is called sabotage, a word that reveals something of the recent history of this activity, in this case intimately connected to wood as a material for shoes and shoe parts. The word derives from the French word for shoe, *sabot,* and the verb *saboteur,* which means "to make a loud clattering noise with sabots."[44] Legend has it that French workers disrupted new machines they considered threats to their livelihood by chucking their sabots into machines, but there appears to be no direct evidence that they did this. The original French meaning of *sabotage* is more nuanced than messing up the works; it implies doing a bad job as well as destroying the machine.

How, then, did making a racket with wooden shoes come to be associated with its meanings of covert malingering and even treachery by peoples who borrowed the word? The answer may well lie in the nature of the artifact and in attitudes towards the people who wore the gear. The sabot was one of the more common forms of footgear among working-class and rural people in many parts of Europe and in some areas of the British Isles. Made of hardy woods such as beech (the genus *Fagus*) or alder (*Alnus glutinosa*), a tough, pliable and water-resistant wood,

FIG. 90

"Manière de faire les sabots" (Manner of making wooden shoes). Plate 26, volume 1 of Denis Diderot, L'Encyclopédie *(1751). Here is depicted work in the woods, much like that of the chair bodger or lumber camp.*

the sabot, or clog, was carved from a single block of wood, hollowed out, and shaped to the wearer's foot size and gait. Starting with a clear log sawed to approximate length and width, the shoemaker dried the wood thoroughly and then roughly shaped it with a large stock knife, a blade with a handle on one end that was attached at its other end to a sawhorse or bench. This arrangement, in which the pivoting end acts as a fulcrum and the blade is both a cutter and a lever, allowed the artisan to exert great force on the block of wood. After roughing out the exterior, the sabot maker fashioned the interior of the shoe with hollowing tools similar to those of the cooper, small rounded bladed scoops (or scorps) attached to a wooden handle. [FIG. 90]

Wooden shoes and coarse leather footwear with wooden soles (common in the British Isles) were associated with the working class because they were inexpensive and less refined than the shoes and boots of the well-to-do, which were made almost entirely of leather. (Even the footware of the wealthy often contained some wood—little pegs used in the bonding of the sole's heavy leather to the soft uppers.) While in our time wooden shoes have become associated with quaint fairy tales of people in Holland, they once meant "working class" and "poor" and, in the minds of the well-to-do, "untrustworthy" and "sneaky."

The reference to making a loud noise with the wooden shoes, which was embedded in the original meaning of *sabotage,* emphasizes as well the class associations tied to making a racket with this footgear. That clatter announced that one not only was poorer but also had a shuffling gait, a stark contrast to the mannered and graceful bearing of the wealthy, who were trained to walk, sit, converse, and position their bodies with "ease."[45] Meeting this responsibility and expectation was of course easier in leather-soled shoes with leather uppers that fit the feet, and with training in the genteel arts. Noisy and boisterous behaviors were not genteel but were the acts of ruffians, clodhoppers, and boors. Men of station were allowed such social extravagance, but only in the liminal areas they frequented, the saloon, the barn, the bordello, the gambling den, and the blood-sport pit and ring.

Many people still condemn the loud among us as crude, though the social penalties for inappropriately loud behavior are more muted than they once were. Now, ironically, the covert and sneaky acts of saboteurs and other criminals are often depicted as the acts of evildoers who are able to move about without making any noise whatsoever, their feet clad in silent leather or rubber.[46] This may have its roots, in the United States at least, in the legends of Native Americans who were alleged to be able to move—even run—through the forests on moccasin-clad feet that made no noise as the clever warriors closed in on unsuspecting white folks, who had no such powers, especially in wooden-soled boots and shoes. Sabotage, once the act of the loud and the overt, has become the act of the silent and the covert.

By the 1960s, the loud clatter and clomping of the wooden-soled shoe had taken on a more decorous and nuanced meaning. Indeed, while criminals sneaked around on canvas-and-rubber shoes called "sneakers" in the vernacular ("tennis shoes" among the bourgeoisie), "hip," "cool," and "natural" people banged around on clogs imported from Scandinavia, where they had never really gone out of fashion. For the most part worn by women, clogs and their American variant, the contoured wooden-soled Dr. Scholl's sandals, helped define some of the generation of younger Westerners united against the older generation by their music, opposition to the Vietnam War, support for equal rights for all, and burgeoning environmental consciousness. These were small wooden identifiers, at least for a while, like long hair on young men. In time, however, they lost their connections to any particular politics or lifestyle.

The myriad ways in which small wooden things of almost infinite number and variety have altered human existence attest to the significance of wood as a material in everyday life. The large things and events of history concern much of what historians do and say most of the time. The little things, often so mundane that they escape notice or seem not to warrant attention when discovered, are often more difficult to perceive because they are common, inexpensive, and designed to be used and therefore to wear out and be discarded. But shoes, matches, toothpicks, rulers, pencils, clothespins, and the like are essential precisely because they are so used and so common, and their history, form, and shape reveal to us more complicated and unexplored regions of experience, if we know where and how to look at them.

8

Bat and Battle

*I*t may seem odd to link the murder and destruction of war with the gentler and more humane acts of play. Toys and games meant for children are for the most part innocent enough of the taint of war, which is in the end the activity of benighted miscreants whose actions speak to the darkest realm of the human spirit. But many of the sports and games of adults, such as archery, javelin, polo, and rowing, are in some measure derived from the activities of war or the preparation for it. Others have nothing to do with war or violence, but have in some measure taken on the language of warfare, as if the "battle" over the Ryder Cup in golf or the World Cup in football (soccer) is anything like the real thing. It may be that a few of the more violent members of humanity obtain an outlet for their destructive tendencies in competitive adult sport and games; conversely, it may be that this metaphorical equation is an impetus for antisocial behavior, as in the riots that have taken place after a beloved team is victorious or the organized mayhem of football "hooligans." Whatever the case, it is clear, as the Dutch cultural historian Johan Huizinga demonstrated in *Homo Ludens: A Study of the Play Element in Culture,* that play elements exist in such widely different activities as law, war, art, philosophy, and poetry.[1]

The story of warfare begins when humans first figured out that they could use other materials as tools to get food more easily and effectively; that is, when they first picked up a tree limb and figured out that they could use it to better effect than their hands and arms alone, or when they realized that throwing a stone extended their reach. Most people understand that swinging a stick multiplies the force of the hand and arm because the far end travels much faster than the end in one's hands. They also know that a long handle will move faster than a short one, as long as hand speed remains fairly constant. A long axe can deliver more power than a short one; the head of a golf club can travel at over one hundred miles per hour.

Hunting and Trapping

The earliest wooden weapons for war and hunting were clubs, cudgels, and pointed sticks. At the outset the first hunting spear was probably an accident of nature. Initially spears were likely used to stab animals to death, but soon hunters figured out that throwing the spear was more effective and safer, once they understood that a straight stick made for more accurate flight. With the advent of stone working, the idea of binding a tip to the end of a shaft for a more effective and long-lasting weapon was a logical next step in the evolution of this type of weaponry.

Just as the spear and stick were more effective in bringing down game than were the weapons they succeeded, the trap and snare were even more efficient means to obtain game, even after the advent of gun-

FIG. 91

Animal trap. Spruce or pine. Inari, Finland, twentieth century. The Sami made use of several different types of snares and traps to catch everything from birds to bears. Here the victim is lured into entering the trap by bait and crushed or disabled when the log falls on it.

powder and guns. Instead of tracking and stalking animals, hunters could set traps and return periodically to claim the catch. The pit trap, a deep hole dug in an animal's habitat and preferably on an established game trail, often had a battery of sharpened sticks closely implanted in its bottom, onto which the animal was impaled when it fell through the false surface laid over the hole. Wooden traps that lured animals to bait that, when disturbed, caused a heavy log to fall on the animal's head were also part of the hunter's repertoire, as were smaller traps that caught an animal's paw in a slot as it reached for food. Hunters also caught small animals and birds in snares.[2] [FIG. 91]

All of the captured creatures, save for those impaled or crushed, suffered their trapped condition alive, until killed by the trapper. The advent of iron and steel traps changed this not at all, since the leghold method does not kill but merely seizes the animal between metal teeth, further maiming and torturing it as it struggles to get free. These sorts

FIG. 92
*"Roman Balista." Drawing
from Ralph Payne-Gallwey,* The
Book of the Crossbow, *300.*

of traps were uncommon in warfare, since they were too difficult to use in situations where masses of people fought, although iron leghold mantraps were used by some great landowners trying to discourage and prosecute poachers.

Warfare

Wood was still the most important material for war until the nineteenth century, even after advances in metallurgy and the invention of gunpowder had made hunting and warfare more lethal. Wood carried supplies and weaponry overland as armies moved against each other. Wood was the building material for the ships that brought troops and heavy ordnance all over the world, and the major material in the spear, lance, and pike. It was the primary material of the archer and of the crossbowman, and it constituted the stock of the musket, as well as the support structure for the barrel. The earliest massive war machines—catapults, mangonels, onagers, balistas, and trebuchets—were made primarily of wood.

In use since the time of the ancient Greeks and Romans, balistas launched arrows and spears at the enemy. [FIG. 92] By the middle of the third century BCE the Greeks had developed a rock-launching machine, the palintonon, similar to what we know as the catapult. The

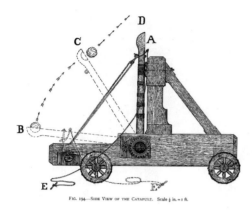

FIG. 194—SIDE VIEW OF THE CATAPULT. Scale ⅛ in. = 1 ft.

FIG. 93
"Side View of the Catapult."
Ralph Payne-Gallwey, The
Book of the Crossbow, *282.*

Romans attached wheels to these machines, enabling the armies to move them rather than having to build new weapons on each battle or siege site. [FIG. 93] They were the primary siege engines of ancient warfare. Trebuchets, probably developed in China in about 300 BCE, differ from catapults and balistas because they do not make use of wound cords to propel the ammunition but use the force of gravity to sling the payload toward the target. The machine employs a long and short arm that pivot on a cross-piece. The short arm is equipped with a large container of rocks on its end and the long arm with a sling or container for the ammunition. Soldiers operated the trebuchet by pulling the long arm to the ground, raising the rock-loaded short arm. Then they loaded its sling compartment, and released it. Once free, the weighted short end plunged to the ground, launching the missile.[3] [FIG. 94]

These siege weapons were deadly but not particularly accurate. They were useful because they could propel arrows, rocks, diseased animal carcasses, and framing materials at great speed over long distances, thereby piercing the armor of foes or merely braining them. They were also important weapons when armies besieged towns, since they could heave heavy and destructive payloads over walls and into crowded and flammable buildings. They altered the balance of power in favor of the attackers, neutralizing the fortifications that surrounded towns by sim-

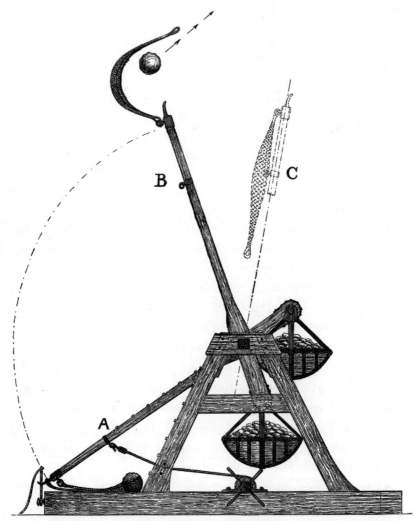

FIG. 212.—THE ACTION OF THE TREBUCHET.

A. The arm pulled down and secured by the slip-hook previous to unhooking the rope of the windlass.   B. The arm released from the slip-hook and casting the stone out of its sling.   C. The arm at the end of its upward sweep.

FIG. 94
"The Action of the Trebuchet." Ralph Payne-Gallwey, The Book of the Crossbow, *310.*

ply going over them without the huge losses in human life that storm-
ing walls entailed. They remained in use well into the medieval era,
when they were supplemented and then supplanted by gunpowder-
based cannonry. They began warfare's assault from the air when aircraft were but a fantasy.

All sorts of simple and complex machines were employed to wind as
tightly as possible the cords and ropes that powered these weapons, to
impart maximum energy to the missiles. Soldiers used large windlasses
(giant cranking mechanisms), human- and animal-powered treadmills,
and multiple pulleys to increase mechanical advantage. Armies often
employed small field machines for open-field or unfortified battle, but
besieging a city or town often meant using more than one hundred great
catapults and at least as many small pieces. Large catapults and balistas
were as much as twenty feet long with slinging arms often greater than
ten feet. Many components of these contraptions were of solid wood at
least one foot thick. These were massive machines because the slinging
action of the catapult and trebuchet, for example, only worked if the
long arm was stopped at a point just short of its apex, while the missile
continued onward. The great force generated would shatter the cross-
pieces into which the arm crashed unless they were thick and strong;
likewise the tremendous potential energy in the wound cords of the
balista would wrench apart a lightweight support structure and frame
when tightly wound for the launch.

Armies continued to use catapults, and particularly trebuchets, long
after gunpowder and cannonry had been developed for warfare. Shortages of cannonballs and other ammunition at siege points allowed the
older form of weaponry to retain its attractions, especially since its missiles could be had from the ground itself and most of it could be constructed on site. A large catapult could launch a fifty-pound weight
roughly three hundred to four hundred yards.[4]

Hand-to-hand combat involved swords, cudgels, and an assortment
of lethal metal and wood implements. Avoiding that stage of warfare
was the ideal way to win on the battlefield or in the town, but it seldom

happened so neatly. For centuries archers and crossbowmen were the long-range killers, in concert with operators of catapults, trebuchets, and balistas. While the latter sought to kill or burn by landing large and dangerous objects somewhere near the enemy, the former tried to pick them off one by one.

The Long and Short of the Bow

Arrowheads discovered in Bit-El-Atir, Tunisia, date archery to around 50,000 BCE, and archaeological remains unearthed in Germany in the 1930s suggest that they may have been in use in that region since around 7500 BCE. The bow and arrow were common to nearly all cultures in

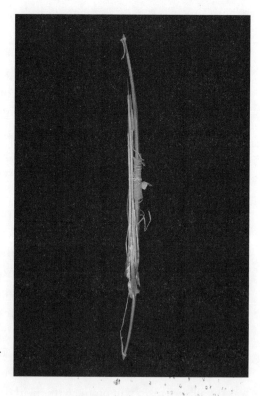

FIG. 95
Bow and five arrows. Wood, metal, feathers, cat-gut, flannel, steel. Plains culture. Late nine-teenth century. Gift of Reverend Herbert Probert. © Museum of Fine Arts, Boston.

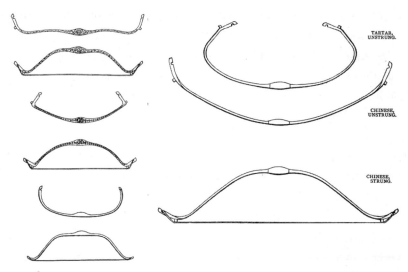

FIG. 96
"Comparative Dimensions of Reflex Composite Bows . . ." Ralph Payne-Gallwey, A Treatise on Turkish and Other Oriental Bows . . . *Appendix to* The Book of the Crossbow, *16–17.*

the world, and making the weapon gradually became a sophisticated process in which knowledge about the flexibility and strength of wood species and the best materials for the line (or string) was critical.

The various world cultures developed distinctive shapes for bows and philosophies that turned around the practices of archery. Native American bows were most often relatively straight pieces of wood when not under tension, and bent in a single arc when strung. [FIG. 95] More common in India, Central Asia, China, Iran (Persia), and the Ottoman Empire were reflex bows, so called because in their unstrung state the bow wood bends in the direction opposite to that when put under tension. In other words, stringing the bow bends the wood past straight and in the direction *away* from its natural bend, thus creating great tension in the wood and on the string, and consequently great power (as well as difficulty in drawing the string or sinew). [FIG. 96] Stringing such bows usually required two people and often a simple machine such as a pulley or winding mechanism.[5]

Species choice for the bow was critical and bowyers eventually gravitated toward those local woods that were resilient (so that they would not take a permanent curve when used) yet difficult to bend, since the range and power of the weapon relied on increasing the potential energy in the bow when pulled. Prime woods also had to be grained in such a way that they did not separate along the boundaries of the annual rings when stressed, and the ideal wood had a minimum of new wood in each year's growth. Sapwood in some woods (Osage orange, for example) was too weak for bows of any sort, except toys, but in other species (ash, hickory, and elm) it is of great strength. The heartwood of Osage orange, on the other hand, is so strong it is one of the most desirable woods for bows. Other species, in particular yew, have these qualities in both the heartwood and sapwood.

Whatever the variety of strengths in species, the straightest grain was essential. Twisted woods were useless, since when split into billets (the first step in making a bow: a length split from the tree that was wedge shaped in cross-section), they would follow the line of the grain. Shaping a twisted billet into a bow would result in a large number of a tree's longitudinal cells' being cut along the length of the bow, thereby weakening it. With a straight billet, the only cut ends would be those revealed as the bow was narrowed from its widest point (where it was held while drawn) to the tips, where the string was attached, thus limiting the number of points of potential separation and fracture.

Ancient bows unearthed in Denmark (the Holmegaard excavations on the island of Zealand, dating from ca. 6000–4000 BCE) and those found in Asia and North America demonstrate considerable evolution in knowledge of the craft. Unlike earlier bows, the orientation of the annual rings is roughly parallel to the belly, or round side, of the bow, a design that added considerable strength over the perpendicular orientation of rings in the earlier, German bows. Some of the Danish bows were made of elm, and "overbuilt," wider and deeper than required to take great tension. They were, as well, nearly the size of a mature adult male, between sixty and sixty-four inches long. Yew bows found in Germany dating from about 2000 BCE show a similar efficiency in power and construction.

Outside of Europe, particularly in Turkey, Persia, Central Asia, China, North Africa, Korea, Japan, and the Americas, bows were often made of several materials, usually some combination of wood, horn, antler, and sinew. Turkish bows of relatively recent vintage (eighteenth and nineteenth century), for example, are composed of a multisectioned wood core, the pieces joined with glued complex joints such as dovetails or V-shaped saw teeth. Strips of horn were glued into shallow grooves in the front face, and bowyers then applied sinew backing to this joint, glued and clamped it, and dried the bow for several months before they finished it and put it to use. Unstrung Turkish and Crimean bows are roughly C-shaped reflex bows. Indian and Persian bows resembled the Turkish bow, and all were used both in warfare and in sport, particularly for long-flight competitions.[6]

Hunter-gatherers inhabited the Japanese islands from at least 7000 BCE. They used the bow and arrow for hunting game, and eventually in warfare. These early bows were short, with a centered grip, like those of most other early cultural groups. About 2,500 years ago the bow, and particularly an asymmetrically gripped long bow, became a symbol of power in Japan, as well as a devastating implement of war. In China, the bow and arrow similarly evolved from an ancient hunting instrument to a weapon for warfare at about the beginning of the Han dynasty (206 BCE–220 CE).[7] Soon after (between about 300 and 1200 CE) Japanese craftsmen adopted the Chinese composite bow, joining wood (often mulberry) with bamboo. Beginning in 1192, after the appointment of Minamoto no Yoritoma as shogun (military governor), more formalized training and practice in archery were established, and the technique of making laminated bows reached an apex.

Archery in Japan has for about two thousand years been a ritualistic as well as practical activity. Shintoism, the indigenous religion of Japan, and Buddhism, which came to Japan from China in the late twelfth and thirteenth centuries, have exercised a strong influence on archery among the warrior class. In the seventeenth century, after the Japanese civil wars ended, *kyujutsu,* or training for warfare, evolved into the more philosophical *kyudo,* a system that interpreted the actions of archery as part of a larger cultural complex of beliefs centered on individual betterment and harmony with the world and spirit. Like other martial arts, *kyudo* was prohibited in the immediate aftermath of World War II, but by the early 1950s it returned as a ritualized practice.[8]

The history of the longbow in Japan parallels that of the English yew longbow, which dates as far back as 2500–2700 BCE. Yew was ideal for longbows because its heartwood is highly resistant to compression and is therefore ideally suited for the inside of a bow. The sapwood is resistant to the forces of expansion, and therefore perfect for the outside of the bow. Archers with bows of apparently great length are depicted in the Bayeux Tapestry, a graphic record of the Battle of Hastings (1066), though they appear to have played a minor part in the clash. By the thirteenth century, however, the longbow formed an essential part of the British army's weaponry and continued to be an effective weapon until armor plate replaced chain mail as a soldier's protective gear. The longbow had several advantages over other weaponry—even early firearms. It was easier and quicker to reload and fire, unaffected by wetness, more accurate than many early firearms, and could be used in stealth attacks with more effect. But the longbow's great disadvantage was that its power was no match for armor plate's protection. By the end of the sixteenth century it was superseded by improved musketry and handguns, which could be fired from any position (rather than standing) and which were easier to transport and carry.[9]

In North America, Eastern Woodlands Indians usually used single-piece bows, rather than the composite bows favored by the Plains Indians. European records of the earliest contact with eastern Native Americans and archaeological findings indicate that the eastern peoples used bows of over sixty inches long. In use at least as early as the eleventh century CE, bows of Osage orange and other such woods (pecan, sycamore, and black locust) were part of the warrior equipage east of the Mississippi River. Shorter bows were common among the plains and western peoples, and became very popular after the Spanish introduced horses to North America. That they are more widely known among people in general than the longbow of the eastern peoples has more to do with depictions of Native Americans in film than it does with historical evidence.

The best bow in the world is not much use if arrows are not straight. In the modern, industrial era, the obvious tool for making straight wooden arrow shafts is the lathe. But the weapon predates the tool so there must have been other means of straightening shafts. The most obvious solution was rare—a straight tree branch. It is more likely that people shaped shafts using heat, edge tools, and moisture. Once shaped and dried, arrow shafts in North America were finished with a two-part sandstone cylinder with a straight concave depression carved in the matching pieces of stone. The shaft was rotated in the form until it was smooth and straight.

The Lewis and Clark exploring party of 1804–6 encountered western Indians who used compound arrows, in which a long shaft was fitted into and sinew-bound to a short piece onto which the point was attached. The advantage of this arrangement was that, after the initial extra work of joining, the long part of the arrow could be reused if the pointed end had been damaged. In addition, since the main shaft was often made of a softer wood such as pine or red cedar, the two-part construction allowed for the use of a harder wood for a foreshaft that could better stand up to impact. A great many of the surviving bows and arrows used by Native Americans are highly decorated with geometric patterns; some have carved references to spirits and deities. The intricate work of the weapons suggests that these peoples attached philosophical and religious connotations to archery, hunting, and warfare, as did archers in Persia, Turkey, China, and Japan.

Superseded in warfare in the industrialized world by the sixteenth century, and somewhat later among peoples in less industrialized or technologically advanced areas, archery became a popular sport in the West among middle-class and wealthy people, and especially for women in the middle and late nineteenth century. [FIGS. 97, 98] Archery is now a sport and method of hunting for a few, and it is often contested with

FIG. 97
Dinner plate, "Archery pattern." Glazed earthenware. Hercula-neum Pottery Co., Staffordshire, England, ca. 1833–36. Part of an entire set of tablewares, this pattern attests to the popularity of archery among the bourgeoisie and the wealthy.

FIG. 98
"Drawing the Bow." Scribner's Monthly. Vol. 14, no. 3 (July 1877):273. This plate from an article in a popular magazine concentrates on bow hunting, with only pass-ing reference to target shooting. Even as firearms became more deadly and easier to use (after cased cartridges were created), bow hunting remained a modestly popular pur-suit, both for its greater challenges and for its association with Native American warriors.

fiberglass, titanium alloy, and other manmade composition bows. Complex sights and other high-technology gear enhance the performance possibilities of the bow and high-tech arrows. Tradition-minded bowyers who work only in wood and sinew can still be found, but they are scarce.

Crossbows

The earliest records of crossbows indicate that they were first employed in China in about 300–400 BCE. Within about 750 years crossbows were depicted in Roman carvings, and they are noted by the Roman historian Vegetius in 385 CE.[10] The Normans introduced the weapon to England during their eleventh-century conquest, and it became a weapon of choice in part because of its great range, up to five hundred yards. Essentially a short bow and string attached to a mechanism roughly akin to a gunstock, the crossbow could be operated from behind cover and at a kneeling position, unlike the long bow. The bolt (akin to an arrow, but shorter) was let loose by means of a trigger mechanism. The force of the discharge was so great that only dense and hard woods such as oak would work for the stock, while strong and flexible woods such as yew were appropriate for the bow part of the weapon. [FIG. 99]

FIG. 99
"Crossbow Finished." Ralph Payne-Gallwey, The Book of the Crossbow, *118.*

Crossbows had the bow's advantage of silence, which made stealth much easier, as well as an accuracy that powder and shot did not have. Their great power, however, came with a price. To get the string pulled back and locked in a stretched position required great strength or the use of auxiliary tools. One of the earliest such implements was a stirrup attached to the front of the stock, below the piece on which the bolt rested. The crossbowman placed one foot in the stirrup and drew the string back, using his leg as a brace. Some crossbowmen used a hook on their belt, to which they attached the string and then pushed the bow toward the ground by pressing on the stirrup. Others used more sophisticated apparatus, including separate cranks and wheels applied to the crossbow, integral screw mechanisms that brought the string back, or a device called a goat's foot lever. [FIG. 100] The latter was a curved iron lever attached to the stock behind the string hook that when pushed forward engaged the slack string with a double hook. Pulled back toward the crossbowman, it used the increased mechanical advantage of the lever to bring the string into a locked position.[11]

For all its power, the crossbow did not provide an overwhelming advantage in battle. Its long reload time meant an accomplished archer

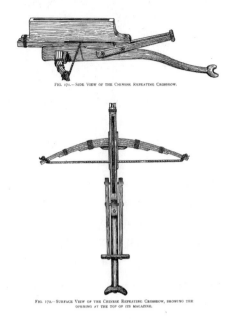

FIG. 171.—SIDE VIEW OF THE CHINESE REPEATING CROSSBOW.

FIG. 172.—SURFACE VIEW OF THE CHINESE REPEATING CROSSBOW, SHOWING THE OPENING AT THE TOP OF ITS MAGAZINE.

FIG. IOI
Chinese Repeating Crossbow.
Ralph Payne-Gallwey, The Book
of the Crossbow, *240.*

could launch several arrows in the time it took a crossbowman to launch his first and second bolt. The goat's foot lever ameliorated this problem somewhat, but too late to counteract the influence of gunnery. By the sixteenth century it was obsolete, at least in Europe. Chinese warriors continued to use the crossbow well into the nineteenth century, employing a repeating crossbow (the bolts stacked in a magazine directly above the groove in the stock) in the Sino-Japanese War (1894–95), which they lost. [FIG. 101] Crossbows now find use in sport and in discharging drugged darts to temporarily disable animals in the wild.

Some crossbows were designed to propel bullets. Instead of a straight trough into which the bolt was set, a small pouch was attached to two small U-shaped pieces attached to the release mechanism. Essentially a slingshot mounted on a gunstock, the first such weapons propelled stones and were used for hunting small game and birds as early as 1500.

More sophisticated ammunition and lever mechanisms were applied to the bullet crossbow by the nineteenth century, and it remains an outré sporting instrument to this day.

Gunstocks and Bullets

The bullet-shooting crossbow probably derived from the handheld gun, which employed a wooden stock and metal barrel to contain the power of the gunpowder charge while sending the ball or load of metal detritus on its way. This progression inverts the trajectory of technology we have come to expect, perhaps because the gun was in some ways a less effective killing machine than the longbow. The technology for casting the perfect spheres needed for straight flight was primitive at best until the nineteenth century and, in any case, barrels with a smooth bore tended to launch balls with irregular or little spin. This lack of movement on the ball produced a flight that wavered and fluttered, rather like a baseball pitcher's knuckleball. Bullets that were not perfect spheres—and usually they were not—added to the deflection from a straight trajectory.

More precise molding of ammunition and cutting regularly spaced spiral grooves on the inside of a barrel—called rifling—changed this, dramatically increasing the accuracy of firearms. The spiral groove introduced a reliable spinning motion to the ball or, better yet, a pointed bullet when fired.[12] Such firearms of great accuracy and range dramatically increased the scale of slaughter in warfare. Generals seem not to have recognized this change, and continued to use tactics appropriate to the older firearms technology long after it was obsolete, sacrificing thousands of soldiers in the American Civil War and other conflicts, as they sent them charging into volleys of bullets.

Wood enters this picture as the material gunsmiths usually used for the firearm's support structure. Gunstocks, except for some of the fanciest weapons of the wealthy, were almost always shaped of a hardwood that could take the succession of small explosive shocks without cracking or breaking. Because wood was lightweight compared to most available metals, and because it did not conduct the heat that constant firing generated, it was the perfect material for the stock. For both decorative and practical purposes the best wood for stocks incorporated complex grain, thought to be a visible manifestation of the wood's strength. The shaped stock, with a concave butt end, was developed to facilitate firing from the shoulder while eying the target down the line of the barrel. For centuries gunstocks were shaped by hand, with drawknives and curved shaves, but by the nineteenth century lathes were developed that could follow a pattern or template and turn out the irregularly shaped stocks. All sorts of exotic woods have been employed for gunstocks, including black walnut, mango, mahogany, Australian coachwood (*Ceratopetalum apetalum*) and, in the United States, curly or tiger maple, figuring that gives the gunstock a luminous appearance that competes with and complements the blued shine of the well-maintained barrel.

Now plastic and other composition materials are used for stocks on many less expensive modern shoulder arms, but wood continues to be in demand. Using real wood for the stock carries with it a sense of closeness or identity with the outdoors for hunters, and a romantic attachment to real or imagined heroes of the past for other gun owners and some sportsmen. Embedded in the wood gunstock is the idea of the frontier, however bowdlerized it has been by television, film, literary fiction, and history. It comprises a powerful enough set of images and myths that many plastic stocks are still colored to look like wood.

Sticks and Stones, Bats and Balls

The connections between warfare and the sporting aspects of hunting seem clear enough. Marksmanship is critical in both pursuits, and in some instances the techniques of stalking cross over as well. Killing game for the sport of it, rather than for food, seems a little too distant from necessity and a little too close to war—or murder—for some. But many of the nonlethal games, competitions, and sports in which people have engaged for centuries were, if not directly derived from warfare, encouraged as a preparation for it by virtue of enhancing physical dexterity and fitness. Such activities also were supported for their potential

FIG. 102

Old-time game sticks. Painted ash softball bat, hickory shafted golf clubs, laminated squash racket.

to develop less tangible personal qualities, such as teamwork, courage, leadership, and ability to function while in pain.

Ball games fall into two basic groups: those in which some tool is used to send the ball (or puck) flying—cricket, field and ice hockey, shinty, hurling, lacrosse, baseball, golf, tennis—and those in which the body alone propels the ball—soccer, American football, volleyball, basketball, water polo, and bowling games such as bocce, boules, and skittles. In the former group the tool used to hit the ball is (or was) made of wood, and is used much as the primitive cudgel or club was used in warfare. The ball in most of these stick games probably began as a stone, then was made of a hard wood. In some cases the wooden ball gave way to a leather version with a cork (cricket), feather (golf), or bran (tennis) core. In a few of the games in which people propel the ball, wooden balls have remained in play, although in most such games wood has been super-seded by rubber, plastics, or other manmade substances. The same gen-eral point holds true for the stick—once entirely made of wood, it is now composed of steel, titanium, plastics, or some other material that increases player performance. [FIG. 102]

The oldest games involving hitting a ball with a stick predate the Christian era. *Camán* (the Irish word for *hurley,* the stick used in hurling), is mentioned in a description of the Battle of Moytura, which occurred in the thirteenth century BCE. The object of hurling is to hit a ball between two goalposts with a curved stick, traditionally made of ash. Shinty (*iomain* and *camanachd* in Scots Gaelic) probably came to Scotland when Irish missionaries arrived. It too involves trying to propel a ball into a goal by means of hitting it with a curved stick, though the shinty stick does not have the flat blade of the hurley, on which players are allowed to carry the ball if they can balance it on the blade while running. The curve in both arose from the obvious need to provide an instrument that a person could wield while running or standing that simultaneously provided a section parallel to the ground with which to whack the ball or, in the case of hurling, pick it up off the turf.

The ideal stick for these games and their offspring, hockey and golf, was a single shaft, bent by nature or the clever hand into the shape needed to hit the ball. A naturally curved stick does not expose end grain at the bend, and it is likely that the earliest players quickly figured out that by cutting and shaping the butt end of a log, where the root system begins to show, they could make an ideal stick. Ash has both the light weight and great strength needed and is easier to work and shape than other woods of similar physical chracteristics. According to the Irish Department of Communications, Marine and Natural Resources,

Ash timber, when grown quickly, is strong and flexible with a good capacity for shock absorbency. For this reason ash has been traditionally used in Ireland for the production of hurleys. Hurleys are manufactured from the butt log (bottom 1.5 metre of the stem) and from trees of a diameter at breast height of approximately 30 centimetres. Only fast grown, straight and branch free ash can be used for this purpose. The same shock absorbing qualities make ash suitable for other sports equipment and tool handles.[13]

While sticks are probably sawed from butt logs in modern manufacturing, it is likely that the traditional ash stick was riven from a log around the radius, in effect quartering out large pieces, much like traditional builders had done when making shingles or even clapboards.

The commitment to a wood species over generations of play can reveal the depths of national commitment to certain games. A rumble of concern about the use of non-Irish ash in hurley production recently became a small-scale cause célèbre in Eire. On September 13, 2002, the *Irish Examiner* revealed that much of the white ash used in the hurley business was in fact imported from Scandinavia and Wales.

> [T]he RDS [Royal Dublin Society] has now revealed that most of the ash timber used in the manufacture of hurleys is imported from Wales and Scandinavia. For the 2002 RDS Irish Forestry and Wood Awards, now in their 15th year, will feature a new special award for hurley ash plantations and forests.
>
> It will cater for ash plantations that are being managed specifically for the manufacture of hurleys. The aim is to encourage the planting and cultivation of ash for the specific use of hurley manufacturing, reducing imports in the process.[14]

Similar articles appeared the following day in the *Limerick Leader* and the *Irish Farmers Journal*. Previous to this exposé, there had been debate about Irish ash supplies in the Irish Parliament, the Dáil Éireann, in June 1975, April 1976, and November 1983, at which times the government was encouraged to aid in the expanded planting and nurturing of ash forests dedicated to the manufacturing of hurleys.[15]

Wood in this instance had deep nationalist associations. Whether Welsh or Scandinavian ash made poorer hurleys was not necessarily the issue; many Irish citizens devoted to the national game were convinced that the quality of their national life and the virtues of the national identity were compromised by imported wood. Thus when the Irish government allocated 20 million pounds to support the development of plastic hurleys in 1998, a small-scale revolt began with the founding of a resistance group, the Irish Guild of Ash Hurley Makers.

Cricket, the stick-and-ball game most closely associated with England, probably emerged from the same stone and stick country games as hurling and shinty, with the added influence of skittle games, in which wooden pins or sticks are knocked down in the course of play. One clue to its origins lies in the language of the game, since a *wicket* is part of a sheepfold fence. Defending the wicket gate of the fence with a stick or bat from the thrown stone or hard piece of wood (perhaps a burl or even a nut) was probably the first manifestation of the game, and the premium placed on driving the missile beyond a boundary or out of the reach of other shepherds probably came later, as players noted that some could drive it farther and more consistently than others. Other linguistic clues to the game's origins are the French *criquet* (club), the Flemish *krickstoel* (a low stool knelt upon in church, vaguely resembling a large wicket) and *krick* (stick), and the old English *cricc* and *cryce* (staff).

Pictorial and written references to a game resembling cricket can be dated to about 1300, when Prince Edward (son of Edward I) was alleged to have played a game called *creag*. The first reliably documented record of a cricket match is that of the contest played at Coxheath, in Kent, in 1646. Bowlers at this time propelled the ball underhand to batsmen who swung their sticks somewhat like they did golf clubs or other game sticks. It does not take a genius to figure out that a flat, wider surface works better for defending the wicket and for whacking the ball. The first county match took place in 1709, and such competitions and the accompanying gambling and carousing helped spur inventive players to increase their advantage by refining their equipment. In 1743 one Shock White's use of a bat as wide as the wicket prompted the beginning of codified cricket rules, adopted as the Code of 1744. Great local and regional variation in rules doubtless resulted from these early attempts, but after the Marleybone Cricket Club published its version, *The Laws of Cricket,* in 1788, its standards were adopted throughout much of England. This was perhaps as a result of the club's elite position and perhaps to make more rational the betting odds among the wagering classes, though the latter explanation seems to presume rather more honor among gamblers than was likely.

Players rapidly adopted the standard of a four and one-quarter inch bat, and its shape evolved into the modern, slightly curved white willow (*Salix alba caerulea*) blade and cane handle. Like Irish ash, French cask oak, or fiddleback sycamore, this wood is purpose-grown for cricket bats. Once felled, the close-grained willow is cut to length, cleft from the pith outward, and dried for at least a year before being shaped into blades. These are attached with the now-traditional V-splice to a cane and rubber-strip layered handle that is held together by tightly wound twine covered with a final layer of rubber. Unlike sticks in most other sports, nearly all cricket bats today are still made of white willow.

How does practicality become laden with tradition? When does technology in the form of new materials supersede it, even bury it? Cricketers appear to have resisted the temptations of non-wooden bats. In American baseball, this is hardly the case. Most wooden bats today are made of northern white ash, associated by older players with two companies, one located in Dolgeville, New York, at the southern border of the Adirondack Mountains, and the other in Louisville, Kentucky. Early baseball players used bats turned from a variety of different woods, including maple, oak, hickory, and elm, reflecting both the widespread and regional nature of the game and the availability of lathes. Ash's physical characteristics and ready availability (and perhaps its limited use in the furniture and building industries) helped it become the standard. It is still that in the highest levels of professional baseball (although hard maple is making inroads), but nowhere else in the game, or its variant, softball. There aluminum superseded wood, beginning in the 1970s. Aluminum bats did not break or splinter, thereby eliminating the danger from a freak accident. More important, that they did not break meant they lasted far longer in the hands of players, though they cost more per bat for exactly that reason.

They also changed the game. Well-struck balls flew farther off aluminum bats than their wooden counterparts, especially those hit slightly off-center. Pitched balls hit off the "handle" of an aluminum bat (near the hands) still flew far enough in many instances to become safe hits. Pitchers could no longer depend on getting hitters out by throwing fastballs on the inside part of the plate and grew up relying on pitching outside, or "away" from the hitter. At the professional baseball level, where wooden bats are mandatory, the inside pitch is still a viable weapon, so prospective major league pitchers find that they have to re-orient their strategies, and scouting of college and high school players must take into account "aluminum" hits.

The rules makers of professional baseball have deemed that wood is the only acceptable material for bats in order to protect existing playing fields and their surrounding accommodations for spectators. Allowing aluminum bats would make existing baseball stadiums obsolete, since the distance hitters could drive the baseball would increase dramatically. But the fences over which hitters must hit home runs cannot be moved back at will. Spectator seating rings most ballparks, and most stadiums are surrounded by urban or suburban real estate development and there-fore cannot gobble up more land for expansion, nor would it make eco-nomic sense to do so. Moreover, the dimensions of the field, especially the infield in which the bases themselves are placed, would prove insuf-ficient to prevent increased player injury from batted balls traveling faster than those struck by wooden bats.[16]

Regardless of the efforts of traditional wooden sporting goods makers, and heedless of their warnings that traditional games will be compro-mised, makers and consumers of sporting goods have leaped with little reflection or remorse into the business of making and buying athletic sticks made of new materials. For the most part these have succeeded because players regard the new materials as superior performers in the field of play. Lacrosse sticks, once made only of bent hardwoods and sinew or gut, are now made of metals, usually aluminum or some special titanium-laced alloy. Field hockey sticks are similarly now made of manufactured materials, as are tennis rackets, first made of bentwood, then laminated bentwood, and now metal, with plastic rather than gut stringing. All the racket sports have followed a similar pattern, allowing a player to swing a lighter racket faster with no fear of breaking it. Bigger and stronger players with bigger and faster rackets now show spectators serves of over 125 miles per hour at tennis's highest levels, though the serve is for the most part a blur in real time, and only really visible to spectators because of the slow-motion replays. Ice hockey sticks are also increasingly made of plastics; the puck flies faster, goalies wear bigger pads, and all must wear helmets, further encouraging the more violent of the players to wield the stick as a weapon.

There is no more obvious example of the alteration of a game than in golf, where technological innovation in shaft and head materials has altered the calculus of the game and the dimensions of golf courses. Originally, the game began in Scotland, along the shores and among the grass-covered dunes of fertile low-lying coastal plains called machair (pronounced "mocker"). Sheep kept the grass clipped close to the ground, and the shore winds eroded rabbit holes and sheep-made shelters on the lee side of hills into wide stretches of sand.[17]

As the game evolved and iron heads were attached to wooden shafts for the various shots that had to be played on the irregular turf, players settled on hickory for the shaft. Its light weight, great strength, and flexibility allowed players to "whip" the club head through the hitting area, thereby sending the ball great distances. The advantages of the flex and torque of the hickory shaft carried with it the disadvantages of increased shaft and clubhead deflection and consequent dispersion of the shot when it was ill struck. Alternatives to the hickory shaft were tried in the late nineteenth and early twentieth centuries; alternating strips of bamboo and hickory, coopered along the entire shaft length, were briefly popular, perhaps a derivation of the composite bamboo bows made in China centuries earlier or perhaps derived from the fishing rod industry.

Steel shafts ended the reign of the hickory club forever. In the 1930s many of the first steel shafts were covered with a yellow coating "grained" to look like hickory. But in a few years that concession to the old days was gone as well. By the 1960s fiberglass challenged steel shafts, but these took some time to catch on with some players because they were black (rather than shiny and metallic) and because their light weight made them harder for some players to control. By the 1980s materials for golf club shafts were even more diverse, as graphite composition materials and other plastics competed with steel.

Perhaps more important for the nature of the game was the evolution of the wooden headed clubs intended for the longest shots, particularly the drive from the tee. Originally "woods" were just that—the heads made of a hardwood, often backed with a lead insert and a horn piece inset in the sole of the club (the part that rested on the ground). In the United States persimmon (*Diospyros virginiana*), a wood native to the southeastern region of the country, became the standard material for wooden-headed clubs, owing to its great strength and workability. Laminated maple competed with persimmon by the post–World War II era, promising the strength of end grain without the danger of splitting. With steel shafts and new heads, players drove the ball farther and straighter, and the case for technology as the improver of the game for the professional and the amateur was won, with only the occasional scold warning that the game would be forever changed from its ancestral roots.

Metal heads—some of gigantic proportions—have entirely replaced those of wood, relegating them to the collectors' areas of sporting goods stores, antique shops, and online auction houses. Titanium heads launch the ball so far that professionals now hit the technologically improved golf ball distances people only dreamed of in the not-too-distant past. Players are stronger and more athletic, and the combination of these factors has transformed the game utterly. Older golf courses, which were even in their infancy consumers of a considerable amount of real estate, and maintained at considerable expense, are now easy marks for the accomplished player. New courses are built to greater lengths than ever before, often as much as fifteen hundred to two thousand yards longer than their predecessors. Woods are now called "metal woods" (a strange misnomer) and increasingly, simply "metals." The nomenclature is not surprising; it indicates some guilty discomfiture, although critics might do well to remember that the irons are no longer made of iron, but steel, titanium, and other materials.[18]

The transition from the hickory shafts and wooden heads of the long-distance clubs is not merely one of reducing the effective length of golf courses. For most players, lightweight and more flexible hickory shafts mandated a different swing—slower and longer, with less torque applied to the player's back and legs during the backswing. Older courses, designed for the distance that players could hit the ball with these clubs, generally placed their emphasis on flatter and slower greens and on the player's ability to manage the bounces on hard fairways that were more akin to mowed native grasslands than green carpets of dense grass that needs daily watering and mowing and copious fertilizing.

Modern nonwooden clubs make golf an air game rather than a ground game, far from its roots in the fields or on seaside courses. The qualifier to this characterization of modern courses is putting, which was once akin to billiards, croquet, and bowls, but is now a contest of wits with the designer of the humps, hollows, and slopes of the closely shaved grasses and lightning fast surfaces. Courses that do not sufficiently lengthen holes to keep up with the acceleration of distance the new clubs (and balls) provide often turn to longer carries over hazards and tricked-up greens so slick the surface is closer to polished stone than grass.

In the end the cries that new materials in lieu of wood compromise the game are overwhelmed by players' delight in the greater lengths they can hit the ball and the longer life of the metal or plastic stick. Moreover, the "tradition" invoked as an argument against changing the material of the stick or club is itself often a product of late nineteenth century Romantic nationalism, rather than of an unbroken line reaching back ages. A closer look at hurling, shinty, lacrosse, and many other ancient games reveals that they fell out of favor, even among their acknowledged founders, until the folk and craft revivals of the nineteenth century. Even golf, which was continuously played in Scotland from at least the seventeenth century onward, only caught on in the United

States and much of the rest of the British Isles in the late nineteenth century. The "revival" of "traditional" sports, music, crafts, and cuisine characterized and underpinned independence movements in Ireland, Scotland, Finland, and many of the national groups straining under the yoke of the Austro-Hungarian, Russian, German, and Ottoman empires.[19] Folklorists, such as Elias Lönnrot in Finland, collected ballads and tales handed down through oral tradition; Sweden's Artur Hazelius directed the building of the first outdoor folk museum, Skansen, in Stockholm. Composers sought inspiration in folk dances and folk music. Traditions were recaptured, sometimes rethought, and sometimes invented in the broad-based enthusiasm for the ordinary folk, their history, and their sport.[20] Nonetheless, dramatically altering the calculus of an old game—whatever its roots and traditions—by means of technology renders transgenerational customs, mores, and comparisons at best problematic and at worst impossible or meaningless, further alienating generations and reducing the pool of experiences that they might have in common.

The Racket in the Snow

Indigenous peoples in the polar and wintry regions of the world developed a panoply of snowshoe forms to make their way through the variety of snow surfaces of a winter season. Those unfamiliar with human history and the artifacts of alien cultures might well link the tennis, squash, or badminton racket in its wood-and-gut form with this common and ancient tool for walking on snow. Some snowshoes bear a striking resemblance to early sport rackets, but it is unlikely that northern-based Native Americans, the Araucanian peoples of Andean Argentina, the peoples of northern Russia and Siberia, or the Sami of Scandinavia played racket sports. If they played ball games at all, some form of ancient stickball was the likely version of their game.

Snowshoes are ancient tools, at least six thousand years in use, as are their kin, skis. Whereas skis are meant to slide over snow, snowshoes are clearly intended for those conditions of deep snow in which walking is at best a trial and at worst nearly impossible. The four most common shapes of snowshoes made by North American indigenous peoples are the Maine, Ojibwa, Alaska, and bearpaw. The Maine (also called the Michigan) shoe most resembles a tennis racket, the long tail piece serving to keep the shoe in line with the foot while walking. The Ojibwa has the Maine shoe tail and an upturned nose, the better to slip over exposed roots and other brush in the woods. The Alaska shoe has a broad oval form with a tail, useful characteristics for tramping in the cold and often fluffy snow of the Far North. The bearpaw is a shorter shoe, but wide, designed for close environs in the woods.

Making snowshoes required the skills of the leather worker, the weaver, and the woodworker. Most snowshoes were made by wetting and heating a pliable species such as pine and then bending it around a frame, clamping it until it had dried and cooled, and then binding the ends together at the base, or tail. Weavers then constructed a mesh of gut or leather and attached a thong or harness to link the shoe to the wearer's boots. Long narrow snowshoes, such as the Ojibwa form, were often made of two sections of wood, bent around framing stretchers and joined at both front and tail ends.

Snowshoeing as sport or competition probably originated with Native Americans. The earliest written records of competitive snowshoeing, however, are those of the Montreal Snow Shoe Club, an organization founded by English émigrés in 1843. Previous to this formal organization, well-to-do members of the Montreal English community engaged in informal "tramping" on snowshoes, having learned the skill from the native peoples and French-speaking Quebecois.[21] Organized winter snowshoeing events continued throughout the nineteenth century in Canada, although the organizations soon began to exclude native peoples from "open" competitions because they continually outraced the immigrants. By the late nineteenth century tramping was somewhat popular in the United States, especially in the western mountain states and to some extent in New England. In recent years snowshoeing has undergone something of a revival in the United States, particularly since

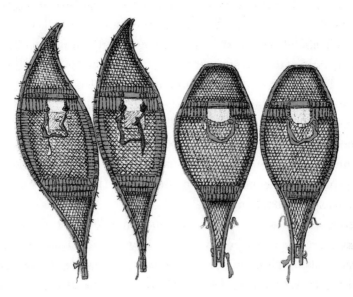

FIG. 103

Ojibwa and bearpaw snowshoes. Plate from George Catlin, The Manners, Customs and Conditions of the North American Indians, *2 vols. (London: by the author, 1841).*

aluminum-tube snowshoes with teeth on the base have been introduced. They are easier to manage in the snow than the longer and wider traditional wood-frame shoes, which are now for the most part consigned to the walls of lodges and bars as decoration.[22] [FIG. 103]

Skis are about as ancient as snowshoes, dating back at least 4,500 years.[23] The earliest skis were often fur-covered planks or boards, the grain of the fur providing both a slicker surface for gliding and resistance for skiers trying to proceed uphill. Skis probably first emerged among the Arctic peoples of Siberia and Fennoscandia, and experts have identified regional differences in the method by which the binding thongs were attached.

Skis operate on an entirely different principle of locomotion. While not particularly useful in deep powdery snow, their length makes them a fast and efficient mode of transport over more dense frozen surfaces.

The glide of skis comes from a combination of factors. The skier's weight compresses the snow slightly, even melting a microscopic layer, which provides a small amount of lubrication. Forward pressure on the skis or gravity propels the skier forward. The upturned end of the ski (like that of the Ojibwa snowshoe) helps prevent the front from digging into the snow and facilitates the glide. Early skiers eventually figured out that smoothing the bottom of the ski with scraping implements further aided locomotion, as did the application of lubricants such as animal fats and, ultimately, waxes. Just about any strong and flexible wood makes good skis, and in Europe and North America, ash and hickory were the choice for many ski makers and users.

Both snowshoes and skis played important roles in the history of war. During the Seven Years' War (1757–63), French and Indian forces soundly defeated the English near Fort Ticonderoga in a battle that has come to be known as the Battle of the Snowshoes, because the undermanned and underestimated French and Indians wore them. Snowshoes have since been used in nearly every war in which battles or reconnaissance took place on the ground during the winter. During World War II Canada's Hudson's Bay Company produced over seven thousand pairs of snowshoes for Allied use in northern Europe. Snowshoers also fought in the Italian campaign and the Finnish Winter War of 1939–40.

Skiers have had a more storied history during warfare. Russian troops, themselves no strangers to war on skis, encountered relentless resistance from white-clad Finnish forces on skis after Stalin blundered into the Winter War.[24] Officially created in 1945, the United States Army's Tenth Mountain Division fought in Italy. The idea for a division of skiers came to several members of the National Ski Patrol Association several years before, when they took note of the "phenomenal job the Finns were doing on the Karelian Isthmus in crucifying the Russians. A perfect example of men fighting in an environment with which they were entirely at home and for which they were trained."[25] Soon after the formal entry of the United States into World War II, the Eighty-seventh Mountain Infantry also was formed to train soldiers for mountainous and winter weather campaigns. As George F. Earle reported in his *History of the 97th Mountain Infantry in Italy* (1945),

> In January of 1942 all manner of ski and winter equipment was issued, and ski calisthenics were given in the mud of Fort Lewis, Washington. Late in the month, Companies A, B, and C moved up to Tatoosh and Paradise Lodges on Mount Rainier for ski training. . . .
>
> At this time, the unit was made up of world-famous skiers, mountaineers, forest rangers and trappers, lumbermen and guides, and a group of cowboys, muleskinners and horsemen, in addition to the regular army cadre.[26]

Eventually these troops participated in the Italian campaign, fighting their way through the mountains of northern Italy south of Bologna and north of Verona.

There is little evidence that skiing was anything but a practical activity until the nineteenth century, although the Norse sagas of about 1000 CE make note in passing of recreational skiing. Competitive and recreational speed skiing began in early nineteenth-century Norway, especially in the Telemark region north of Oslo. Telemark skiers discarded the more ancient "pole riding" method of skiing—implanting a pole in the snow and turning—for a new method in which they turned by working the skis alone. Sondre Norheim, a farmer from Telemark, is credited with inventing a new binding in about 1850 that made the style easier. By using flexible birch roots to link the heel to the toe binding, skiers were able to turn by shifting their weight without losing control of the skis. In 1868 Norheim and two other Telemark skiers surprised spectators and other participants at a skiing festival in Oslo (then called Christiania) with their sweeping turns and quick stops. Norheim refined the shape of the downhill racing ski, paring away the width in the center, an improvement that still shows in contemporary skis.[27]

Skiing became an integral part of Norwegian national identity and an important cultural element in the Norwegian quest for independence from Sweden in the latter nineteenth century. Norheim himself left his farm for the plains of North Dakota in 1884, but continued to be venerated as a Norwegian pioneer and nationalist. By that time Norway was exporting both single-plank and laminated skis (pine over ash), which had been developed in 1881 to meet the demands of the sport's growth. The first factory for making skis in the Telemark pattern opened in Norway in 1886. Telemark and Norwegian skiing got further boosts from the exploits of skiing explorers such as Fridjtof Nansen, who was the first to cross the mid-Greenland glacier in 1888 and the polar ice cap in 1895. (Nansen used sledges and kayaks as well as skis to negotiate the eighteen-month trip.) Perhaps more important for popularizing skiing internationally was Roald Amundsen's successful trip to the South Pole on dogsleds and skis in 1911, in a race in which his competitors, the English explorer Robert F. Scott and his team, perished.

Skiing on the western mountain slopes of the United States and Canada took a different form, somewhat in keeping with the wilder environment of mining and frontier life in those parts. Norwegian and other Scandinavian immigrants to the West pioneered long-board skiing, in which racers strapped on skis as long as twelve feet, which they then rode down slopes at speeds that were reported to approach ninety miles per hour. In the Alpine sections of Europe skiing was grafted onto an already thriving resort life that revolved around "taking" mineral waters and breathing the clearer air of the mountains. Herein was born the swanky and expensive version of what began as an economically accessible and athletic activity.[28]

By the late 1940s aluminum skis had reached the market, but did not sell well. In 1947 Howard Head's aluminum-over-plywood skis did succeed, since they enabled the skier to turn with much greater ease than did wooden skis.[29] Plastic ski bottoms ended the chore of waxing and rewaxing, and in the 1950s fiberglass skis, pioneered by the European makers Rossignol and Kneissl, were gaining a share of the market. Traditionalists for a time hung onto wooden skis, particularly those who participated in cross-country, or Nordic, skiing. The latter was an activity that many of its strongest adherents thought separated the rich and crass from the "genuine" and the "authentic," and it became the favored skiing sport of the "counter-culture" in the 1970s. Some of that edge has worn away from Nordic skiing, but the bushwhackers, back-country gliders, and even the resort cross-country devotees still seem of a culture different from that of the snowboarders and the Alpine skiers.

The Racket on the Lawn

Tennis—and most of the other racket games—probably evolved from ancient games played across an arc of the globe that stretched from North Africa to India and China and Japan. Exploration, trade, and empire-building brought the games and the goods to Europe. In the West, historians locate the origins of tennis in France as early as the ninth century, when monks adapted a Moorish game in which they tried to hit a ball to each other using their hands or a stick of some kind. By about the fifteenth century, that activity had been superseded by an indoor game, a hybrid of modern tennis and squash played off walls and over a net with a wood-and-gut racket. It spread quickly among the European wealthy by the sixteenth century, but declined in popularity until, by 1800, it was a quaint anachronism. Within seventy-five years, however, it was again popular, resurrected as lawn tennis. It may be derived from the Indian game of poona, a relative of badminton, and from an ancient game played in China, Japan, and India for around two thousand years—battledore and shuttlecock.[30]

The lacrosse stick in a sense links the snowshoe and racket with the trebuchet. The webbed end of a lacrosse stick is slack, in order to catch, carry, and sling the ball, rather than hitting it back at an opponent. Originating among Native American peoples, lacrosse became popular in North America, and especially in Canada, where it is still one of the nation's official national games.[31] Hickory is the preferred wood for making lacrosse sticks in the traditional manner because of its strength and flexibility. After shaving and smoothing dried hickory, traditional makers (most of whom are Native Americans) steam-bend one end of the stick into a shape resembling a shepherd's crook. (One form, called the Great Lakes stick, is often made of ash and bent completely back on the long shaft, to which the end is attached and bound.)[32] As in racket sports, wooden sticks are now rare among most organized lacrosse players, who favor the lighter-weight aluminum or titanium gear.[33]

Casting upon the Water

Fishing involves several means and devices to get the slippery creatures out of their element and into human cuisine. Traps and nets rely on the fish being swept into them by currents or on the fish's meandering into the mechanisms of their doom. The advantage of these methods is that these devices sometimes catch many fish at once, and in many instances no bait or lure is needed. As a practical matter these are the most desirable modes for catching fish for those who must have them for food. What they lack for sport fishermen is the trickery and the thrill of outwitting the fish, based on knowing its habits, and the challenge of finding a lure or bait that will fool the creature into biting the hook, battling to get free, and ultimately being landed.[34]

The fishing pole also has the advantage over the net and trap in that it is easier to make, at least in its simplest form—a stick and a string, to which a hook and bait are attached. The earliest fishing lines were probably woven tendrils or vines thick enough to withstand a strike. Hooks of the same vintage were fashioned from bones. Anglers using pole and line are depicted in Egyptian art of about 2000 BCE, and the Greek philosophers Plato, Aristotle, and Plutarch note fishing as a pastime as well as a method of gathering food. Dame Juliana Berners's *A Treatise on Fysshynge with an Angle* (1496), part of her *Booke of Saint Albains,* is probably the first work that provides details on fishing for sport along with instruction on making rods, lures, flies, tying knots, and information on the ways to study the eating habits of the quarry. Berners's work both formed the basis of fishing knowledge and established parameters for instruction books to follow.[35] The most famous of those books is Izaak Walton's *The Compleat Angler, or the Contemplative Man's Recreation* (1653). Part philosophical treatise, part instruction manual on the technology and techniques of fishing, and part natural history and science tract, *The Compleat Angler* remains the most important book on sport fishing ever published.

While it is possible that ancient fishermen used carved hardwoods for hooks, the wooden element most important in angling remains the pole. Usually made of a flexible yet strong wood that would resist fracture, such as ash, alder, hazel, willow, or yew, serious European sport fishermen's rods from as early as the seventeenth century were approximately fifteen to eighteen feet long. The line (eventually made of braided horsehair) was generally about twice that length, since there were no reels until the latter decades of the eighteenth century.[36]

By the nineteenth century reels and line were mass-produced and fly fishing had become a popular pastime. Many sporting anglers of means bought lures and flies from artisans and manufacturers, rather than making them themselves. Running rings, through which the line is threaded along the rod, had been in sporadic use for more than a century, and became standard equipment as better methods were developed for lashing them to the rod, and reels provided the opportunity to carry and cast more line. Tackle makers experimented with species of woods other than those of the earliest rods, such as lancewood (*Oxandra lanceolata* and *Calycophyllum candidissimum,* also known as lemonwood), bulletwood (*Manilkara bidentata*), moso bamboo (*Phyllostachys edulis*), eventually settling on greenheart (*Chlorocardium rodiei*), an evergreen that can grow to a height of 125 feet and a diameter of three feet. Native to northern Brazil, Suriname, French Guiana, and Guyana, it is more than twice as strong as oak for bending purposes; that is, it can withstand great compression along the grain without breaking.[37] But greenheart is difficult to work, dulls edge tools rapidly, and is relatively scarce; rod makers therefore experimented with other species. A Pennsylvania violin maker, Samuel Phillippe, developed an alternative in 1845, using lengths of split bamboo that he glued together. Lighter than greenheart and still very flexible, glued-together bamboo ribs, carefully coopered and shaved, became the standard for fine fly casting rods, although there was considerable argument about their worth among fishermen, especially when hunting for larger fish, such as salmon.

Frank Forester's Fish and Fishing (1858) was one of the mid-nineteenth century's most popular books surveying the sport in North America. Along with a brief survey of the natural history of fish, Forester (whose real name was Henry William Herbert) offered his opinion of the ideal fishing rod: "The best rods I have ever seen were those made by country fishermen. . . . These rods were all of English ash, butts and middle pieces, and lancewood tops."[38] Genio Scott's *Fishing in American Waters* (1875) was a popular expansion of the Walton model for angling books and offered a counterpoint to Forester's opinions about rod materials, as well as some indication of the passion with which fishermen regarded the question of wood species:

For the butt, 5¾ feet of well seasoned, selected memel [oak] with the fibre of the wood running straight in the direction of the rod. . . . For the middle piece, 4⅞ feet of selected ash. For the top, 4⅞ feet of lancewood. . . .

Francis Francis [sic] states that "the best wood is unquestionably greenheart, and next to it hickory;" adding that they in the British Isles have tried bamboo, and found it a failure. . . .

I have seen the same rod [split bamboo] used in Canada, where it was pronounced . . . the best . . . ever seen in use. . . . I frankly confess that the split bamboo is vastly . . . their superior in delivering a fly at a great distance, and retrieving the line; in playing a large fish while the angler is on the shore of a wide, rapid river, . . . The rod is twenty feet long and not more than three-fourths the weight of a greenheart or hickory of the same length.[39]

In the end, split bamboo emerged as the favored material for rods, until replaced by plastics, graphite, and titanium.

The history of the fishing rod reads like the history of golf club shafts, sporting rackets, bows, and skis, equipment in which flexibility and resistance to breakage are paramount qualities. Bows require great spring and stiffness to send an arrow great distances, while golf clubs and fishing rods need "whip" to impart greater speed to the club head or the line than could be accomplished with a stiff shaft or rod. Both varieties of lancewood were tried in the fishing rod industry, as well as by bowyers. The woods (which are biologically unrelated) ultimately proved less desirable as rod material because they were too stiff, but were successful among bow makers for whom they had just the right flex. For larger game and food fish, heavier and stiffer rods were popular since a strike could put tremendous pressure on the rod. The trade-off was the lack of precision in casting, but in many cases—especially when fishing for some species from a boat—that was incidental.

Wood also entered the angling repertoire in the form of the woven splint creel, the catch net, and a device known as a fisherman's priest. This instrument is a turned length of hardwood, usually about ten inches long, with a short cylinder of brass attached to one end. [FIG. 104]

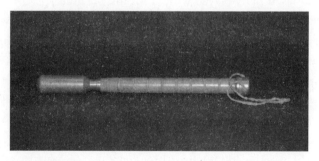

FIG. 104
Fisherman's priest. Beech and brass. England or Scotland, ca. 1900.

With it the angler could quickly dispatch a caught fish by smacking it on its head, thereby preventing the long and presumably gruesome death by suffocation that fish normally endure. Anglers seldom use the priest today, perhaps because they do not want to damage the skull of their catch, or perhaps because they do not think the fish suffers, or perhaps because they do not want to more directly participate in the death of the creature.

Bentwood and net forms seem to be logical, natural, or even mundane solutions to their tasks, but they are anything but that, since making them requires both great skill and a deep theoretical knowledge of ways to solve a complex problem—whether it is trying to catch, run with, and propel a ball, netting a fish, or making your way across deep snow. By conceiving of such artifacts as the "natural" products of people without formal education, we demean the accomplishments of indigenous peoples and cultures. By asserting an intuitive or instinctive—and therefore nonintellectual and unscientific—character to cultures other than our own, we presume that those in an industrial and postindustrial society could easily have developed these solutions. There is no evidence to suggest that this is true.

Skittles, Bowls, and Billiards

Skittles games, usually played with nine pins, are at least 750 years old. Most often played in inns in Britain and Europe, the size of the wooden pins, alleys, and balls varies from region to region. The American variant, ten-pin bowling, may have originated as a ruse to evade laws forbidding nine-pins. Bowls, boules, and bocce, games in which a ball is rolled toward a target ball, at first may have been played with rounded stones.[40] Ancient Egyptians played a game similar to bowls with stones, and the Romans eventually began using olive-wood balls. In many parts of the world, wood is still the first choice for these games.

These games may well have been the antecedents of ground billiards and croquet.[41] The latter require expanses of smooth turf and may have been upper-class alternatives to poorer people's entertainments. Ground billiards players struck their ball around the court, at a stake at one end and a hoop at the other. Fourteenth-century woodcuts depict players at the game, and in Italy it was known as *biglia,* in France as *billhard,* and in Spain as *virlota.* Both billiards and croquet appear to have come from this game, and both were originally aristocratic pastimes that achieved popularity throughout much of western Europe by the sixteenth century.[42]

Croquet was played on a greensward with mallets and balls made of lignum vitae or perhaps ironwood. Billiards most likely developed as the indoor version of the game, played with turned sticks and smaller balls on expensive wooden furniture with a flat wooden or, by the fifteenth century, slate bed covered with green cloth, presumably a reference to the outdoor origins of the pastime. A 1470 list of goods purchased by Louis XI of France includes "billiard balls and billiard table for pleasure and amusement." Mary, the ill fated queen of Scots, brought a table back from France and allegedly had it in her cell as she awaited the headsman's blade. The earliest printed explication of the rules of billiards in English is included in Charles Cotton's *The Compleat Gamester* (1674).[43]

The billart (stick or mace) and balls were made of wood, and it was especially important that the balls could be turned into smooth, perfect spheres that could withstand powerful impact. Lignum vitae was a favorite choice as the game evolved into one in which people struck, rather than pushed, the ball. Striking the cue ball with a paddle-ended stick was difficult, but hitting it with the now-familiar leather-tipped cue was much simpler and allowed players to impart spin to the ball, dramatically increasing both the number of successful shots and theatrical possibilities of the game. By the second decade of the nineteenth century, the mace had largely disappeared from billiards.

The elite origins of billiards and the aristocratic pretensions of flashy display are revealed in cue sticks. The turned stick became a visual field for the show of wealth and sophistication, through the clever combination of exotically colored woods such as purpleheart (*Peltogyne panicu-lata*), Burma padauk (*Pterocarpus macrocarpus*), ebony (*Diospyros* spp.) and goncalo alves (*Astronium fraxinifolium*), species that contrasted dramatically with finely grained white woods such as hornbeam and maple. Wooden billiard balls by the 1600s were replaced by those made of ivory and, by the latter nineteenth century, industrially manufactured composition balls.[44]

The competitive aspects of billiards and bowling games, the ample opportunity for gambling on them, and the relatively inexpensive equipment needed for some games played on a small pitch or table helped make them immensely popular—so much so that they were banned or discouraged by social critics, religious leaders, and other scolds in many parts of Europe. Monarchs thought they distracted their subjects from the more important (to them alone) practice of archery, which obviously had greater value for warfare than did pitching or hitting a ball at a target. Stern religious leaders saw gaming—and play in general—as the entering wedge of licentiousness and other godless behaviors. Secular social critics saw billiards as interfering with the development of "habits of industry" that they were certain were what the lazy and morally challenged underclasses needed. And all critics decried the gambling that took place routinely. None of it worked. Eventually the bans were lifted (first for the nobility, then for all).

The historical and cultural linkages between war and play seem odd—even bizarre—to twenty-first-century observers. They are the stuff, it would seem, of some soldiers who survived, commanders above the fray, and those who have never served but who long for the glory of victory within the confines of the stadium, court, or pitch. Fervent believers in the expanded metaphor of sport as war have the weight of at least some historical experience upon which they can draw, if unknowingly. Kings and other rulers did indeed sponsor and support certain sporting activities, seeing them as ways to keep their subjects ready for battle and distracted from the possibilities of rising in rebellion against them. The shrewdest of the lot, such as Scotland's James I, figured out that he had better let Lancashire men play at sports if he wanted them on his side against his numerous opponents. Thus in the Book of Sports (1618) he declared:

> The common and meaner sort of people . . . [had need of] such exercises and as may make their bodies more able for warre, when wee, or our successors, shall have occasion to use them. . . . That after the end of Divine Service, Our good people be not disturbed, letted or discouraged from any lawful recreation, Such as dancing, either of men or women, archery for men, leaping, vaulting, or any other such harmless recreation, nor from having May games; Whitson Ales, and Morris-Dances, and the setting up of Maypoles, and other sports therewith used.[45]

James saw a connection between sport and play in general and his need for soldiers. Succeeding rulers and functionaries have likewise appreciated the confluence of sport, physical training and military preparedness, from the overt associations such as the Victorian phrase, "Wars are won on the playing fields of Eton," to the "President's Council on Physical Fitness," organized by John F. Kennedy and headed by the University of Oklahoma football coach Charles "Bud" Wilkinson in the early 1960s. Some have equated war-planning capabilities with the strategic aspects of board games such as chess, which traditionally has "men" made of carved and turned light and dark woods such as box and ebony, an obvious military reference in the knights, and state references in its castles, kings, queens, and bishops. The most numerous and least powerful pieces are the front-line pawns, presumably those folk allowed their games and amusements so that the back line of pieces can "use" them.

Wood is no longer the material of war or play. It has been superseded by higher-technology materials that have advanced human physical capabilities without a concomitant advance in human consciousness. Plastic and titanium may have altered games utterly, but the cost-benefit relationships are still those of recreation, rather than destruction. Many sports and games may have had their origins in preparation for warfare, but the gulf between these two activities could not now be wider.

9

Fire, Smoke, and the Costs of Comfort

W ood and fire have a complex—even peculiar—relationship. Fire can destroy a forest and consume a house, turning into ash what was once alive or fashioned into practical and beautiful things. But forest fires can also be the engines of renewal in mature woods and the fires of the hearth have kept us warm and cooked our food. Smoke is both an irritant and a preservative, a choking cloud and a holy environment. Wild fires can kill us; charcoal made the first phases of the Industrial Revolution possible.

Humans probably first encountered flame by accident, the result of a lightning strike or (rarely) in lava flows. A fire out of control was a fearsome event, but a small blaze accidentally ignited that soon went to embers perhaps gave people the idea that this strange occurrence might well have uses.[1] Fire became a magical as well as essential element in people's lives, its history and nature sources of wonder for millennia. The Apache trace the advent of fire to the machinations of the fox, who stole it from the fireflies and brought it to the rest of the world by burning his tail in their fire. He was punished for this by losing any chance of using it for himself. The Cherokee credit the water spider with bringing fire from a lightning-struck hollow sycamore tree, after the raven, three owls, and two snakes failed.[2] The ancient Greeks settled on the myth of Prometheus as the fire-bringer to the humans, and other cultures saw fire as a gift from deities and properly revered them for the largesse.

In the West, fire was once considered one of the four "elements" that constituted the building blocks of matter on the planet, joining earth, air and water. This ancient idea was based on a set of binary oppositions that defined the nature of the elements. Fire was hot and dry, earth was cold and dry, air was hot and wet, and water cold and wet. Apparently based on the sense of touch (the other senses perhaps thought to be insufficiently trustworthy), this conception of the elements maintained a strong hold on both scientists and philosophers well into the medieval era.

Scientists and alchemists gradually undermined these theories by experimentation, in particular on the effects of heating metals. The former sought information about, among other things, the nature of matter, and the latter sought to understand the properties of metals so that they might change the base (lead, for example) into the precious (gold). In their analytical processes early scientists and alchemists noticed that wood and other flammables in large part disappeared when burned, and the residue of combustion seemed substantially different from the fuel. This eventually, if accidentally, focused the attention of some scientists on the basic physical character of matter and the nature of ignition and combustion. Ultimately the chemists Robert Boyle, Antoine-Laurent

Lavoisier, J. B. Richter, and others completed what the alchemists wittingly or by chance began—the collapse of the four-elements theory of matter.

Joseph Priestley's discovery of oxygen and Lavoisier's experiments identified burning as a chemical reaction of oxygen and (in the case of wood) hydrocarbons that is initiated by heat and that produces energy in the form of more heat and light. Since wood is almost entirely composed of cellulose (a complex hydrocarbon), oxygen, water, and trace amounts of minerals, when it burns the result is the combination of oxygen and carbon to form carbon dioxide and oxygen with hydrogen to form water vapor, in addition to energy in the form of heat and light.

Outside the laboratory, people learned from observation and information handed down through generations about how different woods burned. They knew that green wood burns with difficulty and produces voluminous smoke and that some woods left sticky and smelly substances when they burned. Others learned that smoke, if carefully controlled, could be used to preserve and flavor meats and fish. As the nineteenth century began, European and American inventors, amateur scientists, agricultural experts, and sylvaculturists conducted experiments to ascertain which woods performed best for heat, light, ease of combustion, and quality of smoke and ash. They discovered, among other things, that the dense forests of deciduous trees such as hickory and oak produced the most heat per standard volume than did softwoods such as pine and cedar, and more than some hardwoods that they thought would be great heat producers, such as maple.

Cords of Dependence

The United Nations Food and Agricultural Organization's *State of the World's Forests* reported that in 1992 54 percent of wood consumed internationally was burned for fuel.[3] Historically the percentage has been declining from nearly 100 percent since coal and other fuels became available in the nineteenth century. For most of human history, people living in small, widely scattered settlements in areas in which there were abundant forests gathered fuel from the woods, relying whenever possible on deadfall, especially smaller pieces that did not have to be split or that could be broken into manageable length for the fire. As towns and cities began to grow, and agriculture expanded into the hinterlands to feed the densely populated areas, firewood increasingly became a precious crop to be harvested for use in cities.

Population expansion in a region has had a historically consistent effect on regional forest resources. In China the population grew from about 70 million in 1400 to 270 million by 1770, prompting massive deforestation, both to provide firewood and to open land for agriculture. Internal dislocation and warfare notwithstanding, the population expansion of China during this period appears to have stimulated a crisis in firewood that lasted for approximately 450 years. Agricultural expansion led to the clearing of woodlands with no provision for reforestation; once the forest was burned (both as waste and for fuel), ever-increasing distances to haul firewood doomed ordinary people to lives of scarcity and scavenging to meet their fuel needs.[4]

Firewood shortages had become a problem in larger European cities (especially London and those of the Low Countries) from about 1300 on. The population catastrophe wrought by the mid-fourteenth-century bubonic plague forestalled the fuel crisis for about one hundred years, but by the beginning of the sixteenth century European populations had rebounded, especially in cities. In about 1500 there were twenty-six cities with populations between 40,000 and 200,000. By 1700 there were forty-eight, with Amsterdam, Naples, Constantinople, Paris, and London peopled by more than 200,000 inhabitants.[5] Such population centers wiped out vast amounts of woodland to feed cooking and heating fires. In London the price of firewood between 1450 and 1700 rose at twice the rate of commodity prices as a whole, as the population surged from about 45,000 to 530,000. The historical geographer Michael Williams estimates that a town with a population of 40,000 required 1,200 to 1,500 *square miles* of woodland to provide a sustainable source of fuel annually. But most forests were not managed for the long duration. By 1800

Parisians were reaching more than one hundred miles into the countryside to obtain firewood, often competing with the demands of the iron industry and its voracious appetite for wood in the form of charcoal. Throughout Europe, the vast majority of the people were too poor to afford the fifteen cords burned by the moderately well-off in central Europe (where the towns were smaller and the forests larger), but their minimal demands still accounted for 8 percent to 10 percent of their meager annual incomes.[6] Those forest-poor areas blessed with coal deposits (central England, for example) gradually turned away from wood, accepting the greater cost of extraction as a net gain over the high cost of transporting wood.

In areas heavily forested and sparsely populated, such as the United States and Canada, firewood consumption was enormous on a per capita basis, and wood outpaced all other fuel sources until well into the nineteenth century. In the eighteenth century, Americans burned about forty cords per family annually.[7] Between 1630 and 1930, historians estimate that Americans burned 12.5 billion cords of wood for industrial and domestic uses.[8] As late as 1850 wood still provided more than 90 percent of the energy used in the United States, and approximately 50 percent of all wood harvested was burned. (Lumber accounted for nearly all of the rest.) By about 1885 coal supplied about as much energy as did wood, with a relatively small amount of energy coming from water, oil, and coal gas. The decline of wood as an energy source continued apace in the United States until by 1960 less than 5 percent of the total produced came from the forests.[9]

Even with the bounty of seemingly endless forests, fuel shortages plagued American cities, which grew rapidly as the Industrial Revolution proceeded in the nineteenth century and immigrants came across the oceans looking for work, land, and wood. The United States was blessed with several superb natural ports, a river system that reached far into the continent's interior, and, eventually, a web of canals and railways that linked the hinterlands with the coasts and the Great Lakes; soon the port cities of New York, Baltimore, Boston, Philadelphia, New

Orleans, San Francisco, and Charleston were joined by inland giants such as Chicago, Cleveland, Detroit, St. Paul, and Buffalo as manufacturing and trading centers and gargantuan consumers of wood for energy and buildings. New York, for example, grew from 79,000 persons in 1800 to 2.5 million in 1890 and drew upon the forests of the northern parts of the state served by the Hudson River and the western lands opened up to trade by the Erie Canal, which was completed in 1825. Coal superseded wood as a heat source in many cities by the latter half of the nineteenth century, especially near the centers of mining in Pennsylvania, West Virginia, and Kentucky; it was also used in the parts of the Great Plains that were virtually devoid of trees but had access to rail lines.

Home fires were not the only ones burning up the forests. Wood was also the major source of energy for two of the most important forms of transportation in the United States and, to some extent, the rest of the industrializing world. Wood-fired steam locomotives consumed enormous quantities of the fuel, as did the steam engines that powered factories and other sorts of transportation, such as large riverboats. Although this was a small percentage of the total consumption of wood for fuel (3.5 percent in 1879), it was still more than 2.5 million cords in that year.[10]

The basic unit of firewood is the cord. *The New Shorter Oxford English Dictionary* simply notes that it is "a measure of cut wood, esp[ecially] firewood usu[ally] 128 cu. ft."[11] Other sources reliable and suspect elaborate slightly, agreeing that the size is somehow related to a rectangular stack of wood four feet by four feet by eight feet.[12] The word may be rooted in the Latin verb *accordare,* meaning "to agree," from which *accord* derives. A few definitions refer to the word's possible linkage to a method of measurement using a rope of standard length. Four feet seems to be a basic unit for firewood, but it bears little obvious relation to other traditional lengths, such as the cubit, the measurement associated with the length of the human forearm, or the foot, whose association is obvious. There does not appear to be any connection to the dimensional traditions in vernacular housing.[13]

Perhaps the tradition had to do with the size of the fireplace in housing built before the nineteenth century. In large and small houses fireplaces were wide and deep, often as wide as the room, since they also included ovens in the chimney block. Cooking in these cavernous areas did not take place as romantic images of the "old hearth" suggest—in a large iron pot suspended from a crane that swung out over a roaring fire. In reality the cooking hearth usually included several small fires, actually piles of embers that cooked food in smaller kettles, pipkins, and roasters. Even the spit-roasted fowl or joint of meat was cooked in a special apparatus that sat on the hearth in front of the fire. There was a large fire built and tended in the fireplace; it provided heat for the room and the embers for cooking. The mythic, traditional image confuses food preparation on the interior domestic hearth with cooking over an open fire outside the house—an activity more correctly associated with military campaigns and herdsmen who often lived in the field for extended periods.

Starting a fire from wood and tinder was arduous and time-consuming, so most people thought it imperative to keep at least one fire going continuously. The larger the logs, the better; such a fire had great longevity and less chance of going completely cold. Perhaps four feet became the traditionally accepted length based on the experience of building and keeping a fire lit while allowing for the customary practices of the hearth. Maintaining a fire in these early Euro-American houses usually meant setting a large backlog on andirons over a hot bed of embers sitting on the hearth, placing a smaller log near the front of the andirons, and piling other logs between the two. With the proper draft the big logs ignited from the glowing fragments, eventually dropping more embers below the andirons.

Cooking with an open fire or on the large hearth was a tricky and dangerous business. Managing the several small fires and knowing the combustion traits and tendencies of wood species were important skills, and it required years of experience to produce foods for the table that were even minimally edible by twenty-first-century standards. A good cook's reputation therefore went a long way in the marriage market. But these fires posed real dangers to women because of the ease with which their skirts could ignite. Between that hazard and the strength it took to hoist heavy cast-iron pots (even when empty), cooking was drudgery and required both culinary skill and considerable upper-body strength. This was not a job for fashion-runway models. It is easy to see why women of considerable heft were thought the best partners in marriage.

Large fireplaces such as those built in early modern houses often drew well because their chimney flues were large. Chimneys "draw" largely for two reasons. First, warm air rises, so when a fire is lit on the hearth, the smoke and warmed air drift upward. Second, and probably more important, air blowing across the top of the chimney produces a lowered air pressure in the flue, and air in the fireplace below then rises to equalize the pressure. The stronger the wind, the more the pressure is lowered, and the more active is the draft.

The chimney flue was in almost all ways an improvement on the traditional smoke hole at the top of the roof, since it concentrated the soot, creosote, and other deposited residues in the chimney rather than on the interior of the roof. This was less of a concern in moveable housing, such as yurts and tepees, and more troublesome in, for example, Native American longhouses and hogans, or the wooden and woven grass buildings constructed in many other parts of the world. Using a chimney to direct smoke also enabled people to construct second floors and garrets to isolate storage and sleeping areas from public spaces (parlors, halls), a living arrangement that became increasingly more important as people began to value privacy. Great manor houses surrounded their large central public spaces with an assortment of small rooms, but ordinary people had not the luxury of such great open areas and smaller chambers.

There was a price for these benefits. Soot and other deposits in the flue could eventually clog it and cause a chimney fire. People used brick, stone, and mud chimneys as anchors for the wooden framing members of the building. This provided greater structural stability but also could lead to burning the house down if the flue got very hot, a chimney fire erupted, or flawed masonry allowed sparks to escape into the structure. Over time beams nearest the chimney became extremely dry, and the perfect environment for an escaped spark to become a flame.

Even chimneys with competent masonry did not always draw as they were planned. Modern observers with some knowledge show great respect for surviving houses and other aged buildings, but these are in the main the work of the most skilled artisans of the past. Remarkable as these craftsmen were, they were more than likely a small minority in a great sea of people with minimal skills, knowledge, or intelligence who combined daring, arrogance, and obliviousness to produce structures that collapsed or burned to the ground, leaving little or no trace. Masons made mistakes and built hearths that were too large for the flue; they turned flues (called *corbling*) in ways that constricted air flow, or they used mortar that was ineffective. Though a great improvement over the open flame in the middle of the room, the open fireplace was nonetheless an inefficient heater and often ineffective at conveying smoke out of the house. Not until the turn of the eighteenth century, when Benjamin Thompson (Count Rumford) developed the shallow chimney design that still bears his name, did the heating and cooking areas reach some measure of efficiency. By the twentieth century only poor people regularly burned wood—until the oil shock of the 1970s.

The first of these recent shortages and fuel-oil price spikes inspired many people to reconsider heating with wood. Sales of air-tight wood-stoves skyrocketed, especially in the United States, which seems to specialize in fuel-guzzling automobiles and trucks as if they were a birthright mysteriously omitted from Jefferson's triumvirate of "Life, Liberty and the pursuit of Happiness" in the Declaration of Independence.[14] "Back-to-the-landers," poor folks, and environmentalists hopped on the woodstove wagon, though the latter expressed some concern about the air pollution from the wood burners. By the 1980s the first of the oil crises eased in the United States, as oilmen and their supporters took over the White House. Oil furnaces returned to homes, gasoline became more plentiful, and automotive behemoths with low miles-per-gallon rates returned to the road.[15] In this context woodstoves have become more a fashion accent in the suburban or country house than the alternative source of heat they had been for the previous generation. The labor of hewing, cutting, splitting, and stacking—and then hauling the wood into the house to load the stove—got tiresome. The Iraq war spiked the price of oil again in the early 2000s, and cordwood prices took off to parts unknown and unimagined, perhaps explaining why in this instance far fewer people abandoned their oil and natural gas furnaces for the joys, aromas, work, and smoke of the woodstove.

The Glories of Smoke

Smoke was a nuisance in the parlor or kitchen, but a godsend for preserving food. With no refrigeration and little ice to use for this purpose until the later nineteenth century, salting and smoking meats and fish was an important part of nearly every meat-eating culture on the globe, and the low-burning fires required for smoking were tended with great care. Smoking, drying, and salting kill bacteria and provide a barrier against many forms of contamination in the future, although the process is not foolproof, nor does smoking indefinitely preserve foods. Smokehouses were often among the first outbuildings constructed on farms once people built houses and barns.

Smoking foods was a tricky operation, and great importance was attached to a smoker's skills at managing the process; even on the large slave-labor plantations of the American South and Caribbean, the slave who demonstrated superior smoking skills was placed in a special class, closer to that of a house servant than a field hand. One of the ways slaves managed some small measure of resistance and revenge for their condition was to appropriate food from the smokehouse for themselves, although the penalties for that were severe.[16]

There are two basic smoking methods. Hot smoking involves the use of a fire that actually cooks the food as the smoke flavors it. Cold smoking, while still a warm process, involves maintaining a smoking environment at a much lower temperature so that the meat is not cooked but flavored after several days with the aromas of the embers as the meats dry. Drying eliminates one of the conditions critical for the growth of bacteria that can rot the meat, and the cold-smoking process introduces some chemicals that can impede (but not wholly prevent) spoilage. Both types of smoked meats require further steps for preservation, save those entirely cooked by the process and then quickly consumed.

The difficulty in satisfactorily curing meats with salt, sugar (or molasses), and smoke elicited advice in American agricultural and other periodicals throughout the first half of the nineteenth century. In March 1837, a correspondent to the *Genesee Farmer,* a Rochester, New York, weekly newspaper, boasted that he had "discovered the whole secret," which was to reopen a small ventilation hole in his "tight brick smoke house," a process that provided him with hams "dry and seasoned, and as good in September as in April."[17] Fifteen years later a correspondent to the *Michigan Farmer* magazine advised readers to "pack them down in coarse rock salt in a box" to secure them "from the flies and bugs during the summer."[18]

These two writers attest not only to the continuing importance of smoked meats but also to the practical scientific method employed by ordinary people—experimentation, observation, and a trial-and-error method that are the building blocks of improvement in the mechanic and domestic arts. In short, it is another manifestation of Henry Petroski's observation that innovation moves forward because of failure. The first correspondent had been smoking meats for twenty years in his brick smokehouse, and while he had a measure of success, the results were not entirely to his liking, so he didn't stop tinkering with the process. The second writer had his smoked meats spoil before he could eat them all, a situation that could be dire once winter had set in.

Small brick or wooden smokehouses dotted the rural landscape of Anglo-European North America, as well as that of Europe, China, Russia, and central Asia. Smoked wildfowl were important and special parts of the diet in China and other parts of Asia, and smoking pig and wild boar in central and southern Europe was an important part of the annual cycle of farming and preparing for the lean months. Smoked fish were a significant part of the diet along the coasts of countries in cold climates; Native Americans on the northwest coast burned alder to preserve salmon, and the Baltic peoples smoked and salted huge quantities of herring and salmon. By the later nineteenth century giant industrial meat-processing and packing companies gradually took over the food processing and preserving trades.

Refrigeration made smoking and drying obsolete as preservation processes, but today smoked meats, many treated with nitrites and other chemicals to preserve them after the smoking process, still occupy a considerable amount of space in the meat and fish cases of grocery stores and butcher shops. Warnings about the possible carcinogenic effects of eating nitrites and the residue of smoking have prompted some enthusiasts of smoked foods to opt for smokehouse products free of them, but culinary trends show little chance that smoked goods will disappear, whatever the mode of preparation.[19]

Sweat and Belief

Several cultures treasure the ritual of heat and smoke-induced sweating. For centuries Native Americans have incorporated sweating in a small structure as a way toward religious knowledge and bodily purification. Early settlers and travelers to America remarked on the practice of sweating among the native peoples as early as 1643, when Roger Williams of Rhode Island colony noted the practice. Later travelers such as George Catlin observed that Mandan, Omaha, Pawnee, and other tribes partook of the sweat lodge, both for religious and health reasons. Using either hot rocks (onto which they tossed water) or, less commonly, burning logs, native peoples of the Americas from the far northern rim of the continent to at least as far south as Mexico engaged in the activity.[20]

The exact nature of the religious ritual and its meaning vary among indigenous peoples, but the sweat lodge is most often connected in some way with knowing and reaffirming the creation history of the earth and of the Native Americans. Some peoples believed that Manitou resided in the hot rocks; others that the lodge itself represented Mother Earth, while still others link knowledge of secret and sacred wisdom to the sweat lodge experience or allow its transmission to initiates to occur only there.

The sweat lodge of Native Americans who lived in permanent dwellings was usually a small, low-ceilinged and close-to-the ground structure into which members of the tribe and their close friends retreated to breathe in smoke and to sweat. Nomadic tribes constructed small temporary structures of flexible wooden branches and overlaid them with hides and blankets. Although missionaries and government agents tried to discourage and even forbade sweating, the practice has continued to this day.

The more secular counterparts to the Native American sweat lodge in Russia, Finland, Sweden, and the former Ottoman Empire are, respectively, the *bania,* sauna, *bastu,* and *hammam*. The written history of the Russian sweat bath reaches back over a millennium, and is noted in the *Primary Chronicle* of 1113. Built of a variety of materials, including wood, mud and stone, the "black" bania was a smoky room stocked with rocks to be heated, water to splash on them, and birch switches known as *vennik,* which were lightly slapped on the skin to encourage better circulation. Russians considered the bania essential for a variety of health and ritual practices associated with birth and death, as well as for other physical therapeutic purposes. Often accompanied by bloodletting to help purge the body of poisons and pressures thought to disrupt its equilibrium, it was an essential part of the medical practice of nearly all who partook of the experience. The counterpoint to the black bania was the urban "white" bania, the concrete and tiled sweat bath, a smoke-free room that was often one part of a series of spaces for sweating, massage, and bathing.

The bania is the Russian counterpart to the Finnish *savusauna,* or smoke sauna. Recorded history of the Finnish sauna is scant before the seventeenth century, but that is in the main the result of the scarcity of recorded history of the Finns before then. Since the early migration of the Fenno-Ugrians likely took place at about the same time that other peoples migrated from central Asia to the northwest of Europe, it is likely that the Finnish sauna and Russian bania emerged from roughly the same roots and at roughly the same time. The traditional Finnish sauna that evolved over centuries is a free-standing wooden building, ideally located near a lake or stream in which hot and sweaty bathers could cool off and rinse their bodies. Those wood-fired structures dotted the agricultural landscape for generations before the Finns left the countryside in droves in the late nineteenth and twentieth centuries. [FIG. 105] In modern Finland the wood-fired sauna is still the preferred form for the sweating experience, and many Finns insist that the quality of its heat is different from that created by electricity. While most urban Finns now partake of electrically heated saunas in their apartment buildings, their attachment to the traditional kind at the edge of a country lake remains very strong. Indeed, there are about half as many summer cottages and small, free-standing saunas as there are people in Finland. Whether in city or country, most Finnish saunas are lined with wood, usually Finnish spruce, with seats of obeche or aspen.

The sauna and its Swedish counterpart, the bastu, became focal points for reformers' attacks in the eighteenth century. Criticized in Sweden for its carnivorous appetite for wood in a country fretting over its forest resources, the bastu—and by extension, the sauna—received the attention of moralists who connected the bastu with the spread of venereal disease, just as reformers had so linked the public baths of southern Europe and the Ottoman Empire two centuries before. The combination of church and state pressures reduced the bastu to an occasional and often Yuletide ritual, while in Finland the customs of bathing, birthing, and healing in the sauna continued in the face of the ruling Swedish opposition. Finns asserted that their sauna was a deeply religious place rather than a location for licentiousness. The place and the practice also

FIG. 105
Sauna. Åland archipelago, Finland. Twentieth century.

became a rallying point for Finnish nationalism, against both the Swedes, who exercised sovereignty over Finland until 1809, and the Russians, who did so until 1917. Finnish national identity and Romantic nationalism of the later nineteenth century in part relied upon the identification of the sauna as a ritualistic place with material characteristics and ethnically specific practices, especially the *savusauna.*[21]

The *hammam,* the Islamic counterpart to the sauna, is also primarily a place for physical and spiritual cleansing, deeply connected to the social and sacred rituals of rites of passage. Often close to or attached to mosques, most hammams were heated with wood fires. Large complex Roman baths may have provided some architectural inspiration to Islamic builders as their faith spread throughout the Middle East, North Africa, central Europe, and Spain. Eventually women were admitted to the hammams (though always segregated by gender), and travelers wrote admiringly and perhaps enviously of the languid joys of the heat.

The convergence of this form of bathing, relaxation, and religious ritual suggests broader questions about the nature of the human relationships to smoke, steam, and heat. The cleansing effects of the hot and steamy room have been well documented by legions of practitioners over centuries. Physiologically, the heat opens the skin's pores, allowing sweat to carry off minor impurities, while inside the body the raised temperature of tissues stimulates a more rapid heartbeat. The after-effects of a sauna or a steam bath are for most people akin to a dramatic easing of tension and stress, a slowing down after the elevation of the senses in the hot room.

The Essential Ingredient

Everyone who has tended and cleaned after a wood fire knows that ash—a useful source of sodium hydroxide, or lye—is the tangible end product of combustion. At some time in the distant past—at least six thousand years ago—someone discovered that slowly and carefully burning wood produced a lightweight black substance—charcoal—that they could use to make fires that burned at a temperature that allowed them to smelt ores. With higher burning temperatures, tin, which requires 1100°C for smelting, could be combined with copper (which can be smelted from ores at 800°C, a temperature realizable without charcoal) to produce bronze. Because it can be cast and withstands use better than either pure tin or copper, bronze was a material central to the gradual transformation of technology in human civilization over several millennia.

In addition to making bronze, charcoal also made possible iron smelting. By about 500 BCE, iron was in production in many areas of the world. It and bronze became the basic metals of weaponry and tools. Those civilizations that had these materials and their technologies conquered those that did not. By the late fifteenth century, so great was the demand for charcoal that the English Crown enacted laws to regulate its production, aiming to prevent charcoal burners from taking trees needed for the navy.

Charcoal was the essential ingredient in just about all metallurgy until coke (first produced in 1735) began to challenge it in the larger ironworks in England by the mid-eighteenth century. The English, because of their chronic wood shortage, made the changeover in the eighteenth century: 95 percent of the pig iron manufactured in 1750 came from charcoal furnaces, but within fifty years only 10 percent was. The advent of coke did not end the use of charcoal in industrial smelting in the rest of the world, however. In the United States 78 percent of iron produced in 1865 was made in charcoal-fired furnaces, and the abundant forests of the country assured that making charcoal (also called colliery) continued to be an important part of the rural industrial economy throughout much of the nineteenth century.[22]

Making charcoal was a long and taxing process that required both skill in organizing the burning stack and patience and attention to monitor the operation for a period of several days. For centuries, instruction in the techniques of colliery was likely passed down from one generation to the next, keeping the mysteries of the process within a small network of people. By the nineteenth century, however, the secret was out; books such as Frederick Overman's *The Manufacture of Iron, In All its Various Branches* (1850) laid out the process in detail, complete with diagrams.[23]

A successful charcoal burn required a large round pile of vertically stacked wood, about ten yards in diameter, which was lit in the center at ground level and then the whole pile covered with small branches and sod to seal it. By carefully controlling oxygen access to the fire, it smoldered for days, eventually producing charcoal. Overman exhorted farmers to "char the [pine] wood that is now altogether being destroyed in clearing new lands," rather than hauling it "by thousands of cords to the southern cities, . . . especially since mineral coal is now brought at such a low rate, by rail-roads and canals, coming every day more into use, as fuel in families as well as in factories."[24] [FIG. 106]

Charcoal burning was viewed as a more efficient means of realizing profits from forests, and in some measure a way to reduce the voracious

FIG. 106
Charcoal burn. Illustration from The Plough, the Loom and the Anvil. *Vol. 3, no. 8 (February 1851): 513-16.*

and inefficient use of the woods that had come to characterize American forestry. In the temperate climates of Europe, where charcoal burning and timber harvesting had been ongoing pursuits for centuries, the problem was more dire, as the interests of industry, agriculture, forestry, and shipbuilding collided, at least until steel began to replace wood as the building material of choice. In part this explains the use of coppiced wood to produce charcoal, particularly in England, since coppicing produced young saplings and branches within a few years. Farmers also benefited from colliery, discovering that areas where charcoal kilns had once been built were more fertile. "The spots where charcoal pits were burned 20, some say even 30 years since," wrote one William H. Trimble in a letter adjoining Overman's article, "still produce better corn, wheat, oats, vegetables or grass, than the adjoining lands."[25]

In addition to its uses in metallurgy, charcoal was important as a purifier and filter element. In the July 1849 issue of *The Plough, the Loom and the Anvil,* a periodical of early-nineteenth-century America, correspondent "J.T." wrote of his discovery that pulverized charcoal, when added to a previously foul-smelling mixture of Pategonian guano and water (a popular fertilizer then), rendered the mixture absent of "offensive, or indeed scarcely any scent whatever."[26] In the twentieth century, the absorptive properties of charcoal have made it an essential element in filtration processes, particularly for purifying water and filtering bourbon whiskey. Modern charcoal is for the most part produced in metal ovens in which temperature and oxygen levels can be carefully monitored and adjusted without having to climb atop a smoldering kiln, thereby risking life and limb. Colliers no longer preside over the ancient craft, living rough in the woods; technicians and engineers handle the operation in factories. Much of the product no longer is bound for metallurgical purposes but to fuel people's penchant for grilled, charred, and smoked cuisine made in the barbecue, smoking pit, or metal cooker and smoker sitting in the backyard.

Using small hot coals for cooking is hardly a practice solely of the American suburbanite. The Japanese hibachi has been in use for at least two centuries. Originally a ceramic container in which embers were placed, it was part of a larger cooking area commonly known as the *kamado* in Japanese houses. In the United States, outdoor cookery over hot coals became a favorite method of "leisure cooking" in the suburbs following World War II. The favorite facility for this sort of activity, called barbecuing or grilling, was the brick, stone, or cement-block grill, a small chimneyed structure over which (usually) the male of the household was supposed to preside, as if the outdoor location was in some way connected to camping, hunting, and other manly pursuits in which residents probably had little interest, know-how, or talent.

This outdoor grilling area fell into disrepute in the United States, a victim, perhaps, of the younger generation's ambivalence or outright hostility toward the culture of the 1950s. In the 1960s, a small cast-iron version of the portable Japanese hardwood- or charcoal-fueled hibachi became popular, especially among urban dwellers and younger people more mobile than those living in the suburbs. Eventually the hibachi itself was superseded by another charcoal-burning apparatus, the kettle grille. Mobile in that it was small (eighteen to twenty-two inches in most forms) and wheeled, it had neither the limitations of the hibachi's small size nor the retro associations of the outdoor fireplace. Eventually even the fuel changed, from raw, or lump, charcoal to the compressed and molded briquette.

Most histories of the charcoal briquette attribute its discovery as a marketable product to the car manufacturer Henry Ford, who was looking for a way to use the scrap wood left after timber harvesting and milling on one of his Michigan forest properties. (Ford used many wood parts in his early cars, including wheel spokes, running boards, and dashboards.) The process of pulverizing charcoal, mixing it with a binder, and pressing it into briquettes was patented by Orin F. Stafford, and Ford latched onto it as a way to further his profits and reduce waste. His factory opened in 1921, employing one of his early sales agents, Edward G. Kingsford, a distant relative and lumberman. Ford Charcoal

Briquettes were produced until 1951, when the Kingsford Chemical Company bought the firm.

Most briquettes contain all or some of the following ingredients: wood char, mineral char, mineral carbon, starch, borax, limestone, sawdust, and sodium nitrate. Raw or pure charcoal contains only what remains from the wood after charring, and there is great dispute among aficionados of grilling about the differences in flavor of foods cooked with various fuel forms, as well as about the health benefits or threats from the processes. In more recent years many people have switched from charcoal or briquettes to propane for their outdoor cooking, and some have used "liquid smoke" products to obtain the smoky flavor in their cooking. Some chefs have ignored all of the these trends and use wood as a grilling fuel.

Camping and Cooking

Armies once traveled almost exclusively on foot and operated from temporary quarters in which the cooking (such as it was) was done over the wood-fired flame. So too explorers, trappers, traders, lumbermen, miners, hunters, and herdsmen traditionally lived significant portions of their working lives in the wild, or at least far from home and permanent settlements; their cuisine was for the most part kettle cooking—stews and pottages heated in a single pot suspended from a tripod over a wood fire.

For most people in the industrialized world, however, the open fire outside the hearth and yard is an experience connected to leisure or childhood experiences in organized youth groups such as the Boy Scouts, Girl Guides, and Camp Fire Girls.[27] The latter were organized into three divisions: Wood Gatherers, Fire Makers, and Torch Bearers, names that clearly exhibit not only gender distinctions but also the central importance of wood and the campfire in popular culture. The outdoor wood fire was the overarching symbol for this experience and the Camp Fire Girls appropriately used it in their logo.

Children in the Scouts and Guides learned some of the simpler skills of the Indians, for which they earned badges, sang camp songs, and otherwise connected with their peers if not with real Native Americans or other "early" peoples.[28] A few may have emerged from these activities as budding naturalists, but most did not. Their camping was more likely a bonding experience among children of similar classes. The bourgeoisie constituted a larger percentage of the scouting ranks than did the wealthy or the poor, though not entirely to the exclusion of the latter groups, whose children often joined the preadolescent scouting groups, such as the Brownies and Cub Scouts. But as well-off children advanced to their teenage years, they characteristically abandoned the Scouts for the experience of private camps they attended for several weeks of the summer, much to the relief of their parents. Poorer children often left the Scouts to go to work.

Out of Control

The "gift" of fire is also the bane of wooden buildings, especially those in cities and towns. While ancient and early modern peoples protected themselves from marauders by building on hills and surrounding themselves with stone walls, earthworks, and wooden palisades, within the confines of these walls the housing stock was often made primarily of wood. Later cities, built in a more expansively optimistic environment, or because trade and economic growth pushed their boundaries outside of the original settlement, were almost always built of wood—until they burned to the ground, as nearly all cities eventually did.

The relative ease and lack of expense of building in wood (as opposed to stone and brick) greatly aided the expansion of cities and towns that has occurred during the past two thousand years. But land pressures and poverty quickly led to extremely dense urban areas. Because the only available source of heat and light was the open flame, the catalyst for incineration was ever near to a ready supply of fuel—the housing and its furnishings. Once fires were kindled by accident, arson, or forces of nature, they easily got out of control. Nineteenth-century firefighting was primitive and inefficient, relying on bucket brigades and private fire departments that often competed with each other before working to extinguish the flames.[29]

Large areas of nearly all of the major cities of the world have burned at one time or another. The great fire of Rome (64 CE) is probably the most famous example (owing to the Emperor Nero), but blazes also swept through London (1666), Edo (now known as Tokyo, in 1657 and 1772), Stockholm (1697, 1759), Moscow (1812), Turku, Finland (1827), Toronto (1849), and Paris (1871). In the United States, almost no cities escaped conflagration. New York burned in 1776, 1831, and 1835; Washington, DC, burned when the British set it alight during the War of 1812, when York, Ontario, and Buffalo were also torched. Burned also were Pittsburgh in 1845, most of the cities of the South during the Civil War, Portland, Maine, in 1866 (when ignited by a firecracker), Chicago in 1871, Boston in 1872, and San Francisco in 1851 and 1905, the latter after an earthquake fractured natural gas lines that subsequently ignited. Nearly all the cities in the conflict areas burned during World War II—Dresden, Tokyo, Hiroshima, Nagasaki, London, Hamburg, Helsinki, Moscow, Stalingrad, Kiev, Leningrad—the list of destruction from that catastrophe seems endless.[30]

Smaller fires were a common part of daily life in cities. The 1839 volume of the American newspaper *Niles' National Register* reported fires in St. John's, Nova Scotia; New Orleans (2); St. Louis; New York (5); Natchez (2); Philadelphia; Brooklyn; Aiken, South Carolina; Mobile, Alabama (3); Brooklyn; Constantinople; Norfolk, Virginia; Königsburg, Prussia; "Quibdo, in the province of Chaco S. A."; Yazoo City; and "since the beginning of this month [October], no less than twenty-four fires that we have accounts of" with the destruction of six hundred buildings valued at $4,040,000, in addition to the list above.[31]

Even stone and brick buildings were also no match for a conflagration. They did not burn, but the flames of buildings burning nearby heated them until they ignited their own internal wooden elements. Fires also create convection currents that can carry burning embers—even boards and beams—hundreds of feet up and outward, landing on roofs of other buildings, igniting them immediately if the roof is flammable and dry. Barn fires are especially dangerous because they are usually full of ready fuel—straw, hay, and exposed beams—that often send flaming parts a quarter of a mile away.

One result of repeated conflagrations was that city politicians began to require people to build in stone and brick, making wooden buildings a rarity in many urban areas. Wood became the irresponsible, untrusted material for central-city building, consigned to the factories and residential areas in which there was more space between structures. In tightly packed urban areas, brick, brownstone, marble, sandstone, and other such materials replaced the charred remains of what was once a wood-dominated space.

Anyone who has ever seen the remains of a city after a large fire or bombing knows that the standing rubble presents more problems than the total destruction of a fire in wood-built areas. Walls remain, but the buildings are gutted, owing to their incinerated wooden interiors. The remaining stone or brick structures are rendered useless in many cases by the heat of the flames and must be torn down and carted away. Air power, chemistry, and physics have enabled the human race to destroy places with sophisticated weaponry so that the fire resistance of stone and brick, while substantial, is in the end no match for the darker side of our nature.

Outside of the cities, fires have repeatedly occurred in relatively uninhabited woodlands and on prairies during dry summer months, ignited by lightning or careless people.[32] Farmers still burn off fields in some areas and indigenous woodland peoples regularly set fire to the edges of forests to stimulate new growth and attract game in the next growing season. In modern times some people in tropical areas still burn off trees to get at arable land, and in developed areas debates about forest conservation and management still ignite the passions of forest-service professionals and managers, businesspeople, and the general public when the principal issue is wildfires. Should they be fought or allowed to run their course? Forests regenerate relatively quickly from fires, but is that worth the sacrifice of natural resources that took decades and even centuries to grow? What, if anything, should be done to save pristine woodlands and wilderness destined for human use (and not abuse) that fire would render "useless" for a half century or more? Should people be allowed to build in tinder-dry woodlands and then be able to acquire fire insurance when a big burn is nearly inevitable? Should the rates of others living in less fire-prone areas increase to cover insurers' losses?

Most troubling of all fires are those that are deliberately set for no agricultural purpose. Arson has been a weapon of revenge and an act of the deranged for centuries. Fires are often set to provide cover for other criminal acts. Burning the evidence makes conviction for a crime much more difficult; this was especially true in the age before forensic medicine and chemical and physical analyses. As soon as fire insurance surfaced in an economy, those who would seek to profit from the payout for torching their own property emerged from the murky pool of greed, crime, and indifference. Often arson was a weapon of ethnic and cultural conflict, as it remains to this day. Warring gangs of one religion lit up the residences, businesses, places of worship, and recreational halls of people of other beliefs; "reformers" of a vigilante nature burned the hangouts of those they considered vice ridden; African slaves used arson in their sporadic war against their masters; night riders such as the Ku Klux Klan set alight the houses of former slaves long after the Civil War had ended; union arsonists fired urban railroad terminals and other buildings during strikes in the summer of 1877.

Underlying all these natural and deliberately created disasters was the flammability of wood. Whatever its other affective and intellectual associations, wood is inextricably bound to fire and smoke. Both are essential to human civilization, and both can be lethal. Smoke cures and flavors foods; wood in its charcoal form made the metallurgical revolutions without which human civilization would hardly exist. But the metals and the knowledge that they wrought, albeit indirectly, have also served to threaten human existence as well as to enhance it. Smoke burns the eyes and ultimately will suffocate a living creature; fire will consume life and nearly all the matter that composes it. Managing both—and avoiding or controlling the temptation to use them to subjugate others—remains one of the great challenges of the twenty-first century.

Crisis in the Woods

Wood was not only the major building material and fuel source for most people for millennia; it was also *in the way* when farmers expanded their fields to meet an increasing demand for planted crops or to provide grazing land for animals. Since it was also the building material that drove the maritime empires of the world for several centuries, wood supplies were no match for the expansive political appetites of European and Asian politicians and monarchs.

In China, intense pressure on forests came from several directions, as land clearing pushed far into the northern deciduous woodlands. By the tenth century a thriving iron and steel industry and the urban expansion noted earlier in this chapter (there were five cities with more than one million inhabitants by 1100) gobbled up Chinese forests. Further population pressures fueled agricultural expansion, as the number of inhabitants grew from about 300 million to 500 million between the early eighteenth and early twentieth centuries. Technological innovation aided as well in the further denuding of the Chinese landscape, as the timber cutters and farmers used new technology to push into the hills and mountains previously too difficult to log and clear with older tools. The resulting erosion and river siltation in turn exacerbated flooding and reduced the fertility of older agricultural lands, provoking even more expansion into what little forest remained.[33]

The Chinese got a head start on most of the rest of world in forest clearing, but the other participants in the race to strip the woodlands from the planet made up ground quickly. In roughly the same two centuries in which the Chinese virtually eliminated their once-vast woodlands, western and central Europeans more than doubled their croplands and wiped out millions of acres of forests. In their great cutting, however, they were no match for either the Russians or the North Americans. Russia, which had about one-half the total acreage under cultivation as did Europeans in 1700, surpassed Europe in that department by 1920 and cut down about five times as many acres as were hewn and cleared in Europe. Of course, the caveat in those figures is that by the eighteenth century most European woodlands were gone or off-limits to further cutting, either because they had been commandeered by the monarchs, held as woodlands for wealthy landowners, or increasingly managed for long-term productivity.

In North America the change was even more dramatic. Starting with essentially no cleared cropland in 1700, North Americans (for the most part in the United States) by 1920 matched the Russians in the amount of such land, with *sixty times* the number of acres under cultivation as had been the case two centuries before. (Not all of the newly tilled acreage was the result of forest removal: 59 percent of the total came from cultivating grasslands in the Midwest; the rest came from cleared forests. Virtually all of the increase in Russia came from the forests. It was about equally divided in Europe.) In Australia the story was slightly different, since much of the interior of the continent is arid. About 35 percent of all Australian wooded areas has been modified since 1780, with the greatest concentration of change in the coastal areas of New South Wales, Queensland, Victoria, and the southwestern tip of Western Australia, where about 70 percent of the forested lands have been cleared for agriculture.[34]

Pressures to cut the forests and expand agriculture also came from the effects of global trade in tropical foodstuffs and other products that Europeans and North Americans demanded. Throughout the tropical climate zones, Western hunger for tea, coffee, cocoa, spices, sugar, and other commodities not grown in the temperate zones stimulated forest clearing for commercial plantations. The first European traders had brought these goods to the consuming public soon after their explorations had taken them to Africa, the Indian subcontinent, the East Indies, and eventually China; in the nineteenth century native farmers and plantation owners in these areas were joined by Europeans who established their own growing and trading businesses, all leading to the transformation of great swaths of virgin timberlands into farms.

In the Americas and many of the tropical areas of the world, not only was wood plentiful, but the forest was an overwhelming place of dark thickets and giant trees, wild animals, and (in the minds of Europeans) wild human beings who lagged behind what the new settlers thought was their advanced stage of development and enlightenment, yet who were, ironically, also more skilled in the arts of wilderness survival and in certain ways of war. Confronted by the enormity of the woods, both newly arrived immigrants and settlers whose families had lived in the Americas for generations saw an endless wealth and supply, and a colossal task—removing the trees to farm the soil beneath the canopy and around the stumps and somehow transforming the wood into useful products and fuel.

The number of acres cleared and the pace of deforestation in America are staggering. Between 1650 and 1850 Americans cleared 114 million acres of woodlands, but between 1850 and 1890 they cleared 137 million more acres, 78 million of them between 1870 and 1890.[35] All this timber went for a booming national housing market spurred by rapid population growth (7.2 million in 1810; 74.8 million by the end of the century), the agricultural demands of the country's urban population, and the industries that needed lumber for fuel and infrastructure.

The railways in particular were significant in this regard, replacing ships and boats as voracious consumers of wood. The number of miles of track grew steadily from the advent of the railways in the 1830s, and especially quickly between 1870 (60,000 miles of track) and 1910 (357,000 miles); the number of ties consumed annually for new and repaired track by 1910 numbered 124 million. The numbers, impressive though they are, belie an even greater consumption of wood since, until preservation with creosote became a common practice in the last decade of the nineteenth century, railroad builders demanded or were required to use the straight heartwood of oak, locust, and to a lesser extent chestnut and redwood because these species resisted rot and breakage under the heavy loads of railroad engines and cars.[36]

Important as well in this leveling of the American woodlands was the increasing technological sophistication of the lumber industry and the web of internal transportation facilities and resources that made it possible and profitable to mow down forests, mill the lumber, and move it to population centers. Moreover, unlike much of western Europe, the United States met its timber needs from within its borders, and thus traders did not have to contend with the vagaries of transnational shipping and international diplomacy. The industry followed the forests westward, cleaning out the Northeast, the Upper Midwest, the upland South, and ultimately settled on clearing the old-growth forests of the American Northwest.

As early as the mid-nineteenth century, the idea that there was an inexhaustible supply of timber that could be used with no thought of replanting or efficiency was under attack. Reformers thought that Americans' forest and fuel-burning practices were wasteful and reckoned that families not only could live on much less wood but could become self-sufficient and self-sustaining if they worked at it. "Fifteen acres of wood and timber land will furnish a farmer his ordinary timber and wood for two fires," wrote one critic. "*Ten* cords of wood will suffice for any man to keep two fires the year round, provided he has tight rooms and good stoves."[37] Two assumptions are important in this passage. First, the exhortation provides for two fires year-round, attesting to the need for constant fire and to the equation of comfort with fires in two rooms. A room without a fire remained unheated, save for radiation from the chimney if it went through or was adjacent to it. Second, the writers in this mid-nineteenth-century magazine base their analysis on woodstoves and not on the use of open fireplaces. Cast-iron and tin woodstoves had been in production in the United States and parts of Europe since the 1830s and were popular for those who could afford them. The poor, therefore, faced both the daunting specter of inefficient heating and great expense and labor just to survive.

Others writing at about the same time saw the timber and firewood crisis in broader economic and social terms, presaging the environmental concerns of the Progressive movement of the early twentieth century and the fuel crisis of the 1960s. In 1855, the editors of the weekly newspaper *The Country Gentleman* printed the first of a two-part article addressing "the rapid disappearance of our forests."

> The wasting away which was strikingly visible, appeared to be the result only of the home consumption of ordinary fuel. In the same neighborhood, an extensive land-owner informed us in looking over the surface that he had denuded of the original forest for his own firewood, he had discovered that *one hundred acres* [emphasis original] had been cleared for this purpose, during the forty years he had resided there.
>
> The same results in a greater or less degree may be found almost everywhere. It has been estimated that at the present rate of consumption of the lumber districts of the country, the whole region east of the Mississippi will be stripped of everything valuable for this purpose within the next thirty years.
>
> What will *then* be done? Where shall we get the materials for houses and barns—for agricultural implements and machines of all kinds—for fencing—for bridges—and for the construction of all kinds of shipping—to say nothing of the three to five thousand cords of wood consumed annually as fuel in every township in the settled portions of the country? . . .
>
> The answer is most obviously, commence the growing of timber trees *immediately*.

The article then proceeds to examine methods for growing trees, compiling information from various sources about the practices that had succeeded or failed, and concludes with a critique of French Canadian woodcutters in the Concord, Massachusetts, region, who "camp in the woods and [make] havoc ... with our forests.... If these denuded wood lots, could be left to produce another growth, it would in some measure mitigate the evil, but the proceeds of a *burnt-land* crop of rye, can be transformed into dollars quicker than a growth of trees can be reared."[38] The refrain continued throughout the nineteenth century.[39]

Massive consumption of firewood and the use of wood for the ubiquitous fencing of the United States were major contributors to the impending crisis, as were the wasteful and presentist practices of woodcutters everywhere. Nothing was done to check the carnage, despite pleas and warnings. Eventually other sources of fuel supplanted wood, and other materials replaced wood for fence rails. Steel and other metals eventually superseded wood in machines, and brick, metal, plastics, and glass took the place of wood in large-scale buildings. But Americans kept mowing down the woodlands, all the while reveling in their ability to do so and to "conquer" the really big trees they found.

Stripping the forest of trees, planting the cleared land with crops, consuming huge amounts of wood with little attention to the reality of diminishing forest resources, and the testimony of history (in the case of the writer above, merely forty years) sound drearily familiar to the modern observer. In the twentieth and early twenty-first centuries, industrially developed nations have focused on the diminution of the rain forests, where acreage has been steadily declining to meet the agricultural needs of rapidly expanding populations while the amount of forest acreage has stabilized in the world's temperate zones.[40] Inhabitants of tropical areas complain (with considerable justification) that the colder-climate countries are pressuring them to sacrifice their agricultural needs and aspirations, their major fuel source, and their wish to profit from the forest, even as those powers have already laid waste their great forests and, in some cases, bungled the preservation or planned restoration of their own woodlands.

It was not and is not a completely bleak picture, however. As we saw in the history of shipbuilding, there was some glimmer of a realization that the forests would soon be gone if consumed at the rate that the big ships took the trees. Little was done about that, given the long time to recoup the investment in tree planting. But coppicing, practiced in many parts of England, eventually met some of the demand for firewood and charcoal until coal and coke replaced the former as a fuel. In Japan, forest management became the norm rather than the exception long before the conservation movement picked up steam in Europe, Britain, and the United States. Forest management—in particular the idea that timber was a crop to be tended and harvested over a long duration—became part of Japanese governmental policy in the seventeenth century. Similar management practices were instituted in Germany at about the same time, although there is little evidence of communication then between the two peoples. While Americans were consuming wood as if supplies were endless, acreage in forests was gradually increasing in Europe.[41]

Steps to regulate or manage the harvesting of woodlands have met with some success in Europe, Canada, and the United States, in some cases undertaken by private enterprise and in others by governmental agencies. In the United States forest management of the sort championed by Progressives at beginning of the twentieth century is being challenged by antiregulation and antigovernment forces. Since most of the wood products in the United States (plywood and paper) are made of fast-growing softwoods, entrepreneurial reforestation concentrates on conifers that can be most quickly and profitably harvested, rather than the diverse species stock that North America once had.

But if market forces seem to be mandating endless acres of pine and spruce, there is some indication that species diversity is returning, if only by accident, default, or neglect. The transformation of the agricultural economy in the United States since the nineteenth century has resulted in the large-scale abandonment of farming east of the Appalachian Mountains, resulting in a partial reforestation of these lands. The process is a slow one, and initial reforestation is often by scrub species or a dense population of rapidly growing species. It will be centuries before anything like the forests of great oaks and maples returns, if it ever does.

Technological determinists take comfort in the transition of the first industrialized areas of the world from an "age of wood" to ages of steel, plastics, and other industrial materials, and thus the resurrection of some forests because wood has become largely irrelevant to the production or composition of most goods consumed in these countries. This is small comfort to those who envision the forest as something more than a product or an obstacle to be overcome, and it provides no counter-argument to newly industrializing areas that see the forests in those terms. The techno-argument resonates, perhaps ironically, with the early Christian position that God created the world for humans to dominate, rather than to be stewards of Creation, and with the (seemingly) secular faith in "markets," as if the latter were ever in fact "open" or free from the sometimes successful attempts to manipulate and control them.[42] Technological advance and the loss of farming in the eastern United States were in part the product of market forces, but these were substantially aided and abetted by military protection from, and subjugation of, the Native Americans, as well as government subsidies to the railways, land-grant colleges for agricultural research, individual farmers (first as nearly free land, later as price supports for crops), and other enterprises that opened up the Great Plains and the prairies. Deforestation continues in the emerging industrial world as it took place in those areas that industrialized first, the rate of change increasing in large part because of the mechanization of timber harvesting and land clearing. The broader environmental impact on the ecosystems of both the tropical zones and the entire planet is unknown in its entirety, although it seems clear that people still have not learned the lessons of the long-term effects of deforestation that were identified centuries ago.

Epilogue

O n the bottom shelf of one of the bookcases in my home office is a long row of magazines. First in the row is the September 1978 issue of Fine Woodworking. Modest in length (87 pages), printed in black and white, and stapled together, it is aimed at readers already in the profession and those seeking know-how in the craft and art. Forty-eight pages of articles offer instruction on subjects as diverse as methods of clamping workpieces, sanding, sharpening, and spindle turning, as well as information about Scandinavian furniture design and that of the California architects and designers Charles and Henry Greene. There are also advertisements for Japanese woodworking tools, a panoply of home workshop-sized electric power tools, work benches, hand tools, adhesives, veneers, lumber, furniture plans, books on woodworking, and finishes, in addition to a page of classified ads offering services, supplies, and other information.

Nearly three decades later, it is a thriving magazine, still aimed primarily at experienced woodworkers, and it is printed in full color. Since that time, the number and circulation of woodworking magazines have grown consistently, both in the United States and internationally. Furniture-quality lumber and veneers, hardware, tools, and other goods related to the craft today are sold through the mail and on the Internet, and the trade in used and rare tools has likewise expanded. At least once a week a catalog related to woodworking appears in my mailbox.

The expanded interest in this field is not entirely new to the later twentieth and early twenty-first centuries, although it has certainly been aided by new communication methods and easier ways of paying for purchases. "Do it yourself" is an idea at least a century old, as is training in the trades for occupational, therapeutic, or social amelioration. During the heyday of the Arts and Crafts Movement in Europe and the United States around 1900, craft work emerged as an antidote to what critics saw as the evils of industrialism and as a means for tired "brain workers" in the growing white-collar society to "recharge their batteries" and fend off one of the diseases of the age: neurasthenia, or nervous debility. Socialists such as the English reformer William Morris hoped to bring back some of the dignity that they thought had been stifled by assembly-line production; Americans such as Gustav Stickley likewise offered both goods and philosophy in the magazine *The Craftsman*. Other entrepreneurs who bought into the craft ideal but not the economics or the politics of Morris, such as Elbert Hubbard of Roycroft fame, were enthusiastic about the beneficial effects of craft work on the "jangled nerves" of the middle class and the elite.[1]

The Mill in the Suburbs

By the 1920s American popular magazines focusing on technology, tools, and the crafts were sprinkled with advertisements for tools and education for the consumer who was hoping to save on the costs of carpentry or home decorating, or who was interested in starting a new hobby interest that could be both practical and therapeutic. Many advertisements in these new periodicals resonated with their readers' dreams and desires to find new sources of income to supplement what already arrived home in the weekly pay envelope. Ads promising readers that they would "make big money at home" were common in magazines such as *Modern Mechanix* and *Popular Mechanix,* especially during the Great Depression of the 1930s. This alluring pitch continues to this day. That September 1978 issue of *Fine Woodworking* contains a full-page advertisement the promises "Your home workshop can pay off . . . BIG!"[2]

While magazines or their advertisers promoted ways to make money out of woodworking and other workshop activities, the United States Department of Commerce issued a series of ten-cent publications aimed at making better use of the country's wood resources, as well as providing information on the handling and care of lumber and wood products. The National Committee on Wood Utilization's Subcommittee on the Uses for Secondhand Boxes and Odd Pieces of Lumber crammed 105 project plans into a 51-page pamphlet, from an "Animal target box" to an "X-ray box." The companion publication, "You Can Make It for Camp and Cottage," included 102 projects, from houses for various bird species to beach sandals, as well as instructions on reclaiming lumber from boxes and the proper way to build and extinguish campfires. These publications were issued in 1929 and 1930, respectively, and were likely not quite as popular as the 1931 issue, entitled "You Can Make It for Profit," a compendium of 105 plans for earning extra cash, although it was unclear who was going to buy the products. The cover had an engaging engraving of a little "wayside shop" in which one could sell the fruits of the inventive and desperate. The contents, in addition to the drawings and instructions, also included information on plywood, necessary tools, glues, glue joints, commonly used hardware, and instructions for painting, decorating, displaying, and marketing. A short bibliography on interior decoration, wood decay and degradation, and workshop techniques was included at the end of the pamphlet.[3]

The appeal of these activities was not merely the seductive vision of getting rid of the boss, supplementing one's income in trying times, or even quieting the spouse about the hours spent in the workshop and the money lost on tools and machines, some of which would never make it out of the box. (These publications were directed at men.) Part of the payoff of amateur woodworking came from the pleasures of working with an older, "traditional" material that had physical characteristics (workability, aroma) that resurrected the positive world of young adulthood, the "mechanical arts" training of secondary school that was (then) free from the distractions of girls and the rigors of academic work (or so the hobbyists thought). Woodworking became the antiwork for blue- and white-collar men for much of the twentieth century, even as it accelerated as a craft, trade, and art for the cabinetmaker producing handmade goods for a small market willing to pay the price and wait for the goods.

Central to this transformation was the alteration in the nature of powered tools. Hand tools remained pretty much the same in form through the twentieth century, though there were marginal improvements in design and changes in the materials used to make them. But power tools changed dramatically. The development of the small electric motor enabled manufacturers to produce smaller powered tools, practical for the home shop. Smaller stationary floor tools such as the table saw, planer, jointer, radial saw, and drill press were joined by even smaller electrically powered hand tools such as the router, jigsaw, circular saw, sander, and drill to produce miniaturized versions of the woodworking implements of the Industrial Revolution. Changing international trade relations and weakness in the organized-labor sectors of many developed nations altered the environment for the small-time craftsman even further. In the late twentieth century Western

manufacturers of these new tools discovered that they could cut manufacturing costs by sending the work to overseas factories (primarily in
Taiwan and China) where labor was cheap and environmental and
worker safety laws either nonexistent or not enforced. Prices for these
products either dropped or remained relatively constant as the prices of
other consumer goods continued upward. In the United States a new
planer or jointer can now be had for about the cost of new brakes on an
automobile.

The woodworking enthusiasm evident in some industrialized countries is thus part of a confluence of factors that ultimately tap into long-
ings and desires that reach beyond the physical characteristics of the
material involved. The functions of wood and woodworking may well
speak to the discomfiture that modern humans experience in the economies that are more global than local, more anonymous than face-to-face,
more "industrial" than "natural," though the opposition suggested by
that last pair of terms is so loaded with irony that it is amusing. Even in
the raw material itself, globalism has had a transforming effect, the results of which we yet know very little. Species native to one area have
been planted in distant places where they never before have grown, and
in some cases they have thrived, only to be shipped elsewhere for manufacturing. A dock in Port Chalmers, New Zealand, recently was loaded
with logs of Monterey cypress, a species native to California; they were
headed to China for processing into boxes. [FIG. 107]

Our comfort with globalism's immediate returns—lower prices for
some goods, near instantaneous communication, and the immense
power of computers—and our discomfiture with some of the intended
and unintended consequences of this sea change—loss of jobs in some
countries, invasions of privacy, and the angst of white-collar jobs—are
grounded in our tendency to see cultural history in simplistic, linear
terms. For at least 150 years children all over the industrialized world
have been taught that the story of the human race is grounded in technology, in particular the development of materials. They learn about

FIG. 107
*Monterey cypress bound for China. Port Chalmers, New Zealand. November, 2005.
This variety of cypress grows well in the New Zealand climate and is now a significant
cash crop there.*

ancient people's use of stone implements, then bronze, then iron, then
steel and then . . . Concrete? Plastic? Fiberglass? Titanium? Notice
how the "material age" idea loses its attraction as we near the present.
Now we think our "age" is all about microchips and materials that we
probably cannot see, much less touch.

   It isn't that simple. It doesn't take much curiosity to discover that not
all people embraced technological change at the same rate and that those
who did so most rapidly were not necessarily more intelligent than those
who stuck with the old ways. Some changes in the materials of tools
often came about as a result of geography and climate (lots of fires to
keep warm and to cook with, a ready supply of fuel, and easily available
materials, such as ores found in outcroppings), the accessibility of healthier foodstuffs, dumb luck, and mere coincidence. Anyone who thinks
that stone tools are necessarily crude hasn't paid much attention to
exhibits in anthropological museums or, more likely, hasn't visited one
since those days in grade school when all we could think about was
getting out of the classroom. In reality some of these tools are both beau-

tiful in their form and, in the case of edged tools, daunting in their sharpness.

It was neither a mark of genius to favor the new over the old nor a mark of stubbornness or stupidity to think that familiar materials were better suited for tools and other goods. We like to think that history's linear progression is marked by sudden and highly dramatic events that change everything for great masses of people. But superstition maintained its grip on Europeans and their émigrés long after the Enlightenment began, and it is still with us today. The Industrial Revolution eventually changed the way nearly all the world's people lived, but it took a long time—centuries—and it is not over yet, just taking other forms—electronics, chips, and so forth.

It is comforting to look at Denis Diderot's *L'Encyclopédie* of 1751. In it are engravings of many of the arts and industries of the eighteenth century. The comfort lies in the simultaneous surprise at the sophistication of the technology of, for example, mills and machines, and in the quaint imagery of eighteenth-century workers going about their business in clean and commodious surroundings, dressed as if they were in a filmed costume drama. It was but thirty-eight years before the French Revolution bade good-bye to all that.

Diderot's work was published in advance of the time most economic and technology historians think the Industrial Revolution began, though not by much. Yet it seems clear that industrialization was ongoing in important ways, as the wealth of industries depicted demonstrates. Charcoal burners are shown in two large engravings, one detailing how to stack the wood and the other how to burn it. And we know how important charcoal was. So if the "Revolution" really wasn't—if it was a much longer process, then it stands to reason that we might see a longer lifespan of the old ways or at least a mixture of the two.

The Mill in the Hinterlands

On the Stevens River, near Barnet Center, a little town in Caledonia County in northeastern Vermont, employees of the Historic American

Engineering Record in 1979 documented a mill that illustrates the amalgam of the old and the new in woodworking.[4] Bartholomew Carrick erected a vertical (up-and-down) sawmill on the site in 1836, and James Goodwillie operated it there until about 1855, when it was demolished, in the record, either by flood, fire, or collapse after abandonment. The site was idle until a Scottish émigré, Alexander Jack, erected a new dye and print works in 1872. Jack, who had patented improvements in dying, printing, and embossing textiles, hoped he would cash in on the woolen industry, which he thought would rapidly expand in Vermont. He was wrong, and when he died in 1887 the inventory of the mill cited several woodworking devices in addition to the dyeing and embossing gear. In 1888 Elmer Ford began operating the cider press he built in the mill; in 1893 James Loren Judkins converted it to a wagon works and wheelwright's shop. He added the old forge from Joseph (father of James) Goodwillie's blacksmith shop in 1895, and his son Fenton installed a hydraulic cider press in 1915. The Judkins shop—equipped with lathes, a table saw, planer, jointer, drill press, band saw, and other machines—made and repaired farm equipment and built other vehicles, including traverse sleds, snow rollers, and wagon bodies. Ben Thresher, a blacksmith and teamster, took over the shop in 1947, after working there for six years. He maintained the cider-pressing function, added a variety of machines to the shop's inventory, and continued to engage in both woodworking and metalsmithing for the rural communities that surround the Barnet area.

Looking in on Thresher's mill and shop illustrates the wide variety of wooden goods still in production. [FIG. 108] A sledge leans against the wall in the front of the image, and the floor is full of nearly all the products of a woodworker. A chair back rests on the floor, while a wooden-spoked wheel in parts and its rim lean against it; a white barrel (perhaps for nails) sits near the wheel. Patterns hang from the exposed second-floor joists; next to them hang a pair of snowshoes, a seeming hybrid between the broad bearpaw and the sleeker Ojibwa. More wheels sit on the floor near the back.

FIG. 108
Ben Thresher's woodworking mill. First floor, looking north-west along the front wall from the doorway to the blacksmith shop. Historic American Engineering Record (VT, 3-BACEN, 1-15). Courtesy of the Library of Congress, Washington, DC.

Thresher's woodworks demonstrate the persistence of old ways of working wood and the enthusiasm for machines and new materials that were developed over the long haul of history. While there is no doubt that technological innovation has changed our lives dramatically, it has not swept away everything in its path. What new materials and machines did accomplish was a leveling of the playing field when it came to the results once achieved by hand work. Relative newcomers to the arts and crafts could approach the level of finish and competence of the wily veterans who had learned their trade on older tools and who could still outwit and outperform machines. New materials and new machines democratized woodworking to some extent but also created the woodworking factory and ultimately the industrial replacements for wood in our everyday lives. In the end, this has also served to cast highly skilled and trained artisans who rely on one-of-a-kind production into the borderland of the specialist and artist whose works seem to be financially accessible only to those with lots of disposable income. (Ironically, the work of a small-scale cabinetmaker often may not be that much more expensive than factory-made goods. But people think it is.)

Just about everything people own today is made of plastic or metal, and nearly all urban housing is made of steel, concrete, and drywall (sheetrock). Given our hankering for seeing history as a succession of "ages," it might be reasonable to conclude that the age of wood is over. But it isn't. If it were, we wouldn't still build houses in the suburbs or the countryside out of timber studs and beams. There is probably a steel split-level or ranch house out there somewhere, but they are certainly rare. Aluminum and vinyl may have replaced wood on the exterior of smaller-scale housing, but steel has not, and not just because it is much more expensive to use. There is virtually no wood in skyscrapers and apartment towers, both because it is flammable and because it does not have the strength to hold them up. But *fireproof* doesn't necessarily mean "safe." Though building materials may no longer actually fuel a fire, a blaze can harm them to an extent that they fail, as they did in the collapse of New York's World Trade Center in 2001. Even if burned buildings still stand, what remains is a hulk that is worse than useless, since it must be torn down and removed before rebuilding can begin.

In large part what is left for wood to do in modern housing built outside of the cities is often invisible—framing and support. Wood is also still the primary material for interior trim and some furniture, in spite of challenges from inorganic materials. The kitchen table (or dinette, as the table and chairs were called from the 1950s on) may be plastic coated or laminated for ease in cleaning, but dining room tables are still, for the most part, made of wood. Television-dining culture and a more generalized fracturing of the dining experience may have taken hold in certain cultures (such as that of the United States), but the table for the special meal is (for those who can afford the space and the furniture) still made of wood.

Wood may not be at the center of human culture as it once was, but I still heat my house (in part) with it; many people living in rural areas where firewood is plentiful heat almost exclusively with it, especially as oil prices jitterbug and politicians and schemers seem only interested in making political hay and lots of money out of that fuel's uncertain availability. When I make my way to the first tee I don't pull out my old persimmon driver (although I still have it). It's titanium for me and I have even gotten used to—and appreciative of—the sweet metallic sound of the ball rocketing off the trampoline face of the club, then sailing as far as I could hit it as a teenager, and just as far off target. No one hits a baseball or a softball with an ash bat any more, save for professional baseball players, who have to do so. Bow hunters use fiberglass weapons and metal-shafted arrows (although some bowyers still use wood in their laminates). New chemical materials are everywhere—from gunstocks to toys and car "floorboards" to church pews. But if all this new stuff is so great, why is so much of it molded and colored to *look like wood*?

Wood matters in the human scheme of things. I suggested some reasons in the introduction: wood is organic and therefore ties us (so we hope) more closely to the benevolent aspects of the natural world, especially the forest in which we now feel safe: that is, the "woods" of parks and managed lands rather than the wilderness. Except for a few dunderheads, everyone probably knows that trees contribute oxygen to the atmosphere, absorb carbon dioxide, and help retain valuable topsoil. No longer do many of the world's people consider trees as merely an encumbrance, to be rid of in order to plant crops or graze animals. One of the persistent themes in Jared Diamond's influential book *Collapse* (2005), an analysis of the rise and fall of civilizations, can be put quite simply: cut down your trees and you will die out—or nearly so.

But the importance of wood in human culture entails more than this. Central to our continuing fascination with and need for wood is its complexity and randomness. We know that there are a multitude of wood species, each with its own special physical and visual characteristics. Oak, maple, purpleheart, cocobolo, wenge, and the whole range of species look different from each other and are endlessly varied. No two boards look exactly alike, even if they are from the same tree. Grain and figure vary for reasons we cannot discover, much less predict, until we cut into the log. Beyond that, most of us do not have any idea how a work of art—a bowl, a sculpture, a piece of furniture—will look until the artist and artisan show us the art in the wood. Wood persists in our lives because it is full of surprise and wonder. It was once a living organism that we can grow again and again, knowing that it will be different no matter how hard some of us try to make it uniform.

About the only way to get uniformity in wood is to slice it or grind it up and reconstitute it as something else—chip board, oriented strand board, and so forth. These materials, like their older cousin, plywood, offer the builder and the consumer of the finished product predictability, increased strength, and a certain amount of comfort borne of the new materials' dimensional stability. That most plywood comes with at least one side covered with a veneer of fancy or finished wood (save for structural or "building skin" plywood and other products that will not be seen by anyone) only confirms my point about fancy and surprise. Even if the wood in a piece of solid-wood furniture today is almost never shaped or otherwise manipulated by the craftsman's hand, consumers still associate this wood with a preindustrial craft culture more "genuine" than the industrial civilization of which we are a part and the furniture a product.

Consciously or not, we associate solid wood with the long history of hand work, and most wooden things with centuries-old forms and designs, even if we are only a little aware of art or design history. Venerable materials, tools, and products surround us, and even if they are reproductions or interpretations of styles once popular, they nonetheless form part of our collective and individual consciousness of the past, albeit a comforting, comfortable, and probably mostly imaginary time to which we can anchor.

We also tend to see solid wood as organic and natural, even though plywood and similar wood products often make more efficient use of the whole tree than does sawing trees into boards. Perhaps this is because we suspect that the adhesives used to hold the layers or particles together are lethal or at least dangerous, especially when burned. Particulate matter in wood smoke may damage people's lungs, and sawdust may clog them, but they both smell better and certainly seem to be less harmful than plywood smoke or dust and certainly more benign than the effluvia from making plastics or other chemicals. Paper mills may in the past have polluted streams and rivers (and to some extent somewhere may still do so), but we don't blame the wood for the damage. And this makes sense. There aren't too many toxic waste sites that were once lumber mills.

Recycling of wood and reforesting of lands have grown apace since the latter decades of the twentieth century. On a small and usually local scale people are disassembling old barns and reselling the timbers rather than letting abandoned buildings collapse or be burned. New Zealanders have been extracting kauri logs submerged in bogs for as much as 50,000 years and selling the lumber at a premium. In North America logs long submerged in lakes and ponds are similarly being extracted and milled. In many areas of the world forests are managed—trees large enough to be milled into useful boards are harvested and the lands replanted. In most cases the industrial tree farmers replant with coniferous softwoods—trees that grow quickly—rather than with the hardwoods that originally grew in the once-great forests. In the equatorial zones of the earth tropical hardwoods are being cut down at a rate that recalls the great tree slaughter of the nineteenth-century United States. There are hopeful signs that the human race has figured out that it must tend to its woodlands, but we are not "out of the woods" yet. Of course there is an irony in that familiar phrase, suggesting that the "woods" are a threatening place and that cleared land is safest for us. Now that contrast does not seem so obvious.

Acknowledgments

The inspiration for this book came from two remarkable people in the publishing world. Stuart Krichevsky first had the idea for a study of wood as a material. He has been an enthusiastic supporter of this project and all that an author could want in an agent. Wendy Wolf, my editor at Viking/Penguin, has been both an encouraging reader and a steadfast and fair-minded critic and adviser. I could not have done this project without their support and guidance. Other members of the Viking staff worked wonders on what I provided them. Carla Bolte turned a manuscript and a bunch of images into a beautiful book. Clifford Corcoran helped me get through a multitude of obstacles and problems, patiently helping me keep the project on schedule.

My colleagues in the History Department at Northeastern University have been enthusiastic and helpful, directing a steady stream of sources and information my way. Three outstanding graduate students, Valerie Okrent, Adriana Maksy, and Paul Blankman, aided me in my research and did so with dispatch and good humor. Paul and Adriana also read initial drafts of the manuscript and provided sage advice and suggestions, especially for illustrations.

I have benefited from the wise counsel of many extraordinarily talented furniture makers, scholars, and artists. Master craftsmen William Thomas and Jon Siegel read chapters and helped educate me on matters of technology and furniture history. They and David Lamb, Jon Brooks, Geoff Ouellette, Ted Blachly, Garrett Hack, and the rest of the New Hampshire Furniture Masters have patiently answered my questions on many occasions, and have pointed me in directions I had not previously considered. I owe Mary McLaughlin a great debt for introducing me to these gifted and generous craftsmen. Ned Cooke, Charles Montgomery Professor of Art and Art History at Yale, read portions of the manuscript and offered important commentary, criticism, and encouragement. Mikko Saiku, Erkko Professor of North American Studies at the University of Tampere, Finland, provided essential criticism on botany and environmental history. Julie Nicoletta, associate professor of architectural history and American studies at University of Washington–Tacoma, critiqued my work on architecture. William Fowler, professor of history at Northeastern, brought his considerable expertise in maritime history to chapter four. Mark Sfirri, professor of art and director of the fine woodworking program at Bucks County College, helped me to see things and wood in new ways and was a great aid in my research. He directed me to many sources, the most important of which was *The International Book of Wood.*

This compilation of work by seven researchers, edited by Martin Bramwell, helped me refine my thinking in numerous ways and alerted me to elements of the story of wood that had escaped me. Other works that I have found so important that they merit special mention are: R. Bruce Hoadley's *Understanding Wood,* Bryan Sentance's *Wood: The World of Woodwork and Carving,* R. A. Salaman's *Dictionary of Woodworking Tools,* Garrett Hack's *Classic Hand Tools* and *The Handplane Book,* and *A Guide to the Useful Woods of the World,* edited by James H. Flynn, Jr. and Charles D. Holder. There are many other excellent works on wood as a raw material and as finished form; they are cited in the endnotes to chapters and in the selected bibliography. While Isaac Newton's contention that we stand on the shoulders of giants is a cliché, it is no less true for that.

Yoshio Komatsu graciously allowed me to reproduce six of his photographs, originally published in *Built by Hand* (2003). Aud Eidal Berg of Hallingdal Museum at Nesbyen, Norway, provided me with a superb vintage photograph of the *staveloft* from Åi. Robin Strand provided two photographs from his considerable body of work on Norwegian culture. I am grateful for permission to reproduce images of artifacts and photographs in the collections of the Museum of Fine Arts, Boston, the Library of Congress of the United States, the University of Washington Libraries, and the Yale University Art Gallery.

The late Bob Brecht lent me two of his many fine antiques to photograph for illustrations and Lew Fifield lent me books, encouragement, and information about design. My dogs Charlie, Emmy, and the sweet Dudley, whose big heart finally failed him in July 2005, kept me company and were always encouraging me to revive my mind and spirit by sporting with them. I am also grateful to the many antique dealers, collectors, and curators who for many years have freely shared their knowledge and their enthusiasm for the artifacts that surround us.

As in all my other efforts in history, I remain indebted to my dissertation adviser, the late Warren I. Susman. His example has been a continuing inspiration to me, though I can never hope to achieve the breadth of knowledge and sheer brilliance that was his. Finally I owe more to my wife, Susan Reynolds Williams, than I can adequately enumerate. She has always encouraged and supported me in this and other work for the many years we have been together. She paid for my turning lessons and even let me build the kitchen, tolerating my mistakes and celebrating my victories.

NOTES

Introduction

1. Those doubting the importance of standardization need only try to work on a house, like mine, that does not always abide by the standard.

2. Paper, which until the mid-nineteenth century was made from rags, eventually was made almost entirely from wood pulp. Cheaper to produce, this paper had the unfortunate characteristic of being self-destructive after a time, the result of the action of the acids used in production and the lignin found in wood. Eventually cheap wood-pulp paper turns brown and becomes brittle and breaks. Paper and other such derivatives of wood are not the subject of this work, since their fundamental physical nature has been altered so that they are no longer recognized as wood.

Chapter 1: Into and Out of the Wood(s)

1. The essential source for comprehending wood and its biological and physical structure is R. Bruce Hoadley, *Understanding Wood: A Craftsman's Guide to Wood Technology* (Newtown, CT: Taunton Press, 1980). Figures for the woods noted are from p. 7. Other useful books include Martyn Bramwell, ed., *The International Book of Wood* (New York: Simon and Schuster, 1976) and United States Department of Agriculture, *Wood Handbook: Wood As an Engineering Material* (Ottawa: Algrove, 2002).

2. Hoadley, *Understanding Wood,* 15.

3. Ibid., 8.

4. Michael Williams, *Americans and Their Forests: A Historical Geography* (Cambridge: Cambridge University Press, 1992).

5. Gregory Allen Barton, *Empire Forestry and the Origins of Environmentalism* (Cambridge: Cambridge University Press, 2002). See also Jared Diamond, *Collapse: How Societies Choose to Fail or Succeed* (New York: Viking, 2005).

6. Bryan Sentance, *Wood: The World of Woodwork and Carving* (London: Thames and Hudson, 2003), 34; Raymond Tabor, *Traditional Woodland Crafts* (London: B.T. Barsford, 1994), 13–33.

7. Michael Williams, *Deforesting the Earth from Prehistory to Global Crisis* (Chicago: University of Chicago Press, 2003), 203–7; 271–73; Keith Thomas, *Man and the Natural World: A History of the Modern Sensibility* (New York: Pantheon, 1973).

8. Simon Schama, *Landscape and Memory* (New York: Knopf, 1995) examines this and other important aspects of the place of the landscape and pastoralism in modern European and United States history.

9. In 2002 my friend and next-door neighbor was killed by a leaner, turned loose by a gust of wind. He was in his forties at the time. He had cut a tree that hung up on the way down. Later he was cutting another tree in the vicinity of the first with a chain saw. He never heard the hanger come down.

10. A drawing of such an arrangement—odd though it seems—is in Bramwell, *International Book of Wood,* 40.

11. See, for example, Grant McCracken, *Culture and Consumption: New Approaches to the Symbolic Character of Consumer Goods and Activities* (Bloomington: Indiana University Press, 1990), 104–17.

12. Hoadley, *Understanding Wood,* 111.

13. All this matters not at all with the advent of aluminum bats, which are a scourge on the game of baseball, as are metal "woods" on the game of golf. Both have altered the nature of their respective game in ways that seem to me destructive to their essences.

14. In addition, the term *grain* can be as confusing as it is helpful. Ring arrangement and width can yield close, dense, coarse, or open grain; the plane of the surface can show end or face grain. Some of the terms, such as *close* and *fine,* are also used to describe the effects of pore size.

15. Figure variations are the result of distortions in cells in the grain of trees and can be caused by a multitude of factors, from climate variation to insect infestation to genetic tendencies.

16. An essential text for the geographic distribution of various tree species as well as brief notations of the uses to which the wood and other tree products were put is James H. Flynn Jr. and Charles C. Holder, eds., *A Guide to the Useful Woods of the World* (Madison, WI: Forest Products Society, 2001), from which much of the specific information about medicinal and other functions is drawn.

17. See Harvey Green, *Fit for America: Health, Fitness, Sport and American Society, 1830–1940* (New York: Pantheon, 1986), 122–36.

18. Chemical Specialties, Inc., www.treatedwood.com.

19. See John Crowley, *The Invention of Comfort* (Baltimore: Johns Hopkins University Press, 1999).

20. See Siegfried Giedion, *Mechanization Takes Command: A Contribution to Anonymous History* (New York: W. W. Norton, 1948) for illuminating discussions of not only assembly line versus disassembly line technology but also important analyses of the ways in which wooden technology preceded that of metals, as in the development of the lock.

21. See Karen Halttunen, *Confidence-Men and Painted Women: A Study of Middle-Class Culture in America, 1830–1870* (New Haven: Yale University Press, 1982).

22. McCracken, *Culture and Consumption,* 31–43.

Chapter 2: Shelter

1. "Finished" implies both the quality of smoothness and the application of some form of protective coating or covering to the wood, which is also termed *polish*. Finish and refinement are linked as concepts, both for people and for wooden and other goods. They not only indicate the condition of the thing or person but also imply that human effort went into the process of transformation. People can exhibit refinement and polish; young women used to be sent to finishing schools, which did not, one assumes, include being scraped and sanded, though they were probably taught how to apply protective coatings in a way that deceived the ordinary eye. Smoothness in the texture of interior flat areas— walls, furniture, floors—was a marker of status and refinement for centuries, separating the refined classes from the "coarse" members of the working or peasant classes.

2. Gunnar Bugge and Christian Norberg-Schulz's book, *Stav og Laft* (Oslo: Norsk Arkitekturforlag, 1990), 97–102, contains several photographs and scale drawings of this amazing building.

3. Ibid., 15.

4. See, for example, Robert B. St. George, *Conversing by Signs: Poetics of Implication in Colonial New England Culture* (Chapel Hill: University of North Carolina Press, 1998), 115–205; and "'Set Thine House in Order': The Domestication of the Yeomanry in Seventeenth-Century New England," in Dell Upton and John Vlach, eds., *Common Places: Readings in American Vernacular Architecture* (Athens: University of Georgia Press, 1986), 336–64. Gaston Bachelard's *The Poetics of Space* (New York: Orion, 1964) remains one of the most important works in an effort to understand the ways in which space conveys meaning.

5. The contrary argument might be that rural Norway and Sweden enjoyed relative isolation from the depredations of modern warfare. This is certainly true for World War II, in which Sweden was neutral and not bombed by either side. Norway, on the other hand, was occupied by the Nazis, but most of the bombing was of the coastal cities such as Oslo, Bergen, Trondheim, and Tromsø. Switzerland was also neutral, and surviving buildings there suggest a similar, if considerably more restrained, use of ornamental work. One possible explanation for this may be that the ideas of strict Calvinism may have penetrated the Far North less deeply than they did Switzerland.

6. Terry Jordan and Matti Kaups, *The American Backwoods Frontier: An Eth-*

nic and Ecological Interpretation (Baltimore: Johns Hopkins University Press, 1992) traces the origins and evolution of the horizontal log building to northern Europe and specifically to the Savo region of Finland. Also somewhat useful is Harold R. Shurtleff, *The Log Cabin Myth. A Study of the Early Dwellings of the English Colonists in North America,* ed. and intro. Samuel Eliot Morison (Cambridge, MA: Harvard University Press, 1939), an early study of this notion.

7. It is true that the log cabin came to resonate with a certain disingenuous symbolic simplicity. Whether it carried with it a distrust of luxury that some argued could cripple the democratic "experiment" that was the United States or whether it resonated with a long-held distrust of education is difficult to determine. More than likely it included both of these fixations. Surely the most cynical use of this tangle of attitudes was the "log cabin and hard cider" campaign that helped carry the otherwise undistinguished William Henry Harrison to the presidency in 1840. He no more lived in a log cabin than does the average stockbroker working the exchange in London, Paris, or Tokyo. But he kept quiet and won in what was probably the first multimedia campaign in U.S. history. There were even sets of china one could purchase, entitled "Columbian Star" on the reverse, with a romantic rendition of a log cabin on the plate center, a prominent barrel (of wood) of what one was to assume was hard cider sitting near the front door.

8. In 1985 my wife and I came very close to buying a little house in Macedon, New York, that had at its core a log house, built in the 1840s. Succeeding generations had added to it and had sheathed the interior.

9. Williams, *Deforesting the Earth,* 118, 169.

10. Abbott Lowell Cummings, *The Framed Houses of Massachusetts Bay, 1625–1725* (Cambridge, MA: Harvard University Press, 1979), 40–51, disabuses any reader who might have thought that all settlers built their own houses.

11. Pins are sometimes used in furniture as well. One of the attributes of fine furniture making is, in many cases, the absence of visible means of joinery, thus no visible screws or nails (or the various ploys to mask them, such as color-matched fillers for the holes) or wooden pins through mortise-and-tenon joints. The whole seems to be held together by magic or through complex joints invisible to the beholder. This in a sense is the obverse of the practice in buildings, in which it is very important to show joinery, at least in those regions of the structure suitable for that, such as the attic. Finished spaces in the building—parlor and other living areas in houses, sacred spaces in religious buildings—are inappropriate for viewing strong joinery. The exceptions in furniture are the works of the Arts and Crafts Movement, in which visible joinery was part of an ideological position that emphasized the "honesty" of the work and the "authenticity" of the joinery, as opposed to that of the factory.

12. Several machines are available that enable people to make dovetail joints with a router, a power tool that employs sharp bits that are spun at high rates of speed. The circular motion of the cutting edge pares away wood in a groove. With the proper device the depth of the groove and the distance between the "pins" and the "tails" can be controlled, and once set up, a serviceable dovetail joint can be created. Complex jigs allow for variable spacing and the making of "through" dovetails, or joints in which the edge of the joint is visible on the front of the adjoining piece. This is generally not desirable on drawers, but is so on larger "case" pieces, such as chests.

13. The great California cabinetmaker Sam Maloof has said that it is quicker for him to cut dovetails by hand than to set up and use a machine developed for the task. Of course he has been doing it for years and his organic designs in furniture may not respond well to the demands of a mass-produced jig or template.

14. Talinn, Estonia, is a good example of such a walled city, for the most part untouched by the ravages of World War II bombing.

15. Cecil A. Hewett, *English Historic Carpentry* (Fresno, CA: Linden, 1997), 32. Thick with information for the devoted student of the subject and full of superb drawings of the complex joinery, the book serves novices as well. See especially the appendices (pp. 263–326), which show individual joints in detail.

16. Bugge and Norberg-Schulz, *Stav og Laft,* 31.

17. Sibyl Moholy-Nagy, *Native Genius in Anonymous Architecture in North America* (New York: Schocken Books, 1976).

18. Richard Bushman, *The Refinement of America: Persons, Houses, Cities* (New York: Vintage, 1993); see illustration, p. 284. Currier and Ives, the famous New York City printmakers, produced scores of scenes of genteel people at various indoor activities in which women are seated in the parlor with their feet on a delicate stool, as in "Maternal Piety," a print of circa 1855.

19. See H. Parker James, "Up on Stilts: The Stilt House in World History" (PhD diss., Tufts University, 2001).

20. A variant of the balloon-frame house, more common in contemporary building, is the platform frame. Essentially the same idea as the balloon frame, it differs in the length of the wall studs. Instead of extending from the sill to the head plate, studs extended only one floor, where they were nailed to a header. Thus the frame went up one floor at a time and made use of shorter studs, which were sometimes less expensive and almost surely less likely to be warped than longer planks. This arrangement also provided a platform on which to stand to build succeeding floors.

21. This might be an example of what Grant McCracken, in *Culture and Consumption,* refers to as the "pursuit" of the wealthy by the less well-to-do, the latter gaining access to goods that previously had been available only to the

wealthy because of the cost of producing them. But mechanization and mass production changed that economic circumstance.

Chapter 3: The Rub of the Grain

1. A decorative square from the corner of one of the panels of the cabinet doors recently simply fell off, the glue finally failing after 150 years.

2. R. A. Salaman, *Dictionary of Woodworking Tools and Allied Trades, ca. 1700–1970* (Newtown, CT: Taunton Press, 1990) and John Walter, *Antique and Collectible Stanley Tools: Guide to Identity and Value* (Marietta, OH: The Tool Merchant, 1996) attest to the multiplicity of tools and their variants. Other books on specialized types of tools (hand planes, for example) or studies focused on particular cultures (Japanese, Chinese, or East African tools) or on specific periods in history are also common. See, for example, Garrett Hack, *The Handplane Book* (Newtown, CT: Taunton Press, 1999); Emil and Martyl Pollack, *A Guide to the Makers of American Wooden Planes* (Mendham, NJ: Astragal Press, 1994); Aldren A. Watson, *Hand Tools: Their Ways and Workings* (New York: Lyons and Burford, 1982); Garrett Hack, *Classic Hand Tools* (Newtown, CT: Taunton Press, 2001); Denis Diderot, *Encyclopédie, ou Dictionnaire Raisonné des Sciences, des Artes, et des Métiers* (Paris, 1751; repr. New York: Dover, 1987).

3. Salaman, *Dictionary of Woodworking Tools,* 299. Seventy-two pages of this 535-page dictionary are devoted to planes. No other category has as much diversity or as lengthy a set of entries. Axes and hammers each get 19 pages; saws get 31.

4. The cap iron serves to aid in breaking the shaving that the iron has cut. It bends the shaving into the familiar curls that a sharp and well-tuned plane creates in the hands of a skilled worker. Not all planes employ cap irons. Some Asian planes rely on different blade angles or thinner shavings to accomplish the break.

5. The terminology for the surfaces of the opening is for the most part anthropomorphic. The back side, on which the blade, cap iron (if there is one), and the wedge rest, is called the *bed*. The front edge is the *throat*. The sides are the *cheeks*. The opening in the *sole* of the plane is the *mouth*. The front edge of the plane is the *toe* and the rear end the *heel*.

6. Small specialized bronze planes for delicate work, such as in stringed-instrument making, were in use for many years before, but they were a relatively minor part of the story.

7. The bed surface in older wooden planes was also sometimes called the frog. The nomenclature for the new part of the plane may be the result of the endurance of terminology, however inappropriate, when change occurs, or perhaps a deliberate deflection of the anxiety innovators thought accompanied change, or mere confusion, a condition that characterizes much more of human history than some of us would like to think.

8. In many cultures in the world people commonly take their rest from standing by sitting on their haunches, though there are instances in which seating furniture is the norm, such as for ceremonial purposes. Chairs were uncommon in traditional Native American cultures, and seating furniture certainly does not enter the mosque or the accoutrements of domestic material culture in many Islamic areas of the world. Textiles—blankets, carpets, rugs, and the like—are important in many nonseating cultures, both as comfort suppliers and barriers between the profane surface of the earth and the human body.

9. The best single resource on this type of chair made in the United States is Nancy Goyne Evans, *American Windsor Chairs* (New York: Hudson Hills Press, 1996).

10. Some country chair makers, called *bodgers,* used lathes to make spindles and legs: some used shaves to shape them. But they uniformly used wood split from the "round" of a tree, rather than sticks or branches, which would crack, warp, and split. "Rustic" furniture, made with the bark still on the branch, was generally made from flexible woods that could take the bending, such as willow or hazel. Windsors were a cut above the rustic style in their finish and fit.

11. On mass-production work in small shops, see, for example, Walter Rose, *The Village Carpenter* (Cambridge: Cambridge University Press, 1937; repr., New York: New Amsterdam Books, 1988), 46. Rose wrote in this instance about mass-producing window sash. See also Raymond Tabor, *Traditional Woodland Crafts* (London: B. T. Batsford, 1994).

12. Room sizes were quite modest in seventeenth- and eighteenth-century Anglo-America, as Robert Blair St. George points out in *Conversing By Signs*. It would be surprising to find large quarters for the poor (nearly everyone) and even the middling sort in the rest of North America or the world. Space is expensive to surround and protect, requiring a great deal of labor and materials. For nearly all of human history almost no one enjoyed (or maybe even wanted) large spaces in which to live. The great spaces of the rich and of the sacred were all the more astonishing and awe-inspiring because they were so much larger and more sophisticated architecturally than ordinary people's domestic spaces. When public and working spaces (factories, libraries, and government buildings) assume gigantic proportions, they compete with and demystify the spaces of the rich and the divine. The rich then find other ways to maintain their distance, as both St. George and Grant McCracken (*Culture and Consumption*) demonstrate.

13. *The New Shorter Oxford English Dictionary,* vol. 1 (Oxford: Oxford University Press, 1993), 1452.

14. Ibid., 313.

15. Gustav Ecke, *Chinese Domestic Furniture in Photographs and Measured Drawings* (Peking, 1944; repr., New York: Dover, 1986), 152–55, offers excellent

line drawings of some common and unusual joints. An excellent study of Japanese furniture and joinery is Kazuko Koizumi, *Traditional Japanese Furniture: A Definitive Guide* (Tokyo: Kodansha International, 1986), 185–94. See also Grace Wu Bruce, *Chinese Classical Furniture* (New York: Oxford University Press, 1995), and Sentance, *Wood*, 67, 71, 73.

16. Thomas Worral, a Lowell, Massachusetts, inventor, manufactured a molding plane with interchangeable cutters and soles for producing curved surfaces and various grooves and rabbets. Charles Miller, a Brattleboro, Vermont, toolmaker, patented a metal combination plane that produced rectangular shapes in 1870, three years after Leonard Bailey's famous patent for metal planes was successful. Both patents were acquired by the apparently voracious (or entrepreneurial, depending on one's perspective) Stanley Rule and Level Company, which also had by that time hired the German émigré toolmaker and prolific inventor Justin Traut, who joined Miller and Bailey at Stanley Rule and Level. From their earlier work the "45" and the later "55" evolved.

17. Bramwell, ed., *International Book of Wood*, 112–13. R. A. Salaman notes that bodgers still used pole lathes in the High Wycombe district of England until about 1960. *Dictionary of Woodworking Tools*, 258.

18. Thomas Paine, the author of the now classic treatise that made the case for the American Revolution, probably knew exactly that when he entitled it *Common Sense*. In making the case for the separation from the corruption of Crown and Parliament, Paine created a polemic designed both to describe the crisis and to convince the American public of the "logic of rebellion," as Bernard Bailyn termed it in *The Ideological Origins of the American Revolution* (Cambridge, MA: Harvard University Press, 1967).

19. This happened in my present house. A piano on the first floor began to push down on the carrying beam beneath it, until a quarter-inch gap appeared between the floor joists and the beam. Floor jacks and several joist hangers later, the floor is solid.

20. Sandpaper is graded from low numbers (50) for coarse grits to high numbers (400 and higher) for fine grits. Recommended sanding sequences are something like this: 50, 80 (if it is really rough material), 120, 150, 180, 220, 280, 320, 400; 220 surfaces are smooth in many cases, but sometimes finish coatings reveal irregularities that the raw wood did not, so most professionals recommend sanding to the 320 or 400 level.

21. Smoothing wood against the grain is more difficult, as the common English figure of speech reveals. "Going against the grain" is never easy.

22. For those who want more detail on bending stresses and the methods of calculating them, see Hoadley, *Understanding Wood*, 121–26.

23. Tabor, *Traditional Woodland Crafts*, 107.

24. The Niagara Falls Rustique Manufacturing Company, for example,

made everything from chairs to planters out of branches, in a factory located near the nineteenth century's most popular natural "wonder."

25. On the racial nationalist elements of the "outdoor" life at the turn of the century, see my *Fit for America*, 137–66, 219–58. The work of New Hampshire furniture maker and artist Jon Brooks is a good example of the contemporary system of ideas and art.

26. Marvin D. Schwartz, Edward J. Stanek, and Douglas K. True, *The Furniture of John Henry Belter and the Rococo Revival* (New York: Penguin, 1981).

Chapter 4: The Empire of Wood

1. Williams, *Deforesting the Earth*, 129–30.

2. Celestial navigation was, as nearly every historian agrees, one of the great breakthroughs in the history of technology. The Portuguese are credited with most of the early work in this field, but it is likely that they were not alone, nor that they were the first to figure out that the position of the stars could be used as a guide. Unless we are committed to the notion that other peoples simply got lucky when they sailed great distances, washing up on far shores because the winds or currents happened to carry them there, as it were, then we in the West must admit what seems obvious from archaeological and other tracings in the rest of the world.

3. Eighteenth-century English boatbuilders measured wood in *loads*. A load equaled fifty cubic feet of wood, about what could be loaded onto a wagon of the time. Williams, *Deforesting the Earth*, 193; Peter Goodwin, *The Construction and Fitting of the English Man of War, 1650–1850* (Annapolis, MD: Naval Institute Press, 1987).

4. See Bramwell, ed., *International Book of Wood*, 172ff.

5. The Irish National Museum in Dublin has on exhibit an enormous dugout discovered in one of the country's many peat bogs. Making a dugout requires only one edge tool—some sort of adze to chip away material to shape the exterior of the log and to hollow it.

6. Sentance, *Wood*, 137.

7. Planks were joined along their long sides by sewing them with twisted halfa grass, a flexible and very strong local plant.

8. The dates are approximate, in large part because the categorization is variable across time and space, since it is dependent on the technology in use.

9. Bramwell, *International Book of Wood*, 176–77.

10. Tom Vosmer, "The Durable Dhow," *Archaeology* 50, no. 3 (May/June 1997), abstract.

11. Ya'qub Yusuf Al-Hijji, *The Art of Dhow-Building in Kuwait* (Portland, OR: International Specialized Book Service, 2001); Clifford Hawkins, *The Dhow* (Lymington, UK: Nautical Publishing Company, 1977).

12. Michael L. Bosworth, "The Rise and Fall of Chinese Sea Power," www.cronab.demon.co.uk/china.htm, offers a contrasting interpretation to that of Paul Kennedy, below.

13. Paul Kennedy, *The Rise and Fall of Great Powers: Economic Change and Military Conflict from 1500–2000* (New York: Random House, 1987), 4–5.

14. Bosworth, "The Rise and Fall of Chinese Sea Power," 4.

15. Kennedy, *Rise and Fall of Great Powers,* 7–8.

16. Persia was also a substantial force with which to reckon in this era, but it was primarily a land-based empire that expanded eastward, entering India from the north by the end of the first quarter of the sixteenth century.

17. Kennedy, *Rise and Fall of Great Powers,* 10–11.

18. Simon Schama, *The Embarrassment of Riches: An Interpretation of Dutch Culture in the Golden Age* (New York: Knopf, 1987).

19. Williams, *Deforesting the Earth,* 196. Colbert raised a similar alarm in France at about the same time; it too was largely ignored.

20. The French seem to have made little use of the dense mahogany that grew in the Caribbean, though the Spanish did so to such an extent that the species nearly disappeared from their possessions. See John F. Millar, *American Ships of the Colonial and Revolutionary Periods* (New York: W. W. Norton, 1978).

21. Bramwell, ed., *International Book of Wood,* 184–85.

22. J. T. Kotilaine, "Competing Claims: Russian Foreign Trade via Arkhangel'sk and the Eastern Baltic Ports in the 17th Century," *Kritika: Explorations in Russian and Eurasian History* 4, no. 2 (2003): 279–311. See also Karel Kubis, "Russia, the Northern War, and Baltic Trade, 1690–1730," *Ceskoslovenský Casopis Historický* 27, no. 3 (1979): 380–414.

23. Markku Kuisma, "Green Gold and Capitalism: Finland, Forests and the World Economy," *Historialainen Aikakauskirja* 95, no. 2 (1997): 144–52. Demand for sawn timber by the British and the Dutch in particular was also great in Finland, since there were fewer restrictions than in Norway or Sweden, the technology of sawing was more advanced in the ports along the Gulf of Finland, and there was less demand for charcoal in this area because of relatively little mining. See Sven-Erik Åström, "Technology and Timber Exports from the Gulf of Finland, 1661–1740," *Scandinavian Economic History Review* 23, no. 1 (1975): 1–14.

24. Sven-Erik Åström, "Britain's Timber Imports from the Baltic, 1775–1830: Some New Figures and Viewpoints," *Scandinavian Economic History Review* 37, no. 1 (1989): 57–71.

25. Toshiaki Tamaki, "English Trade with the Baltic, 1731–1780," *Shakai-Keizai Shigaku* (Socio-Economic History) 63, no. 6 (1998): 86–105.

26. Sven-Erik Åström, "Nordeuropeisk Trävauexport till Storbritannien"

(North European Timber Exports to Great Britain, 1760–1810), *Historiallinen Arkisto* 63 (1968): 133–51. See also Sven-Erik Åström, "English Timber Imports from Northern Europe in the Eighteenth Century," *Scandinavian Economic History Review* 18, no. 1 (1970): 12–32.

27. R. J. B. Knight, "New England Forests and British Sea Power: Albion Revised," *American Neptune* 46, no. 4 (1986): 221–29.

28. Åström, "English Timber Imports from Northern Europe in the Eighteenth Century," 12–32.

29. Francis Sejersted, "Aspects of the Norwegian Timber Trade in the 1840's and '50s," *Scandinavian Economic History Review* 16, no. 2 (1968): 137–54; see also David M. Williams, "Bulk Carriers and Timber Imports: The British North American Trade and the Shipping Boom of 1824–5," *Mariner's Mirror* 54, no. 4 (1968): 373–82.

30. David S. MacMillan, "Russo-British Trade Relations Under Alexander I," *Canadian-American Slavic Studies* 9, no. 4 (1975): 437–48; A. N. Ryan, "Trade with the Enemy in the Scandinavian and Baltic Ports During the Napoleonic War: For and Against," *Transactions of the Royal Historical Society* 12 (1962): 123–40; Lewis Fischer and Helge Nordvik, "Myth and Reality in Baltic Shipping: The Wood Trade to Britain, 1863–1908," *Scandinavian Journal of History* 12, no. 2 (1987): 99–116. Britain also relaxed some of its restrictive tariffs of 1810 in the middle of the century. For more on the Canadian-English timber trade, see David M. Williams, "Bulk Carriers and Timber Imports: The British North American Trade and the Shipping Boom of 1824–25."

31. Anne K. Bang, "Il Consolato Norvegese a Zanzibar, 1908–1928" (The Norwegian Consulate in Zanzibar), *Storia Urbana* 26, nos. 98–99 (2002): 137–50; Jeremy Prestholdt, "East African Consumerism and the Genealogies of Globalization" (PhD diss., Northwestern University, 2003).

32. Raymond E. Dumett, "Tropical Forests and West African Enterprise: The Early History of the Ghana Timber Trade," *African Economic History* 29 (2001): 79–116.

33. James Pritchard, "Fir Trees, Financiers, and the French Navy During the 1750's," *Canadian Journal of History* 23, no. 3 (1988): 337–54.

34. Sjersted, "Aspects of the Norwegian Timber Trade in the 1840's and 1850's," 137–54.

35. Michael L. Bosworth, "The Rise and Fall of 15th Century Chinese Sea Power," www.cronab.demon.co.uk (1999).

36. The Danes had claimed parts of the Virgin Islands chain in the Caribbean, which they sold to the United States in 1917.

37. Goodwin, *Construction and Fitting of the English Man of War,* 3; Bramwell, *Wood,* 187.

38. Ibid., 239.

39. Jouko Tossavainen, "Dutch Forest Products Trade in the Baltic Until 1648" (master's thesis, Department of History, University of Jyväskylä, December 1994).

40. Per Eliasson and Sven G. Nilsson, "You Should Hate Young Oaks and Noblemen: The Environmental History of Oaks in Eighteenth- and Nineteenth-Century Sweden," *Environmental History* 7, no. 4 (October 2002): 659–77.

41. Tim Shakesheff, "Wood and Crop Theft in Rural Herefordshire, 1800–1860," *Rural History* 13, no. 1 (April 2002): 1–18.

42. Tim Shakesheff (ibid., 2–3) found instances in which the convicted tree-napper was fined one pound, seven shillings, and sixpence in 1847, a considerable sum when farm laborers earned about eight shillings weekly, when there was work to be had. After 1844, crop theft convictions surpassed those of wood theft.

43. Ibid., 13.

44. James M. Lindsay, "Some Aspects of the Timber Supply in the Highlands," *Scottish Studies* 19 (1975): 39–53.

45. Peter Sahlins, *Forest Rites: The War of the Demoiselles in Nineteenth-Century France* (Cambridge, MA: Harvard University Press, 1994).

46. Michael Williams provides a table of the number of great (exceeding 18 inches), middling (12 to 18 inches), and small (8 to 12 inches) masts imported to England between 1706 and 1785. (*Deforesting the Earth,* 199). In every period and category except for middling masts between 1766 and 1770 and small masts between 1781 and 1785, the number of masts from the Baltic is greater than that imported from North America. Masts from Norway (accessible from the west without contending with the narrows between Denmark and Sweden) first appear in the table in 1766–70. Few of the great masts imported are Norwegian, but they consistently constitute the vast majority of middling and small masts imported throughout the latter half of the eighteenth century. Not surprisingly, great masts imported from the Baltic far exceed those from America between 1776 and 1780 (17,371 to 289), and Norwegian middling masts outdistance those from either the Baltic or America in the same period (134,474 versus 1,722 and 596, respectively). About six times as many Norwegian as Baltic small masts were imported in this period as well. The extraordinarily large number of masts imported at this time suggests a massive shipbuilding effort to counter the loss of vessels during the wars (since the American Revolution was part of a world war) as well as Britain's push for naval expansion and domination.

47. Patricia Crimmin, " 'A Great Object with Us to Procure this Timber . . .': The Royal Navy's Search for Ship Timber in the Eastern Mediterranean and Southern Russia, 1803–1815," *International Journal of Maritime History* 4, no. 2 (1992): 83–115.

48. World events took the Dutch out of the New World equation early in the scramble for power. By the 1660s the British had supplanted them in New York and New Jersey. The Swedes similarly faded from the imperial picture in North America, perhaps as a result of their hunger for the Ukrainian agricultural riches or the persistent mistake European leaders make about their ability to conquer Russia. Spanish explorers and settlers traveled through the southern Great Plains, Mexico, and what is now the American Southwest, while the Russians were confined to a few outposts along the northwestern coast of the continent. The French, who had penetrated far into the interior of Canada, were defeated in the Seven Years' War in 1763.

49. Russell R. Menard, Lois Green Carr, and Lorena S. Walsh, "A Small Planter's Profits: The Cole Estate and the Growth of the Early Chesapeake Economy," in Robert B. St. George, ed., *Material Life in America, 1600–1860* (Boston: Northeastern University Press, 1988), 185–201.

50. Surveyor General of His Majesty's Woods in America, "December 2, 1729. Whereas by an act of Parliament . . . ," broadside (Boston: Printed by Bartholomew Green), Massachusetts Historical Society.

51. Surveyor General of His Majesty's Woods in America, "Portsmouth, New Hampshire, December 27, 1733, Whereas Paul Gerrish, Esq . . ." broadside ([Boston?, n.d., 1734?]), Massachusetts Historical Society.

52. Surveyor-General of the King's Woods in North-America, "Whereas some persons have formerly gone . . . " broadside (Portsmouth, NH: January 1, 1770), Massachusetts Historical Society.

53. Michael R. Snyder, A Victim of Circumstance: The Timber Bill of 1772 and the East India Company. *Past Imperfect* 1 (1992): 27–47.

54. It is unclear whether this figure refers to square, or "board," feet one inch in thickness or simply linear feet of varying thicknesses. It would seem likely that the measurement would include some provision for thickness and width, however.

55. Commissioners of Customs in North America, "An Account of the Imports into the Several Ports of America...between the 5th day of January 1771 and the 5th day of January 1772," and "Exports to Great Britain between the 5th day of January 1771 and the 5th day of January 1772," log book, British North America Customs Papers, 1765–74, Massachusetts Historical Society.

56. The ports of New York and "Piscataqua" (Portsmouth, New Hampshire) exported more than 2.5 million staves to the "West Indies," and several million feet of boards and plank to that area, perhaps to be carried to further ports of the world. "Exports West Indies," in ibid.

57. Michael Mann, "Timber Trade on the Malabar Coast, c. 1780–1840," *Environment and History* 7, no. 4 (2001): 403–26.

58. George A. Hall, Collector, "General Exports from the Port of Charleston,

South-Carolina, from November 1786, to November 1787" broadside (December 1, 1787), Massachusetts Historical Society. Rice, however, was the major commodity in South Carolina at this time; the state shipped more than 60,000 barrels and nearly 7,000 half-barrels to other ports.

59. "Letter of the Secretary of the Treasury of the United States Transmitting a Statement of the Exports of the United States during the year ending the 30th of September 1815" (Washington: William A. Davis, 1816): 2.

60. Department of State, Statistics of the United States of America, *Sixth Census* (Washington: Blair and Rivers, 1841): 408.

61. During the winter in some years the Bering Straits could be crossed on the ice sheet, but that route was of limited utility for those with imperial designs.

62. Of course the language of the nay-sayers' thought would not have been English; it was more like a dialect spoken in Eastern Africa.

63. Henry Petroski, *The Evolution of Useful Things: How Everyday Artifacts— From Forks and Pins to Paper Clips and Zippers—Came to Be As They Are* (New York: Vintage, 1994).

64. David O. Whitten has written about the global efforts to reduce noise in some cities by employing wooden paving blocks. See David O. Whitten, "A Century of Parquet Pavements: Wood as a Paving Material in the United States and Abroad, 1840–1940" (parts 1 and 2), *Essays in Economic and Business History* 15 (1997): 209–26, 16 (1998): 161–78.

65. The publication of John Evelyn's *Sylva* in 1664 can be seen as something of a stimulant toward the culture of forests, but the extent of its influence seems to have been modest in the actual landscape, however significant it was in the theoretical sense.

66. Patricia K. Crimmin, "'A Great Object With Us to Procure this Timber . . .': The Royal Navy's Search for Ship Timber in the Eastern Mediterranean and Southern Russia, 1803–1815," 83–115.

67. United States Department of the Interior, *The Statistics of the Wealth and Industry of the United States, Ninth Census,* vol. 3 (Washington, DC: Government Printing Office, 1872), 394.

68. Barbara Freese, *Coal: A Human History* (Cambridge, MA: Perseus, 2003), 15ff.

69. Ibid., 212–13.

70. Prestholdt, "East African Consumerism."

71. Figures from tables in Williams, *Deforesting the Earth,* 257, 258.

72. Jon Arno, "Ekki," in Flynn and Holder, eds., *Guide to the Useful Woods of the World,* 328–29.

73. Lydia Sigourney, "Horticulture," *Godey's Lady's Book* 21 (October 1840): 179.

74. Why are there so few of these in the United States? The answer lies in a

combination of factors. First, Americans tended not to build these structures for the long term and did not cover many of them, figuring the cost of replacing them did not merit the initial expense of building the roof and walls. Second, technological advancement in materials led to rebuilding bridges in steel rather than wood, especially as the traffic volume increased. Third, population shifts took people away from the areas in which early covered bridges had been built, thereby consigning them to the status of relics no longer much used by citizens. Since these structures were usually built with tax moneys, maintaining them was a hard sell to the citizenry.

75. www.cprr.org/Museum/Ephemera, a web site connected with the Central Pacific Railroad, shows a cross-section of a surviving pole fragment with 130 annual rings.

76. It is a strange notion that has been afoot in the United States for some time—that its expansion to the Pacific and to Alaska, as well as its seizure of Hawai'i and occupation of the Philippines and other islands nearer by was unlike the activity of imperial powers such as Great Britain and France, which commandeered vast sections of Africa, the Middle East, and India. This odd conviction flies in the face of the historical reality that the United States was itself created out of a set of colonies of the British Empire and that its land mass after 1783 was but a part of North America. The rest of the continent was under the domain (albeit weakly administered and controlled) of, at various times, Spain, France, Russia, and Great Britain, all of whom ignored the claims of the inhabitants who had preceded them—the Native Americans.

Chapter 5: Artifice: Furniture, Faith, and Music

1. *The New Shorter Oxford English Dictionary,* 2747.

2. Sentance, *Wood,* 86, 112–13.

3. John E. Crowley, *The Invention of Comfort: Sensibilities and Design in Early Modern Britain and Early America* (Baltimore: Johns Hopkins University Press, 2000), 3–78.

4. Bruce, *Chinese Classical Furniture,* 1.

5. Bramwell, ed., *International Book of Wood,* 116.

6. Koizumi, *Traditional Japanese Furniture,* 149.

7. Schama, *Embarrassment of Riches;* Neil McKendrick, John Brewer, and J. H. Plumb, *The Birth of a Consumer Society: The Commercialization of Eighteenth-Century England* (Bloomington: Indiana University Press, 1982); Chandra Mukerji, *From Graven Images: Patterns of Modern Materialism* (New York: Columbia University Press, 1983); McCracken, *Culture and Consumption;* Adrian Forty, *Objects of Desire: Design and Society from Wedgwood to IBM* (New York: Pantheon, 1986).

8. See two important articles by Jules Prown, "Style as Evidence," *Winterthur*

Portfolio 15, no. 3 (Autumn 1980): 197–210, and "Mind in Matter: An Introduction to Material Culture Theory and Method," *Winterthur Portfolio* 17, no. 1 (Spring 1982): 1–19.

9. Both are on the oddly named street, Vick Park A. I lived in the first between 1978 and 1985. The floor was buried under three layers of linoleum for decades. The house had a curious element in its layout. A massive single door opened from the outside directly into the dining room, in spite of the fact that the house had an even larger double front door arrangement and a back door.

10. An illustration of this wonder of the world is in Bramwell, ed., *International Book of Wood,* 94–95.

11. David Pye, *The Nature and Art of Workmanship* (Cambridge: Cambridge University Press, 1968; repr., Bethel, CT: Cambium Press, 1995), 20–25.

12. As do fanatics everywhere, these zealots assumed that they alone have the answer, and they have been willing to sacrifice the rest of us for our own good, never realizing that the very act of deciding the faith and fate of others contravenes their own religion, or at least most belief systems.

13. Some art historians may complain about this. Seurat lived long after the *intarsia* artists began their work in Italy. And they worked in wood and he worked in paint. He was concerned with having the eye create the connections in his work. And so it seems to me were the *intarsia* artists. On the other hand, the latter did not use *only* tiny pieces of wood if they had a larger surface of one color or got lucky and had the modulation they wanted in a single piece of wood. That point, so to speak, is taken.

14. David Esterly, *Grinling Gibbons and the Art of Carving* (London: V&A Publications, 1998) is an exceptional study of this gifted carver and artist, in no small part because the author is an accomplished practitioner of the same art and craft as the subject.

15. That figure is of course the average life expectancy. It is low because so many children died. Once a man made it to the age of twenty-one, his chances of a long life grew dramatically, unless he volunteered for or was forced into the military. For women, making it past childhood diseases was not the end of the threat of imminent death, since the end of childhood susceptibility coincided with the age of fertility, and hence the threat of death in childbirth.

16. There are some good examples of this in Esterly, *Grinling Gibbons*; see pages 85, 87, 91, 94, 96, and 98.

17. Portions of the above two paragraphs were presented in "The Modern Mind and the Simple Heart," a paper presented at a conference at the Canterbury Shaker Museum, Canterbury, New Hampshire, August 9–10, 2004.

18. Thomas Vaughan and Bill Holm, *Soft Gold: The Fur Trade and Cultural Exchange on the Northwest Coast of America* (Portland, OR: Oregon Historical

Society, 1982). See pp. 88 and 93 for images of the bear crest hat and a shaman's mask including a carving of the otter.

19. Bramwell, *International Book of Wood,* 216.

20. A similar affect can be seen in the Spanish *vargueno,* a piece of mobile furniture similarly clothed in elaborate metal fastenings. In this case the metal-work may bear a closer relationship to matters of security.

21. Plastics are replacing wood in some instruments, especially as supplies of ebony are becoming more limited and the cost of wooden instruments is prohibitive for some players.

22. American hornbeam, or blue beech, and Eastern hophornbeam do not seem to be much in demand for piano making, because as Jon Arno notes in Flynn and Holder, eds., *Guide to the Useful Woods of the World,* the first tends to warp and the second is very difficult to dry (pp. 122–23; 388–89).

23. Some historians of music trace the marimba to Asia and then to Africa; some theorize that the Guatemalan marimba is not related to the African version of the instrument, though that seems to me unlikely.

24. www.lafi.org. "The Marimba." Latin American Folk Institute, Washington, DC.

25. Sam Maloof, *Sam Maloof, Woodworker* (Tokyo: Kodansha International, 1983).

26. Roderick Nash, *Wilderness and the American Mind* (New Haven: Yale University Press, 1982).

Chapter 6: Thinking Inside the Box

1. The Edsel and New Coke easily accomplish this. There are plenty of other examples.

2. A competent brief history of locks is in Giedion, *Mechanization Takes Command.*

3. An inventory of a farmer and lawyer from the late eighteenth century in the collections of the American Antiquarian Society implies something of this behavior. Clearly a room-by-room inventory, the list of farm implements and tools from the barn is interrupted by a listing of the many pieces of silver flatware and a watch. Were they hidden to avoid paying taxes on the event of the death of the owner or from thieves? Inventory of E. Freeman, circa 1800. American Antiquarian Society, Worcester, Massachusetts.

4. McCracken, *Culture and Consumption,* 3–31.

5. In *The Evolution of Useful Things* (New York: Vintage, 1992) Henry Petroski argues persuasively that the history of technology is in fact the history of failure. It is only when something does *not* work that we look for something better to do the job.

6. *The New Shorter Oxford English Dictionary,* 301.

7. What would an anarchist's desk look like? Certainly it would be a mess, but would it have drawers? And could one find anything if it did? Libertarians might well have no drawers, but perhaps a shredder set into the desk surface.

8. See Koizumi, *Traditional Japanese Furniture.*

9. It seems an appropriate comparison—at least in part—to link the urge for organization embedded in the tansu with that of the Shakers, that small group of radical communitarians known for, among other things, superb woodworking, design, and entrepreneurial acumen even as they sought to be apart from what they termed "the world." Shaker storage furniture and furnishings are often combinations of cabinets and drawers, though unlike the Japanese, the Shakers hung clothing on hooks and pegs, a practice unthinkable in Japan until the twentieth century.

10. Perry originally intended to force his way into Japan by gunboat diplomacy, but he did not arrive in Tokyo Bay with his full complement of ships. He was much more cooperative in his straitened circumstances.

11. Koizumi, *Traditional Japanese Furniture,* 76.

12. Susan Stewart, *On Longing: Narratives of the Miniature, the Gigantic, the Souvenir, the Collection* (Baltimore: Johns Hopkins University Press, 1984); Walter Benjamin, "Unpacking My Library: A Talk about Book Collecting," in Hannah Arendt, ed., *Illuminations* (New York: Schocken Books, 1969), 59–68.

13. *The Maltese Falcon* is probably the best cinematic representation of the thrill of the chase.

14. For a lucid study of the court cupboard, see E. McClung Fleming, "Artifact Study: A Proposed Model," *Winterthur Portfolio* 9 (Charlottesville: University Press of Virginia, 1974): 153–74.

15. Many people still think of museums as places for the weird and strange and get annoyed when they are confronted with natural and human history as teachers and scholars envision it, rather than as something more akin to the offerings of carnival barkers and Barnumesque hucksters. In this sense, "thinking inside the box" means something rather different and more menacing than thinking about the box. The segment of the museum and education professions that has adopted the marketing paradigm of discovering consumer demands and desires and fulfilling them is as dangerous as it is damaging. Education is about expanding knowledge, not about diluting it to the lowest common denominator.

16. See, for example, Jim Tolpin, *Building Traditional Kitchen Cabinets* (Newtown, CT: Taunton Press, 1994); Jere Cary, *Building Your Own Kitchen Cabinets* (Newtown, CT: Taunton Press, 1983); and Tom Philbin, *Cabinets, Bookcases and Closets* (Saddle River, NJ: Creative Homeowner Press, 1980).

17. There are many sources on evolving conceptions of death and mourning. One of the best is Janice Gray Armstrong and Martha V. Pike, eds., *A Time to*

Mourn: Expressions of Grief in Victorian America (Stony Brook, NY: Museums at Stony Brook, 1980).

18. "Exports to Ireland from the Several Ports . . . in North America . . . 5th day of January, 1771 & the 5th day of January 1772," British Customs Papers, 1765–74, Massachusetts Historical Society. The large number of staves may have been for transport only, then to be directed to other parts of the British Empire, although many were used in the rum and molasses trade, since barrels were the containers for those goods. A similar detailed account of exports of 1769–70 reveals a thriving North American–Caribbean import and export trade in wood in general, and in particular the trade in staves and shooks.

19. *Ninth Census of the United States, vol. 3: The Statistics of the Wealth and Industry of the United States* (Washington, DC: Government Printing Office, 1872), 394, 396. Packing boxes took less skilled labor to produce and far less time, so the figures probably represent a similar productivity when considering the volume in cubic feet of storage produced. But barrels also lasted longer than boxes, since they were usually made of higher-grade hardwood than the pine and other less hardy wood species used for boxes that were nailed together rather than held together with iron bands. In 1997 there were 318 establishments making "nailed wood boxes and shook" and 257 making "wood containers." The 1997 statistics are from www.census.gov/epcd/ec97/industry/E321920.

20. These were blacksmithing, boots and shoes, carpentering and building, wagons, men's clothing, flouring and grist mill products, furniture, leather, sawn lumber, saddlery and harness making, tin, and copper and sheet-iron ware. *Ninth Census of the United States,* vol. 3, 394–98.

21. Kenneth Kilby, *The Cooper and His Trade* (Fresno, CA: Linden, 1989) provides an excellent short history of the trade written by a cooper.

22. Ibid., 78–84. There are eighty-seven entries, four for Scottish terms for tools already listed.

23. Ibid., 15–41.

24. Ibid., 45–46, 50–53, 61–64.

25. As wine expert Jancis Robinson points out, "Chips are not necessarily a bad thing; they provide consumers with the sort of flavours they seek for a fraction of the cost of a real oak barrel, but they cannot provide the physical properties of barrel fermentation and maturation." Jancis Robinson, *Jancis Robinson's Wine Course* (New York: Abbeville Press, 1996), 93.

26. United States Census Bureau, "2002 NAICS Definitions: 321920 Wood Container and Pallet Manufacturing," www.census.gov/epcd/naics02/def.

27. Janet Berlo and Ruth Phillips, *Native North American Art* (New York: Oxford University Press, 1999); Robert N. Shaw, *American Baskets: A Cultural History of a Traditional Domestic Art* (New York: Clarkson N. Potter, 2000).

28. This information comes via Per-Olof Johansson, descendent of one of

Denmark's three major splint basket manufacturing families, www.home3 .inet.tele.dk/johansso/splint.

29. Shaw, *American Baskets,* 102–5.

30. Sarah Hill, *Weaving New Worlds: Southeastern Cherokee Women and Their Basketry* (Chapel Hill: University of North Carolina Press, 1997), 113–15. See also John R. Irwin, *Baskets and Basket Makers in Southern Appalachia* (Exton, PA: Schiffer, 1982); Sue Stephenson, *Basketry of the Appalachian Mountains* (New York: Prentice Hall, 1977); Rachel Nash Law and Cynthia W. Taylor, *Appalachian White Oak Basketmaking: Handing Down the Basket* (Knoxville: University of Tennessee Press, 1991).

31. Ethnic cleansing is ethnic cleansing, whether it is engineered by Slobodan Milošević or Andrew Jackson.

32. The reservations were broken up into the 160-acre plots imagined by the Homestead Act of 1862. But by the time the native American peoples had been granted pieces of land for their use, the best farmland had been snatched by white settlers. Thus the reservations were for the most part unsuited to agriculture of the Anglo-European sort, and in many cases the tribes and bands of native peoples were more attuned to hunting than to the more sedentary pursuits of crop and livestock raising. Even though the federal government's own surveyors had argued that 160 acres were hardly enough for farming in the arid West (John Wesley Powell had argued for about 1,000 acres), the Dawes Act (1887) stuck the Indians with 160 acres per family, and the rest was sold off to whites.

33. Hill, *Weaving New Worlds,* 169.

34. Ibid., 257–58.

35. By the 1930s the Shakers had declined so precipitously in numbers that nearly all of their industries had ceased production.

36. David E. Whisnant, *All That Is Native and Fine: The Politics of Culture in An American Region* (Chapel Hill: University of North Carolina Press, 1983); Jane Becker, *Selling Tradition: Appalachia and the Construction of an American Folk, 1930–1940* (Chapel Hill: University of North Carolina Press, 1998); Henry Shapiro, *Appalachia On Our Mind: The Southern Mountaineers in American Consciousness, 1870–1920* (Chapel Hill: University of North Carolina Press, 1978); Garry G. Barker, *The Handcraft Revival in Southern Appalachia, 1930–1990* (Knoxville: University of Tennessee Press, 1991).

37. Siegfried Giedion, *Mechanization Takes Command,* offers challenging interpretations about the mechanization of food.

38. Henry Glassie, *Pattern in the Material Folk Culture of the Eastern United States* (Philadelphia: University of Pennsylvania Press, 1968).

39. Excellent examples of the enclosed farmstead unit can be seen at Seurasaari,

the Finnish national open-air folk museum near Helsinki, and its Swedish counterpart, Skansen, in Stockholm. See also Bugge and Norberg-Schulz, *Stav og Laft.*

40. John Worlidge, *Systema Agriculturae, Being the System of Agriculture Discovered and Layd Open* (London: Printed by J.C., 1669).

41. "Making Fence," *The Cultivator* 2, no. 3 (March 1845): 107–8.

42. "Cultivation of Forest Trees," ibid., 111, 122–23; 117.

43. J. S. Skinner, "Cost of Fences in the United States," *The Genesee Farmer* 10, no. 7 (July 1849): 165. The article first appeared in the journal *The Plough, Loom and Anvil,* a remarkably prescient journal that, as its title indicates, sought to respond to the joined interests of industry and agriculture.

44. *The Pictorial Cultivator and Almanac for the United States, for the Year 1851* (Albany: Luther Tucker, 1851): 21.

45. See, for example, *400 Wood Boxes: The Fine Art of Containment and Concealment* (New York: Lark Books, 2004).

Chapter 7: Little Things with a Point

1. John Kasson, *Rudeness and Civility* (New York: Farrar, Straus and Giroux, 1990), 13, 183–84 ; Susan Williams, *Savory Suppers and Fashionable Feasts: Dining in Victorian America* (New York: Pantheon, 1985), 13–14; Margaret Visser, *The Rituals of Dinner: The Origins, Evolution, Eccentricities and Meaning of Table Manners* (New York: Grove Weidenfeld, 1991), 98, 201, 323–25.

2. Christy G. Turner II and Erin Cacciatore, "Interproximal Tooth Grooves in Pacific Basin, East Asian and New World Populations," *Anthropological Science* 106, (1998): S85–S94.

3. Quoted in Visser, *Rituals of Dinner,* 324.

4. Sue Hubbell, "Let Us Now Praise the Romantic, Artful, Versatile Toothpick," *Smithsonian Magazine* 27 (January 1997): 77–79.

5. www.foodreference.com/html/ftoothpicks.html.

6. Given that peoples of the ancient worlds were at times unaware of the long-term effects of some substances, it is entirely possible that people enamored of flavored toothpicks may have been slowly killing themselves if they used the wrong stuff. But since life expectancy was so short anyway, it probably did not matter.

7. See Roland Marchand, *Advertising the American Dream: Making Way for Modernity, 1920–1940* (Berkeley: University of California Press, 1986), for an excellent discussion of the ways in which advertisers preyed upon middle-class social insecurity.

8. There are in fact good reasons to keep the toothpicks, floss, and brush handy and to use them. Mitral valve prolapse and other heart-related ailments can be caused or abetted by bleeding gums; early stage type 2 diabetes can be

related to gum disease as well. And of course there are the problems of dental caries and gingivitis with which to contend. See www.cdc.gov.

9. My father, an émigré from Bessarabia (now part of Moldova), freely used toothpicks at home, after dinner. When I went on my first long-distance trip off the farm, to Saratoga Springs, New York, I brought home a souvenir for him— a small slab of wood onto which was attached a small wooden cylinder, forming a toothpick holder with the doggerel inscribed on the face of the slab: "A pick in time is sure to be / The answer to a dental fee." It became a treasure to him.

10. http://centerstage.net/music/whoswho/waynekusy.

11. The last term is commonly in use in Finland, and probably other parts of Europe.

12. Oddly enough, they also liked Roosevelt, who was hardly of the stature of, for example, Sam Houston (six-feet six inches) or Abraham Lincoln (also well over six feet tall).

13. The brand is Touch Cure-Dents pour Sandwiches *Club Picks,* made for L. Tanguay Inc. of Sherbrooke.

14. Most archaeologists, anthropologists, and historians maintain that the distinctions between Anglo-Americans and Europeans in the handling of knives and forks is a function of the era in which the North American colonies and England were culturally and politically distant, between 1690 and 1760. See James Deetz, *In Small Things Forgotten: The Archaeology of Early American Life* (Garden City: Anchor Books, 1977), 122–23; Claude Lévi-Strauss, *The Origin of Table Manners* (London: Jonathan Cape, 1978); Henry Petroski, *The Evolution of Useful Things,* 3–21; and Visser, *Rituals of Dinner.*

15. Glass rolling pins were occasionally used, but they were more expensive than those of wood, and they broke much more easily.

16. See my *Fit for America: Health, Fitness, Sport and American Society, 1830–1940,* and James C. Whorton, *Crusaders for Fitness* (Princeton: Princeton University Press, 1982).

17. This of course does not apply to those peoples whose beliefs prohibit the consumption of certain meats and their fats.

18. There is some debate on the date among match historians.

19. A complete, if somewhat aged, history of the match is M. F. Crass Jr., "A History of the Match Industry," *Journal of Chemical Education* 18, nos. 3, 6, 7, 8, 9 (March, June–September, 1941); reprinted, 24 pages.

20. Sweden is the largest producer of matches in the world.

21. Clay pipes were popular for centuries as well, since they could be made en masse in molds. But they were fragile and the bowl got quite hot in use. Meerschaum is a white porous mineral (hydrous magnesium silicate) that is easily carved and polished, and popular with smokers who can afford it. It gradually changes color from white to brown with use, as the mineral absorbs nicotine and

other effluents from the smoke. Corncob pipes were especially popular in the United States in the nineteenth and twentieth centuries because they were easily obtained and either cheap or free. Their longevity is suspect, however.

22. Max Kline, "Mountain Laurel," in Flynn and Holder, eds., *Guide to the Useful Woods of the World,* 306–7.

23. See, for example, the Peabody Museum web site for information and objects, www.peabody.harvard.edu/Lewis_and_Clark/objects. Detailed drawings of calumets can be found in the American painter George Catlin's *Letters and Notes on the Manners, Customs and Condition of the North American Indians* (1841).

24. Not all Native Americans made their pipes of wood. Pipestone and antler, for example, were also used.

25. Max Kline, "Snakewood," in Flynn and Holder, eds., *Guide to the Useful Woods of the World,* 84–85.

26. The great middle-class enthusiasm for the mountains in the United States began with "Murray's Rush," a phenomenon brought about by bourgeois concern for the alleged ill effects of urban life and the publication of a guidebook for the camps and experiences of the wilds written by William H. H. Murray, entitled *Adventures in the Wilderness,* and first published in 1869.

27. Grant, one of the first African Americans to graduate from Harvard, never made much money out of the tee, but it seems a safe assumption that he played golf. While this may not necessarily be the case (he may have merely been interested in the problem or had applied his mind to it to aid an acquaintance), if he did play the game, then golf in some sense was perhaps integrated, and Grant's activities would have been taking place at almost the moment the United States Supreme Court was declaring that segregation in public and other accommodations was constitutional. The Court decision that established this principle was *Plessy v. Ferguson* (1896).

28. There is some debate about whether the leading edge of the shallow cup on which the ball rests acts as a retardant to flight. Alternatives such as the upturned brush are on the market, and if there is ever hard evidence that this tee adds even one yard to the players' tee shots, the wooden tee will be gone with the persimmon head and the hickory shaft.

29. Balsa's common use in the structural members of early airplanes, and even to some extent in mid-century steel aircraft, had the additional benefit of flotation, handy in water crashes.

30. See, for example, www.toydirectory.com for references and pictures of new wooden toys.

31. Henry Petroski's brilliant *The Pencil: A History of Design and Circumstance* (New York: Knopf, 1989) is the only source a reader will ever need for the history of this artifact. Much of what follows about pencils is drawn from this book,

though the concentrated discussion to follow cannot do justice to the research, interpretation, and wisdom of this seminal work in cultural history.

32. Ibid., 36–41.

33. Ibid., 60–61.

34. There is some debate about the exact year of discovery. Ibid., 45–46.

35. Conté was evidently a scientist of considerable breadth and depth, experimenting with balloons and hydrogen. Ibid., 71–72.

36. Petroski cites an entry in the second edition of the *Encyclopaedia Britannica* (1787–94) that defines *pencil* as "an instrument used in drawing, writing, &c. made of long pieces of black-lead, or red-chalk, placed in a groove of cut in a slip of cedar; on which other pieces of cedar are glued, the whole planed round, and one of the ends being cut to a point; it is fit for use." Ibid., 64.

37. Ibid., 203–7.

38. Salaman, *Dictionary of Woodworking Tools,* 397–99.

39. For most variants, the distinguishing characteristic is length. The six-foot folding rule (also called the zigzag rule or surveyor's lath), often with sliding brass rule for interior measurement, solves the problem of length limits and handles most measurements in the workshop; the bench rule does the same and is an unjointed rule up to three feet or one meter; the glazier's rule is up to six feet long for measuring glass panes; the coachbuilder's rule, up to four feet long.

40. Giedion, *Mechanization Takes Command,* 628–713.

41. www.smithsonianlegacies.si.edu.

42. Electric and other powered dryers certainly have their place in the laundry chore, especially in wintry times of the year. What I am suggesting is a more balanced solution, one that is environmentally more responsible than the present practice.

43. www.laundrylist.org/education/laundryhistory notes the closing of the factory and the change to an import and distribution business. The concept of a plastic "straw" is worth noting, if only for purposes of irony.

44. *The New Shorter Oxford English Dictionary,* 2660.

45. See Richard Bushman, *The Refinement of America: Persons, Houses, Cities* (New York: Vintage, 1993), 30–99.

46. Note the term *sneaker* for the rubber-soled shoe, a name now uncommon but once in everyday use for the canvas-and-rubber athletic shoe that had the unintended consequence of providing the larcenous, murderous, and destructive members of society with a vital tool in their businesses.

Chapter 8: Bat and Battle

1. Johan Huizinga, *Homo Ludens: A Study of the Play Element in Culture* (Boston: Beacon, 1955).

2. The best examples of the variety of traps used that I have encountered is in the outdoor museum of the Sami people, located in Inari, Finland. Virtually all indigenous peoples used traps and snares to take game, at least until accurate bows and arrows were common and rifled barrels and better bullets made firearms useful for hunting at long range.

3. An aged and very useful source is Ralph Payne-Gallwey, *The Book of the Crossbow* (London: Longmans, Green, 1903; repr. New York: Dover, 1995).

4. Ibid., 296–97. Payne-Gallwey built and tested models of these weapons and calculated ranges from them. He also analyzed both written and pictorial records, offering criticism of the workings illustrated and the often fanciful claims made by other writers. The book, while over a century old, may indeed have weaknesses, but it has a great strength in its critical eye toward written sources, which have many more compromises than many who study history would like to admit.

5. Steve Allely, et al., *The Traditional Bowyer's Bible,* 3 vols. (Guilford, CT: Lyons Press, 2000) provide a wealth of information on the history, technology, mechanics, and making of bows and arrows.

6. David Gray, *Bows of the World* (Guilford, CT: Lyons Press, 2002). See also Payne-Gallwey, "A Treatise on Turkish and Other Oriental Bows of Mediæval and Later Times," bound with *The Book of the Crossbow*; Charles Grayson, M.D., "Composite Bows," in Allely, *Traditional Bowyer's Bible,* vol. 2, 113–54. Reports of the great lengths Turkish archers could send their arrows vary, but three to four hundred yards appears to be a conservative estimate.

7. Stephen Selby, *Chinese Archery* (Hong Kong: Hong Kong University Press, 1999) is the authoritative source in English on this topic.

8. See Hideharuo Onuma, Dan Deprospero, and Jackie Deprospero, *Kyudo: The Essence and Practice of Japanese Archery* (Tokyo: Kodansha International, 1993).

9. Robert Hardy, *Longbow: A Social and Military History* (Cambridge: Patrick Stevens, 1976); Malcolm Vale, *War and Chivalry* (Athens: University of Georgia Press, 1981).

10. Payne-Gallwey, *The Book of the Crossbow,* 43–44.

11. Ibid., 85.

12. The French introduced the minié ball in the mid-nineteenth century.

13. www.agriculture.gov.ie/forestry/publications/irish_forest_species/Ash_low.pdf.

14. Ray Ryan, "Bid to Stop Clash of Scandinavian and Welsh Ash," *Irish Examiner,* September 13, 2002; archives.tcm.ie/irishexaminer.

15. www.oireachtas-debates.gov.ie for June 5, 1975, April 8, 1976, and November 29, 1983.

16. On the other hand, the increased velocity of the batted ball did not prevent the nearly wholesale installation of artificial turf in many baseball fields, some of them not of the domed variety. Games in such stadia rapidly gained the sobriquet "rugball," a disdainful reference to the plastic-over-cement surface on which balls jackrabbited over fielder's heads or rocketed along the surface.

17. The best example of a natural formation that reveals this landscape evolution that I have seen is on the south end of Harris, in the Outer Hebrides. The area lies west of Leverburgh, near a headland on the southwestern tip of the island.

18. Technological innovation in balls has also changed the game, also increasing the distance achieved and amount and type of spin players can impart to the shot. The international rule makers of the game, the United States Golf Association and the Royal and Ancient Golf Association, have placed limits on the amount of "spring" or resiliency that a metal head of a club can have, as well as the launch velocity of balls at a set impact speed, but they have failed to regulate the size and material of the shaft and head of the club to sufficiently honor and protect the older and more land-restricted golf clubs in the world. In this sense they failed to protect the complex heritage of the game, unlike professional baseball.

19. Hurling offers a good example. Although it was played more or less continuously since its beginnings, and enjoyed a brief revival in the eighteenth century, the present game gained more attention as Irish nationalist and independence movements gained force in the latter nineteenth century. The Gaelic Athletic Association was founded in 1884, and hurling was officially declared the national game. The same sort of thing happened in the United States. Albert G. Spalding and a few other baseball officials formed a commission to investigate the origins of the game and concluded that the West Point cadet, and later Union general, Abner Doubleday was the "inventor" of the game in Cooperstown, New York, in 1839. They reported the results of their "research" to the United States Congress in 1907. Their conclusions, however, were the stuff of national mythmaking, rather than sound historical research. The New Yorker Alexander Cartwright is now credited with inventing the American game, a descendant of the British game of rounders.

20. A brilliant and often amusing study of this latter phenomenon is Eric Hobsbawm and Terence Ranger, eds., *The Invention of Tradition* (Cambridge: Cambridge University Press, 1983). It is probably less amusing to some of the groups who discover that some of their "traditions" date back about one century.

21. Don Morrow, "Montréal: The Cradle of Organized Sport," in Don Morrow, et al., *A Concise History of Sport in Canada* (Toronto: University of Toronto Press, 1989), 7–8.

22. Snowshoeing remains an important element of cultural identity for the

First Nations peoples of Canada, and to some extent among descendants of the early French voyageurs.

23. Remains of ancient skis have been found in bogs in northern Scandinavia. Carbon 14 dating established their approximate age.

24. Stalin figured that socialist and communist veterans of the Finnish Civil War, who had fought on the losing side against the "whites," would ally with invading Soviet troops. The Russian leader had garnered "influence" over Finland as part of the Nazi-Soviet pact of 1939. Like other imperialistic tyrants and misguided political leaders, he failed to acknowledge the power of nationalism in Finland, fighting to a Pyrrhic victory in 1940 and earning the enmity of the Finns until 1943, when they signed a peace treaty with the Russians that included aid in driving the Nazis out of Finland and reparations at the war's conclusion.

25. Earl E. Clark, "The Tenth: How It All Began," www.10thmtndivassoc .org/chronology.html.

26. Idem, *History of the 87th Mountain Infantry in Italy*, www.10thmtndivassoc .org/chronology.html.

27. Morten Lund's "A Short History of Alpine Skiing, From Telemark to Today," *Skiing Heritage* 8, no. 1 (Winter 1996), is an excellent short history of skiing.

28. After World War I and the collapse of the Hapsburg empire, Austrians similarly sought to pin some of their national identity (for the rest of the world's consumption) on recreational and competitive skiing of a more diverse type, combining the new plow, or stem, turn with the British innovations of the slalom race. Johann Schneider was their hero in this endeavor.

29. Head also helped revolutionize racket sports, and especially tennis. He was among the first to market oversized metal rackets for tennis in 1976, when he introduced the Prince aluminum racket.

30. Badminton gets its name in the West from the Duke of Beaufort's estate, Badminton House, where the duke is credited with introducing the game in 1873.

31. Lacrosse grew to become one of the national games of Canada (ice hockey is the other) owing to the efforts of Montreal dentist George Beers. He published a set of rules for the game in 1867.

32. Thomas Vennum Jr., *American Indian Lacrosse: Little Brother of War* (Washington, DC: Smithsonian Institution Press, 1994), 321–22.

33. http://www.nw.wnyric.org/tuscarora/tuscaroraschool/sticks.htm offers a concise, illustrated, step-by-step disquisition on how a traditional stick is made. See also Don Morrow, "Lacrosse as the National Game," in Morrow, et. al., *Concise History of Sport in Canada,* 45–68; and Bruce Kidd, *The Struggle for Canadian Sport* (Toronto: University of Toronto Press, 1996).

34. There is great argument about whether fish experience pain. The chief

combatants are Dan Sharron, of People for the Ethical Treatment of Animals, and James D. Rose, a professor of zoology at the University of Wyoming. Rose argues that fish brains are too small and undeveloped to experience pain; Sharron dismisses that argument and notes that Rose is an angler.

35. Berners was the prioress at an abbey near Saint Albans, and one of the first women to write on sport of any sort.

36. Fishing rods were often made by joining two or more pieces of wood, securing the joints with metal fastenings.

37. Flynn and Holder, eds., *Guide to Useful Woods of the World,* 158–59.

38. Henry William Herbert, *Frank Forester's Fish and Fishing* (New York: The American News Company, 1858), 458.

39. Genio C. Scott, *Fishing in American Waters* (New York: Orange Judd, 1875), 210–11.

40. There is a wealth of information on bowls, bocce, and boules. See, for example, www.laboulebleue.fr, www.bocce.org, www.bocce.baltimore.md.us.

41. The Brunswick Company's book of 1909, *Modern Billiards: A Complete Text-Book of the Game* (New York: Brunswick-Balke-Collender, 1909) dismisses the ancient Egyptian, Greek, and Roman antecedents some had claimed for the game, as well as supposed references by Saint Augustine, which are described as "translator's errors," and settles on French and English roots of the game, while admitting that it could have come from the Middle East.

42. *Billiards* probably is derived from the French *bille,* or ball, and *billart,* or mace (a stick with a flat square head for propelling a ball). The term *cue* is likely a derivation of the French *queue,* or tail of the billart, which some players began to use to strike the ball in the seventeenth century.

43. The full title is: *The Compleat Gamester: or Instructions—How to play at Billiards, Trucks, Bowls and Chess—together with all manner of usual and most Gentile Games either on Cards or Dice—to which is added the Arts and Mysteries of Riding, Racing, Archery and Cock-Fighting* (1674).

44. John Wesley Hyatt, of New York, is credited with developing the first composition billiard ball in 1868, although he was not able to counteract the tendency of these balls to split on hard impact until 1893. The number of elephants that died for billiards, which by then had become a popular working-class game as well as an elite pastime, must have been staggering, almost as daunting a slaughter as that of the American bison, which had the misfortune of living in a geographic area that was flat and open and into which hunters with sophisticated weaponry were able to gain access by means of the railroad.

45. King James I, "The Kinges' Majesties Declaration Concerning Lawful Sports," in L. A. Govett, ed., *The King's Book of Sports* (London: Elliot Stock, 1890), quoted in Steven A. Reiss, ed., *Major Problems in American Sport History* (Boston: Houghton Mifflin, 1997), 23.

Chapter 9: Fire, Smoke, and the Costs of Comfort

1. Margaret H. Hazen and Robert M. Hazen, in *Keepers of the Flame: The Role of Fire in American Culture, 1775–1925* (Princeton: Princeton University Press, 1992), 18, 246, note that the archaeologists C. K. Brain and A. Sillen discovered fossilized remains of charred animal bones and hearthstones in South Africa that date to at least one million years ago. The original article cited is Brain and Sillen, "Evidence from the Swartkans Cave for the Earliest Use of Fire," *Nature* 336 (December 1, 1988): 464-66.

2. Hill, *Weaving New Worlds,* 36–37.

3. United Nations Food and Agricultural Organization, *State of the World's Forests.* The figures are 1,870 cubic meters of fuelwood and 1,600 cubic meters of industrial roundwood, www.fao.org. "In developing countries 80 per cent of wood is consumed as fuel. Fuelwood accounts for 58 per cent of energy use in Africa, 15 per cent in Latin America and 11 per cent in Asia."

4. Williams, *Deforesting the Earth,* 233–35.

5. Roger S. J. Mols, "Population of Europe, 1500–1700," in Carlo Cipolla, ed., *The Fontana Economic History of Europe,* vol. 2 (Brighton: Harvester Press, 1977): 15–82; statistics quoted in Williams, 181.

6. Williams, 181–84.

7. For statistics on firewood consumption, see William Cronon, *Changes in the Land: Indians, Colonists and the Ecology of New England* (New York: Hill and Wang, 1983), 120–21; Hazen and Hazen, *Keepers of the Flame,* 159; Jane Nylander, *Our Own Snug Fireside: Images of the New England Home, 1760–1850* (New Haven: Yale University Press, 1994), 82–87.

8. Brooke Hindle, ed., *America's Wooden Age* (Tarrytown: Sleepy Hollow Press, 1981). Noted in Hazen and Hazen, 159.

9. Williams, *Deforesting the Earth,* 251, 311.

10. Ibid., 315, compiled from Charles Sprague Sargent, *Report on the Forests of North America (Exclusive of Mexico),* volume 9 of *Tenth Census of the United States (1880)* (Washington, DC: Government Printing Office, 1884), 489.

11. *The New Shorter Oxford English Dictionary,* 510.

12. *The American Heritage® Dictionary of the English Language*; Fourth Edition (Boston: Houghton Mifflin, 2000); the North Carolina Cooperative Extension Service Glossary of Terms for Private Landowners, http://www.ces.ncsu.edu; The University of North Carolina (Chapel Hill)'s Russ Rowlett, in *How Many? A Dictionary of Units of Measurement* (© Russ Rowlett), notes that in the United States a cord is by law 128 cubic feet, usually in the form of a stack four by four by eight feet, and in Maryland, stacked so tightly that "a chipmunk cannot run through it"; accessible at www.unc.edu/~rowlett/.

13. Henry Glassie, *Folk Housing in Middle Virginia* (Knoxville: University of Tennessee Press, 1975), 24–33.

14. Perhaps the appetite for gasoline is considered part of the "pursuit of happiness." If so, it would manifest a fundamental misunderstanding of another of the basic tenets that underlay the intellectual grounding of Jefferson (and others of the founding generation): the social contract.

15. Automobiles produced in the United States and most of the rest of the world now routinely fail to achieve even 25 miles per gallon of gasoline. Hybrid vehicles—part electric motor and part gasoline engine—are the exception to this, but they comprise a tiny minority of the vehicles in production. Miles-per-gallon rates have actually fallen over the past twenty-five years, as manufacturers and consumers opt for greater power and speed over fuel efficiency. Technological know-how in the car and truck manufacturing sector now seems devoted to these inefficient engines, even as automobiles contribute more and more to greenhouse gases and global warming. Irresponsible is too kind a word for this trend.

16. The penalties for all sorts of behavior considered miscreant by owners were severe. Slavery in the Americas operated on the principles of terror, the white master class maintaining power even when their numbers were small by virtue of their control of weaponry and the brutal nature of their punishments. In *Letters of An American Farmer* (1782), J. Hector St. John de Crèvecoeur describes coming across a slave bound and enclosed in an iron cage suspended from a tree limb, his body slowly being eaten away by birds.

17. I.W., "Curing Hams," *Genesee Farmer and Gardener's Journal* 7, no. 11 (March 18, 1837): 35.

18. A. Y. Moore, "Preserving Hams," *The New England Farmer* 4, no. 5 (May 1852): 241. Reprinted from the *Michigan Farmer*.

19. The carcinogen most often identified in connection with smoked products is 3–4 benzopyrine, which along with nitrosamines (which under some conditions are produced from nitrites) has been linked to stomach cancer. Arguments countering assertions about nitrites and smoked meats can be found at "Smoked Meats Are Safe, Task Force Concludes," University of Wisconsin–Madison, www.news.wisc.edu/3225. The vast majority of the research information, however, advises that people severely limit their consumption of smoked, cured, and salted foods, as well as charred foods.

20. An excellent source on all forms of sweated bathing and experiences is Mikkel Aaland, *Sweat* (New York: Consortium Books Sales and Distribution, 1978); see also "Löyly" (Spirit of Steam) (Helsinki: University of Helsinki, 2000).

21. There continues to be great interest in Native Americans among the Finnish, both in the academy and in popular culture. In part this is due to the similarities both peoples see in the sweat lodge and the *savusauna*.

22. Williams, *Deforesting the Earth*, 292, 316.

23. Frederick Overton, *The Manufacture of Iron, In All Its Various Branches* (Philadelphia: Henry C. Baird, 1850). Instructions for building and maintaining

the kiln were reprinted as "On Charring Wood," *The Plough, the Loom and the Anvil* 3, no. 8 (February 1851): 513–16.

24. Ibid., 513–15.

25. Ibid., 516.

26. J.T., "Charcoal a Disinfector of Manure," *The Plough, the Loom and the Anvil* 2, no. 1 (July 1849): 38. Guano, especially from the Pategonian region, was thought to be a superior fertilizer, though burdened with a stench many found intolerable. Note that the title of the article suggests the equation of foul smell with "infection."

27. Michael Rosenthal, *The Character Factory: Baden-Powell and the Origins of the Boy Scout Movement* (New York: Random House, 1986).

28. See David Glassberg, *American Historical Pageantry: The Uses of Tradition in the Early Twentieth Century* (Chapel Hill: University of North Carolina Press, 1990) for the broader context of these activities.

29. In the United States, at least, fire brigades expected payment on the spot or soon afterward for their services. Sometimes brawls over the opportunity to extinguish the blaze broke out.

30. There were ninety-three major fires in Edo between 1601 and 1866, according to Michael Williams, *Deforesting the Earth,* 240.

31. *Niles' National Register,* 5th ser., 7, no. 1 (August 1839–February 29, 1840): viii, 151.

32. The best study of wild fires in the United States is Stephen Pyne's *Fire in America: A Cultural History of Wildland and Rural Fire* (Princeton: Princeton University Press, 1982).

33. Williams, *Deforesting the Earth,* 140, 326; see also Nicholas K. Menzies, *Forest and Land Management in Imperial China* (New York: St. Martin's Press, 1994).

34. Williams, *Deforesting the Earth,* 277, table 10.1, 330–31. See also United Nations Food and Agricultural Organization, *State of the World's Forests,* http://www.fao.org, and the National Resources Defense Council, *Forest Facts,* http://www.nrdc.org/land/forests/fforestf.asp.

35. Williams, *Deforesting the Earth,* 303. Statistics drawn from Martin Primack, "Farm Formed Capital in American Agriculture, 1850–1910" (PhD diss., University of North Carolina, 1963).

36. Williams, *Deforesting the Earth,* 315, table 10.9.

37. *The Farmer's Monthly Visitor* 12, no. 2 (February 1852): 41–42, emphasis in original.

38. "Timber and Forests," *The Country Gentleman* 6, no. 7 (August 16, 1855): 124; no. 8 (August 23, 1855): 125, emphasis in original.

39. See, for example, J. J. Thomas, "Woodland and the Timber Crop," *Registry of Rural Affairs for 1864-5-6* (Albany: Luther Tucker, 1889), 256–65.

40. Williams's table of net forest change in temperate and tropical zones between 1700 and 1995 shows the dramatic reversal of the geographic pattern of deforestation, with the intersection of the curves of clearing occurring in about 1920. *Deforesting the Earth,* 396.

41. Diamond, *Collapse,* 294–306; Williams, *Deforesting the Earth,* 324–27.

42. Adam Smith's arguments in *The Wealth of Nations* (1776), however convincing, represent a dream vision of a world in which the English monarchs (and others) kept their greedy hands off the economy. They did not, of course, just as contemporary politicians do not. Nor do business leaders want governments to do that, save for regulations that may cramp their style or their profits. The "hand," whether "invisible" or "hidden" was a brilliant concept—unproven and unprovable since it was beyond sight, the product of logic and reason, with scant evidence.

Epilogue

1. In his magazine, *The Philistine,* Hubbard celebrated big business and a person who would seem the archrival of what Stickley called "the Craftsman Ideal," Henry Ford. Stickley began his career more true to the Morris ideal of handmade work but by the 1920s was himself running a big business from New York offices, a long way from the original workshops in Eastwood, New York, a town near Syracuse. For more on neurasthenia, see my *Fit for America,* 137–66, 259–82.

2. *Fine Woodworking,* no. 12 (September 1978): 11.

3. United States Department of Commerce, National Committee on Wood Utilization, "You Can Make It" (Washington, DC: Government Printing Office, 1929); "You Can Make It for Camp and Cottage" (Washington, DC: Government Printing Office, 1930); "You Can Make It for Profit" (Washington, DC: Government Printing Office, 1931).

4. The entire HAER report, complete with 11 drawings, 25 photographs, 37 data pages, and 3 caption pages, can be found at http://memory.loc.gov and then searching for "Ben Thresher's Mill." In 1979 filmmaker John Karol produced and directed a documentary about Thresher and his mill. In 1982 it was nominated for an Academy Award for Best Documentary.

SELECTED BIBLIOGRAPHY

Allely, Steve, et al. *The Traditional Bowyer's Bible.* 3 vols. Guilford, CT: Lyons Press, 2000.

Arwidsson, Greta, and Gösta Berg. *The Mästermyr Find.* Lompoc, CA: Larson, 1999.

Barker, Garry G. *The Handcraft Revival in Southern Appalachia, 1930–1990.* Knoxville: University of Tennessee Press, 1991.

Barton, Gregory Allen. *Empire Forestry and the Origins of Environmentalism.* Cambridge: Cambridge University Press, 2002.

Becker, Jane. *Selling Tradition: Appalachia and the Construction of an American Folk, 1930–1940.* Chapel Hill: University of North Carolina Press, 1998.

Berlo, Janet, and Ruth Phillips. *Native North American Art.* New York: Oxford University Press, 1999.

Bramwell, Martyn, ed. *The International Book of Wood.* New York: Simon and Schuster, 1976.

Brown, Deirdre. *Tai Tokerau Whakairo Rākau (Northland Māori Wood Carving).* Auckland: Reed, 2003.

Bugge, Gunnar, and Christian Norberg-Schulz. *Stav og Laft.* Oslo: Norsk Arkitekturforlag, 1990.

Burke, James, and Robert Ornstein. *The Axemaker's Gift.* New York: Tarcher/ Putnam, 1997.

Bushman, Richard. *The Refinement of America: Persons, Houses, Cities.* New York: Vintage, 1993.

The Coach-Makers' Illustrated Hand-Book. Mendham, NJ: Astragal Press, 1995.

Cooke, Edward S., Jr. *Making Furniture in Pre-Industrial America.* Baltimore: Johns Hopkins University Press, 1996.

Cranz, Galen. *The Chair.* New York: W. W. Norton, 1998.

Crowley, John E. *The Invention of Comfort: Sensibilities and Design in Early Modern Britain and Early America.* Baltimore: Johns Hopkins University Press, 2000.

Diamond, Jared. *Collapse: How Societies Choose to Fail or Succeed.* New York: Viking, 2005.

Diderot, Denis. *L'Encyclopédie, ou Dictionnaire Raisonné des Sciences, des Artes, et des Métiers.* Paris, 1751. Reprint, New York: Dover, 1987.

Ecke, Gustav. *Chinese Domestic Furniture in Photographs and Measured Drawings.* Peking, 1944. Reprint, New York: Dover, 1986.

Esterly, David. *Grinling Gibbons and the Art of Carving.* London: V&A Publications, 1998.

Evans, Oliver. *The Young Mill-Wright and Miller's Guide.* Philadelphia: Carey, Lee, and Blanchard, 1834. Reprint, Almont, ON: Algrove, 2004.

Flynn, James H., Jr., and Charles C. Holder, eds. *A Guide to the Useful Woods of the World.* Madison, WI: Forest Products Society, 2001.

Goodwin, Peter. *The Construction and Fitting of the English Man of War, 1650–1850.* Annapolis, MD: Naval Institute Press, 1987.

Gray, David. *Bows of the World.* Guilford, CT: Lyons Press, 2002.

Hack, Garrett. *Classic Hand Tools.* Newtown, CT: Taunton Press, 2001.

———. *The Handplane Book.* Newtown, CT: Taunton Press, 1999.

Hardy, Robert. *Longbow: A Social and Military History.* Cambridge: Patrick Stevens, 1976.

Hawkins, Clifford. *The Dhow.* Lymington, UK: Nautical, 1977.

Hazen, Margaret H., and Robert M. Hazen. *Keepers of the Flame: The Role of Fire in American Culture, 1775–1925.* Princeton: Princeton University Press, 1992.

Hewett, Cecil A. *English Historic Carpentry.* Fresno, CA: Linden, 1997.

Hill, Sarah. *Weaving New Worlds: Southeastern Cherokee Women and Their Basketry.* Chapel Hill: University of North Carolina Press, 1997.

Hindle, Brooke, ed. *America's Wooden Age.* Tarrytown, NY: Sleepy Hollow Restorations, 1975.

Hoadley, R. Bruce. *Understanding Wood: A Craftsman's Guide to Wood Technology.* Newtown, CT: Taunton Press, 1980.

Irwin, John R. *Baskets and Basket Makers in Southern Appalachia.* Exton, PA: Schiffer, 1982.

Jordan, Terry, and Matti Kaups. *The American Backwoods Frontier: An Ethnic and Ecological Interpretation.* Baltimore: Johns Hopkins University Press, 1992.

Kennedy, Paul. *The Rise and Fall of Great Powers: Economic Change and Military Conflict from 1500–2000.* New York: Random House, 1987.

Kilby, Kenneth. *The Cooper and His Trade.* Fresno, CA: Linden, 1989.

Koizumi, Kazuko. *Traditional Japanese Furniture: A Definitive Guide.* Tokyo: Kodansha International, 1986.

Law, Rachel Nash, and Cynthia W. Taylor. *Appalachian White Oak Basketmaking: Handing Down the Basket.* Knoxville: University of Tennessee Press, 1991.

McCracken, Grant. *Culture and Consumption: New Approaches to the Symbolic Character of Consumer Goods and Activities.* Bloomington: Indiana University Press, 1990.

Maloof, Sam. *Sam Maloof, Woodworker.* Tokyo: Kodansha International, 1983.

Mercer, Henry Chapman. *Ancient Carpenters' Tools.* Doylestown, PA: The Bucks County Historical Society, 1929. Reprint, Mineola, NY: Dover, 2000.

Millar, John F. *American Ships of the Colonial and Revolutionary Periods.* New York: W. W. Norton, 1978.

Moholy-Nagy, Sibyl. *Native Genius in Anonymous Architecture in North America.* New York: Schocken Books, 1976.

Morrow, Don, Mary Keyes, Wayne Simpson, Frank Cosentino, and Ron Lappage. *A Concise History of Sport in Canada.* Toronto: Oxford University Press, 1989.

Nakashima, George. *The Soul of a Tree.* Tokyo: Kodansha International, 1988.

Onuma, Hideharuo, Dan Deprospero, and Jackie Deprospero. *Kyudo: The Essence and Practice of Japanese Archery.* Tokyo: Kodansha International, 1993.

Palardy, Jean. *The Early Furniture of French Canada.* Toronto: Macmillan of Canada, 1978.

Payne-Gallwey, Ralph. *The Book of the Crossbow.* London: Longmans, Green, 1903. Reprint, New York: Dover, 1995.

Petroski, Henry. *The Evolution of Useful Things: How Everyday Artifacts—From Forks and Pins to Paper Clips and Zippers—Came to Be As They Are.* New York: Vintage, 1994.

———. *The Pencil: A History of Design and Circumstance.* New York: Knopf, 1989.

Pye, David. *The Nature and Art of Workmanship.* Cambridge: Cambridge University Press, 1968. Reprint, Bethel, CT: Cambium, 1995.

Pyne, Stephen. *Fire in America: A Cultural History of Wildland and Rural Fire.* Princeton: Princeton University Press, 1982.

Rose, Walter. *The Village Carpenter.* Cambridge: Cambridge University Press, 1937. Reprint, New York: New Amsterdam Books, 1988.

Sahlins, Peter. *Forest Rites: The War of the Demoiselles in Nineteenth-Century France.* Cambridge, MA: Harvard University Press, 1994.

Salaman, R. A. *Dictionary of Woodworking Tools and Allied Trades, ca. 1700–1970.* Newtown, CT: Taunton Press, 1990.

Schama, Simon. *The Embarrassment of Riches: An Interpretation of Dutch Culture in the Golden Age.* New York: Knopf, 1987.

———. *Landscape and Memory.* New York: Knopf, 1995.

Selby, Stephen. *Chinese Archery.* Hong Kong: Hong Kong University Press, 1999.

Sentance, Bryan. *Wood: The World of Woodwork and Carving.* London: Thames and Hudson, 2003.

Shaw, Robert N. *American Baskets: A Cultural History of a Traditional Domestic Art.* New York: Clarkson N. Potter, 2000.

Stephenson, Sue. *Basketry of the Appalachian Mountains.* New York: Prentice Hall, 1977.

Steen, Athena, Bill Steen, Eiko Komatsu, and Yoshio Komatsu. *Built by Hand.* Layton, UT: Gibbs Smith, 2003.

Sturt, George. *The Wheelwright's Shop.* Cambridge: Cambridge University Press, 1993.

Sweeney, James J. *African Sculpture.* Princeton: Princeton University Press, 1970.

Tabor, Raymond. *Traditional Woodland Crafts.* London: B.T. Batsford, 1994.

Thomas, Keith. *Man and the Natural World: A History of the Modern Sensibility.* New York: Pantheon, 1973.

Turnbaugh, Sarah P., and William A. Turnbaugh. *Indian Baskets.* West Chester, PA: Schiffer, 1988.

United States Department of Agriculture. *Wood Handbook: Wood As an Engineering Material.* Ottawa: Algrove, 2002.

Vale, Malcolm. *War and Chivalry.* Athens: University of Georgia Press, 1981.

Vaughan, Thomas, and Bill Holm. *Soft Gold: The Fur Trade and Cultural Exchange on the Northwest Coast of America.* Portland: Oregon Historical Society, 1982.

Wardell, Allan. *African Sculpture.* Philadelphia: Philadelphia Museum of Art, 1986.

Williams, Michael. *Americans and Their Forests: A Historical Geography.* Cambridge: Cambridge University Press, 1992.

———. *Deforesting the Earth from Prehistory to Global Crisis.* Chicago: University of Chicago Press, 2003.

Yale University Art Gallery. *Wood Turning in North America Since 1930.* New Haven: Yale University Art Gallery, 2001.

Yanagi, Sōetsu. *The Unknown Craftsman.* Tokyo: Kodansha International, 1989.

Youngquist, W. G., and H. O. Fleischer. *Wood in American Life, 1776–2076.* Madison, WI: Forest Products Research Council, 1977.

INDEX